Coordinating Technology

Inside Technology

edited by Wiebe E. Bijker, W. Bernard Carlson, and Trevor Pinch

Marc Berg, *Rationalizing Medical Work: Decision-Support Techniques and Medical Practices*

Wiebe E. Bijker, *Of Bicycles, Bakelites, and Bulbs: Toward a Theory of Sociotechnical Change*

Wiebe E. Bijker and John Law, editors, *Shaping Technology/Building Society: Studies in Sociotechnical Change*

Stuart S. Blume, *Insight and Industry: On the Dynamics of Technological Change in Medicine*

Geoffrey C. Bowker, *Science on the Run: Information Management and Industrial Geophysics at Schlumberger, 1920–1940*

Louis L. Bucciarelli, *Designing Engineers*

H. M. Collins, *Artificial Experts: Social Knowledge and Intelligent Machines*

Paul N. Edwards, *The Closed World: Computers and the Politics of Discourse in Cold War America*

Eda Kranakis, *Constructing a Bridge: An Exploration of Engineering Culture, Design, and Research in Nineteenth-Century France and America*

Pamela E. Mack, *Viewing the Earth: The Social Construction of the Landsat Satellite System*

Donald MacKenzie, *Inventing Accuracy: A Historical Sociology of Nuclear Missile Guidance*

Donald MacKenzie, *Knowing Machines: Essays on Technical Change*

Susanne K. Schmidt and Raymund Werle, *Coordinating Technology: Studies in the International Standardization of Telecommunications*

Coordinating Technology

Studies in the International Standardization of Telecommunications

Susanne K. Schmidt
Raymund Werle

The MIT Press
Cambridge, Massachusetts
London, England

Set in Baskerville using Ventura Publisher under Windows 95 by Wellington Graphics.
Printed and bound in the United States of America.

Library of Congress Cataloging-in-Publication Data

Schmidt, Susanne K.
Coordinating technology : studies in the international standardization of telecommunications / Susanne K. Schmidt, Raymund Werle.
p. cm. — (Inside technology).
Includes bibliographical references and index.
ISBN 0-262-19393-0 (alk. paper)
1. Telecommunications—Standards—International cooperation—Case studies. 2. Standardization—Social aspects. I. Werle, Raymund.
II. Title. III. Series.
TK5101.S244 1997
389′.6—dc21 97-26080
CIP

Contents

Acknowledgements

This study of standardization, which is a difficult phenomenon for social scientists to grasp, would not have been possible without the help of many organizations and people. We gratefully acknowledge the assistance of the Bundesministerium für Forschung und Technologie. The Verbund Sozialwissenschaftliche Technikforschung, set up and funded by the aforementioned ministry, accepted our research proposal and covered about half of its costs in the context of its research in *Technikgenese;* the other half was covered by the Max-Planck-Institut für Gesellschaftsforschung. We want to thank Meinolf Dierkes and Herbert Kubicek of the Verbund Sozialwissenschaftliche Technikforschung, who advised us on the research proposal. Renate Mayntz, a director of the Max-Planck-Institut für Gesellschaftsforschung, had the idea of studying standardization as an element of the institute's research activities on the development of large technical systems. She also helped us with the research design, as did our colleague Volker Schneider. The many discussions we had with both of them, and also with Philipp Genschel, Andreas Ryll, Uwe Schimank, and Fritz Scharpf (the institute's co-director), influenced several steps of the research and left their mark on this book as a whole. We would like to thank all of them and our other colleagues at the Max-Planck-Institut für Gesellschaftsforschung for creating such a stimulating atmosphere.

Our undertaking has, no doubt, profited most from the cooperative attitude of the practitioners in the field. The fifty or so interviews we were able to conduct in many European countries, in the United States, and in Canada proved extremely helpful and gave us the confidence to tackle the complexities of international standardization. Most of all, we are indebted to Theodor Irmer, director of the Comité Consultatif International Télégraphique et Téléphonique, who kindly removed potential barriers of access to the archives and to certain sessions of the standardization committees in the CCITT. We cordially thank him and all our interviewees.

We have presented parts of this research project at several workshops and conferences—including the third meeting in the Conference Series on the Dynamics of Large Technical Systems (Sydney, 1991), the Joint Conference of the Society for the Social Studies of Science and the European Association for the Study of Science and Technology (Gothenborg, 1992), the International Conference of the International Telecommunications Society (Sophia Antipolis, 1992), the International Research Seminar on Institutional Change and Network Evolution (Stockholm, 1993), the World Congress of the International Sociological Association (Bielefeld, 1994), the World Congress of the International Political Science Association (Berlin, 1994)—and at several meetings of the Verbund sozialwissenschaftliche Technikforschung and the Sektion Wissenschafts- und Technikforschung of the Deutsche Gesellschaft für Soziologie. We wish to thank all the participants at these conferences who discussed and commented on our presentations.

We also have to thank three anonymous referees and the three editors of the Inside Technology series, who convincingly suggested many improvements and clarifications to the manuscript. The freedom from other obligations that we needed to complete the revision was generously arranged by Wolfgang Streeck, who, after many years at the University of Wisconsin, returned to Germany to succeed Renate Mayntz as director of the Max-Planck-Institut für Gesellschaftsforschung. This is just one of the many publications initiated and stimulated by Renate Mayntz.

We wish to thank Günter Schröder, Iris Böschen, Friederike Botzenhardt, and Doris "Dojo" Oelerich, who assisted our work at different stages—with Dojo at times doing more than the required! We would also like to thank John Booth for translating our English manuscript into an English that deserves the label. Last but not least, we are indebted to Larry Cohen, who smoothly maneuvered us through the formal arrangements necessary for assembling an assortment of ideas, text modules, and empirical data into an obdurate artifact called a book.

Coordinating Technology

Introduction

Standards and compatibility have become increasingly significant components of technical development, both in research and in the everyday lives of users of technology. With the broad diffusion of computing technologies, intricate problems having to do with the compatibility of software programs, operating systems, and hardware elements have become matters of broad concern. At the same time, nations and corporations have become increasingly aware of the political and economic importance of standards within the framework of global competitiveness. Not surprisingly, academic research on standards has expanded (although most of the recent work has been purely economic in perspective, focusing on models and paying little attention to the social shaping of standards or to their social and political implications).

In the first part of the book (chapters 1–4), we provide a comprehensive overview of the social and political issues raised by standards and of the process of standardization. We look at standard setting by firms, by markets, and by committees; at regional and international committees and consortia; at the function of standards in establishing technical compatibility and competitive advantage; and, centrally, at the ways in which social, economic, and political factors influence the standard-setting process and become enshrined in the standards themselves.

In the second part (chapters 5–9), we present three detailed empirical case studies of standardization by committee, following the trail from the realization by certain actors that a standard is needed through complex negotiations that involve many economic, political, and social interests (all couched in technical terms) to final agreement on a standard. The case studies all involve standards for telecommunications services and terminals, including facsimile terminals and transmission (which have diffused rapidly since the mid 1980s), videotex (a service, strongly promoted in the 1980s, that was largely a failure, with the notable exception of France's Minitel service), and electronic mail (the X.400 group).[1] All these standards were approved by the Comité Consultatif International Télégraphique et Téléphonique (CCITT), the most prominent international producer of telecommunications standards. For this reason, our study concentrates on the procedural and institutional aspects of the CCITT, also taking other arenas into account whenever they are relevant to the creation of a standard.

In the third part we interpret and generalize the results of the case studies from various perspectives on standardization and relate them to our general model of this process. Coordination proves to be the key concept. According to our theoretical approach of actor-centered institutionalism, we interpret the outcome of the standardization process as resulting from the interaction of actor-related and institutional variables with case-specific technical factors.

Our decision to focus on telecommunications technology and on committee standards rather than market standards offers several advantages from the viewpoint of science, technology, and society (STS) studies, besides the fact that it highlights a field in which little empirical research has been done. It offers a unique vantage point for studying the generation of technology. Unlike the laboratories of inventors or corporations, committees have institutionalized rules that govern the participation and the work of the actors involved. Because they work in a neutral arena dedicated to the solution of the problems raised by compatibility, by interworking, and by the growth of technical systems, standardization committees are less hindered by secrecy than are those social scientists who venture into the processes of technical change. Furthermore, telecommunications is an area of technology that has seen significant changes and even turmoil in recent years. The ongoing liberalization and privatization of public postal, telephone, and telegraph administrations ("PTTs"), which until recently enjoyed monopoly status, have brought the future of the established institutional venues for standardization into question. Against this background, our study emphasizes the historical embeddedness of a particular aspect of the generation of technology, points to the range of available options for this activity, and sheds light on the interdependence of market forms and organizational forms of technological coordination.

Standards, Compatibility, and Technical Systems

Standards have only recently begun to attract the attention of social scientists and everyday users, but they have long been central to technology. For example, the competition among alternative standards for high-definition television (HDTV) is really the second generation of a conflict that started in the 1960s over standards for color television. Similarly, the struggles to establish corporate standards for videocassette recorders, compact disks, and digital tapes are simply new variations on a drama that was played out around 1900 when various corporations tied their prosperity to the success of alternating or direct current.

For analytical purposes, videocassette recorders exemplify the problem of standardization nicely. The recorder and the tape complement each other and are useless if taken separately. They are components of a larger system that relies on their compatibility. Standards are abstract specifications of the necessary features of a component that make it compatible with the rest of a system—they ensure its "fit." If we look at only the tape and the recorder, the requirement of compatibility is straightforward. But an additional complication is already present in this simple example: the system also includes a market for video tapes, a market for video rentals, and informal exchanges of tapes. With these elements included, establishing a single type of VCR and a single type of tape has clear benefits; in economics these are termed *positive network externalities.* These network effects explain why, even when there are only two components (as in the case of VCRs and tapes), it is highly likely that only one system will survive.

Failure in such contests over standards may be dramatic for the firms involved, but the compatibility problems associated with consumer products such as VCRs are trivial in comparison with those associated with international telecommunications. Aside from the standardization of such straightforward characteristics as plugs and sockets, a telecommunications system requires standardization of bandwidths, transmission frequencies, protocols, codes, signaling conventions, and modulation procedures.

As the prototype of a network technology, telecommunications demonstrates the coordinative requirements of technical systems. Recently, these needs have multiplied with the unfolding of technological opportunities based on digitization and with the growing overlap between data processing and telecommunications. For about 100 years, the installation and the operation of telecommunications networks were coordinated by public PTTs or by monopolistic private corporations such as American Telephone and Telegraph. Regulated or directly controlled by national governments, these organizations developed technical systems that were almost purely dedicated to telephone service, which generated more than 90 percent of their revenues. This monofunctionality limited compatibility requirements, and national containment of telecommunications allowed these requirements to be met through hierarchical control. The growing demand for transborder communication and the possibility of new services, such as the transmission of texts, have created a need for coordination, however. New opportunities for telecommunications services do not automatically lead to integrated global technical systems. Competing commercial interests, together with diverging national and local technological traditions, often preserve heterogeneity and reinforce developments along differing paths. But the increasing internal complexity of systems requires increased coordination during their construction and operation. When these systems are interconnected and cross national borders, the coordinative requirements multiply. Typically, a wide range of technical components must be linked, and this involves or affects numerous actors and interests.

When autonomous but interdependent actors design, build, operate, or use a technical system such as telecommunications, standards allow them to coordinate their actions. Compatibility standards specify the relational properties of individual technical components that are necessary for the overall system to achieve its technical functionality. Whenever linkages are desired between technical systems or between components of systems that have not been designed for interworking, gateways or adapters are an alternative. Although such ex post linkage of previously separate components and systems can realize crucial functionality, this is a much more cumbersome solution than ex ante integration on the basis of standards; furthermore, it is technically inferior. Anyone who has experienced the variety of electrical plugs and sockets in Europe will concede the usefulness of the single North American "standard."

Viewing standards as facilitating coordination among the actors involved in a technical system is an aid to grasping the increasing importance of standardization. The large infrastructure systems of modern society are highly visible examples of the pervasiveness of standards. The increasing international division of labor that transforms simple products such as cars and washing machines into assemblies of components of various origins is another example. Trade and tourism are as much linked to standards as are basic innovations (e.g., microelectronics) that make possible new linkages between previously separate systems, such as transportation and telecommunications.

This does not mean that the degree of coordination needed and the degree actually provided by standards can be determined objectively. Some actors regard standardization of every piece of a system as functionally necessary; others argue that it inappropriately reduces variety, hinders innovation, and helps to create uncontrollable technological "dinosaurs." To both sides, however, it is evident that technical development, besides depending on inventions and innovations that are disseminated in markets and organizations, requires a considerable coordinative endeavor of the kind that standard setting offers.

Where multiple actors are involved in installing, operating, or using a technical system such as a telecommunications network, the pursuit of compatibility is by no means a trivial undertaking. Even when all parties are willing to cooperate, the necessary rules may be missing, unknown, or unavailable because of their character as a collective good; appropriate arenas for the establishment of these rules may likewise be absent. This problem, already present in a national context, is aggravated when international coordination is needed. International standardization organizations can be seen as collective attempts to overcome this problem. More an institutional framework than powerful "production units," these organizations provide a platform for discussions and negotiations whose ultimate goal is agreement on a standard. From this perspective, standardization organizations, at the same time that they are dedicated to solving problems of technical coordination, symbolically reinforce the high value of coordination in technical developments.

The Analytical Framework

Standards are highly visible examples of the social shaping of technology. As solutions to coordination problems, they reflect the multiple social relations involved in technical systems. Moreover, the demands implied by these social relations receive explicit recognition in standardization committees. Unlike the evolutionary processes through which new technologies come to embody societal relations, standardization occurs in a forum where the feedback loop between society and technology is highly explicit. This does not mean that actors are protected from undesired effects; rather, it means that standardization shows actors explicitly addressing multiple aspects of the interaction between technology and society.

This unique characteristic of committee standardization processes has implications for our analysis. The theoretical approaches assembled under the heading of "social construction of technology" have so far focused on capturing the high level of contingencies in the relationship between society and technology. To analyze the highly structured interactions that occur in standardization committees, we need to integrate these processes into a comprehensive perspective on the social shaping of technology. This perspective is provided by the theoretical approach of *actor-centered institutionalism,* which is particularly well suited to linking the meso level and the micro level of analysis because it treats institutions and actors as equally important in the shaping of social processes. This approach, which originated in general social science, provides appropriate tools for describing and explaining stages of technological development in highly institutionalized contexts.

Standardization organizations, and their committee processes of building consensus around technical specifications that coordinate development, are a good case in point. Specific rules govern participation and procedures, structuring actors' negotiations and their attempts to forge agreements. The institutionalization of standardization work—which implies, for instance, that many problems will be dealt with according to pre-defined rules and procedures—suggests that we adopt an institutionalist approach that can capture these traits of permanence. At the same time, we need to take account of actors' perceptions, interests, strategies, and interactions within and outside the institutional framework.

Instead of attempting to add an institutional approach under the heading "social construction of technology," we have decided to adopt and develop an analytical framework that takes institutional and actor-related concepts from general social science theory—concepts partly rooted in political science, sociology, and social psychology and partly rooted in institutional economics and economic history. Thus, although many of the analytical categories we use will be familiar to readers accustomed to social constructivist approaches, we sometimes use different terms and refer to other literatures. Approaching our study with non-technology-specific concepts allows us to draw on and contribute to a large number of social science approaches and theories. More important, a general social science approach to technology can (although this may appear paradoxical at first glance) realize many of the demands made by social constructivist approaches. This is because it will be much easier to establish the generation of technology as a normal social activity if it is possible to analyze it with general social concepts that are just as well suited to other activities than it would be if we were to use area-specific theoretical approaches. In this sense, our study deals as much with institutionalized social processes of consensus building as with processes of coordination of technological development through standardization. It opens the black box of technology, finds social processes that shape technology, and explains how and with what results they operate.

Theorizing Standards

1

Actor-Centered Institutionalism: A Social-Shaping Perspective on Standardization and the Coordination of Technology

Social research into technical development has experienced turbulent times since the early 1980s, when it started borrowing concepts and approaches from the sociology of scientific knowledge. The programmatic basis of this process was spelled out in a frequently cited paper by Trevor Pinch and Wiebe Bijker, who argued that the study of science and the study of technology should benefit from each other and that "the social constructivist view prevalent within the sociology of science, and which is also emerging in the sociology of technology, provides a useful starting point" (Pinch and Bijker 1984, p. 400). Analogous to the well-established empirical program of relativism in the study of science, the "social construction of technology" (SCOT) approach focuses on empirical study of technology's development. Three crucial principles and concepts of this early variant of constructivist theory are "interpretative flexibility," "closure and stabilization," and "relevant social groups." Bijker (1995a) elaborated on these concepts and principles and added further theoretical components. We will discuss and use some of these components when we sketch our actor-centered institutionalist explanation of technical development.

In parallel to the SCOT approach, other variants of constructivist thinking have evolved in what is now called, by proponents and opponents alike, the "new sociology of technology" (MacKenzie 1990, p. 410; Winner 1993, p. 367). The "turn to technology" (Woolgar 1991) by sociologists of scientific knowledge, however, has largely remained a one-way street from science to technology rather than a give-and-take between two subdisciplines. This has found its most visible expressions in the concept of technology as text (Woolgar 1987, 1991) and in the actor-network approach (Callon 1987; Latour 1987, 1992; Law and Callon 1988, 1992), both of which also originated in the domain of science studies. For Woolgar, the concept of technology has no fixed or inherent meaning. Instead, meaning is attributed by the acting subjects and by the analyst. For this reason, Woolgar would have us abandon a priori distinctions between the technical and the social. In the actor-network approach, Callon and others (see also Akrich 1992) understand these networks as sociotechnical ensembles of "actants" (human and nonhuman actors). Actor-networks include researchers, manufacturers, and ministerial departments as well as accumulators, electrons, and catalysts. Callon and others suggest a *general symmetry* between the human and the nonhuman, which they claim to be able to analyze using the same conceptual framework. Thus, they reject distinctions that have been central to Western sociology (Bijker 1995b). For sociologists in the Weberian tradition especially, it would be unacceptable to treat technical artifacts as actors, since they hold no values, intentions, or beliefs. This does not mean, however, that in this tradition artifacts are regarded as irrelevant in the context of social action (Collins 1986, pp. 77–116). They are socially shaped, and they influence behavior and action. But technology can be distinguished from politics, economics, or religion, although they often are intimately interconnected.

Interconnection and interaction of heterogeneous elements have been stressed in studies of the development of large technical systems, which have been less directly influenced by the sociology of scientific knowledge. Research guided by the work of Thomas Hughes, a historian of technology, has concentrated on the role and interaction of economic, political, and organizational factors shaping the emerging technical systems (see, e.g., Mayntz and Hughes 1988 and La Porte 1991). For Hughes (1987), during the early stages of development it is individual "system builders" who coordinate forces and processes, thus allowing technological systems to grow. In later stages systems develop momentum. They do not become autonomous, but various groups of actors (e.g., engineers, scientists, managers, owners, investors, civil servants, and politicians) now have vested interests in the growth and the durability of the technological system (ibid., p. 77). Although Hughes and scholars in his tradition emphasize interaction, interconnection, symbiotic relationships, and synthesis—a *seamless web* (Hughes 1986) of heterogeneous elements constituting technological systems—they uphold a distinction between the obdurate material world and other social phenomena (though more in their practical research than in their rhetoric, which has increasingly been influenced by the terminology of the sociology of scientific knowledge). However, they still appear to be less "radical" when it comes to postulating a general symmetry.

Debates on possible distinctions between the "technical" and the "social" have been multifaceted, but most of them are philosophical and ontological in character and not very useful in guiding empirical research. In these debates the "social" often appears to be even more mysterious than the "technical."

What have been the merits and shortcomings of social constructivism in the analysis of technology? Constructivism, first of all, has convincingly internalized technical development. It has made technological determinism obsolete. Technical change is no longer seen as autonomous or as external to society; it is seen as influenced and shaped by its societal context (see, e.g., the essays in MacKenzie and Wajcman 1985). Technology is the dependent rather than the independent variable. Furthermore, technical development is not a linear process in which—following an inherent logic from the abstract to the concrete—scientists discover, engineers develop, manufacturers produce, and users apply a technical artifact. Typically, as Winner (1993, pp. 366, 367) puts it, "social constructivist interpretations of technology emphasize contingency and choice rather than forces of necessity in the history of technology." The undisputed methodological strength of constructivism lies in its case-study approach. Case studies have helped researchers open the black box of technology and have illustrated processes of social shaping.

In recent years, however, some discontent with the direction in which constructivism has developed has been expressed. The central focus of criticism is constructivism's lack of explanatory power. The constructivists' reluctance to draw analytical distinctions between the "social" and the "technical" (following from the principle of general symmetry[1]) and their hesitancy about differentiating the social sphere into subdomains or segments such as polity, science, economy, and religion (implied by the notion of the seamless web) are seen as ex ante attempts to exclude theories of social differentiation in the tradition of Max Weber, Talcott Parsons, and Niklas Luhmann or other approaches rooted in a single discipline. Gary Bowden pleads for a reintroduction of conceptual distinctions between social and technical elements if science or technology is analyzed. He argues that what appears as a seamless web at a certain point of time has resulted from an historically contingent process in which many elements have been linked and transformed. To explain the emergence of a seamless web as a new entity therefore requires reference to all the elements that constitute this entity. Using Yves Gingras's (1995) example of baking a cake, Bowden (1995, p. 76) illustrates how the combination of distinct elements can in effect produce a new entity:

> The cook begins with a variety of heterogeneous ingredients (eggs, flour, water, and so on), which, combined in the proper proportions and manner, result in the creation of a homogeneous cake. . . . When one focuses upon the resulting system (the cake), it is difficult to distinguish one ingredient from the other. When one focuses on the process of system creation, however, such distinctions become not only possible but necessary.

Another flaw in the prevailing variants of constructivism is that the basic opposition to any form of technological determinism has led to a social determinism. Bijker (1993) calls social determinism "social reductionism" and characterizes it as blind to the role of technical factors in the development of technology. This development is certainly underdetermined by technical factors. But is it misleading to agree with Vincenti (1995, p. 565) that, for instance, "all electrical engineers, if their devices are to work in the real world, have to adapt to the concepts of resistance, voltage, current and power, plus the demands of Ohm's and Joule's laws, as absolute representations of the world"? Next to and perhaps more significant than the constraints "imposed by the physical world" are what Vincenti (ibid., p. 564) calls time-dependent "state-of-the-art" technical constraints. According to Vincenti, these constraints and opportunities, including perceived and "real" compatibility requirements, are socially shaped, but at a given point of time they are taken into consideration as if they were "objective." For a period of time they are invariant because nobody challenges them, or they are protected through institutionalization. In this sociological comprehension of technical factors we can address and use terms such as "technical paths of development" ("path dependency") and "technical trajectories" appropriately—that is, without referring to internal techno-logics. MacKenzie (1990, p. 168) writes:

> It is not simply that people want [a technological trajectory] to take that form—though sometimes they do—but that they build into their calculations that it will take that form. They invest money, careers, and credibility in being part of "progress," and in doing so help create progress in the predicted form. Finding their predictions correct, they continue to make them with confidence. A technological trajectory, we might say, is in this sense an institutionalized form of technological change.

However, these terms, if not used very carefully, can easily suggest technological determinism, as MacKenzie (ibid., pp. 165–239) demonstrates in his critique of "natural trajectory," a phrase frequently employed in the economics of innovation.[2]

Another line of criticism relates to the level of analysis of social constructivist studies. Analysts in the field of constructivism "are definitely micro oriented in their efforts to describe and explain technological innovation in terms of individual actions" (Sørensen and Levold 1992, p. 14). This can have theoretical consequences. The micro view, it is argued, tends to emphasize the contingency of technical development and the variety of social factors linked to the individuals who influence the development. Pointing to the methodological micro bias of constructivism, Misa (1994, p. 138) argues that "micro studies, in the attempt to demonstrate the socially constructed nature of technology, often omit comment on the intriguing question of whether technology has any influence on anything."

The potential alternative to the micro view, the macro approach, also reveals deficiencies. Macro studies are inclined to technological determinism. They typically focus on the repercussions of technological change while leveling historical processes or conceiving of them as largely independent of human awareness or of micro influences.

To repair the "methodological bifurcation" of micro and macro, Misa (ibid., p. 139) suggests that scholars should direct attention to the meso level—"to institutions intermediate between the firm and the market or between the individual and the state." Misa's enumeration of meso-level institutions that ought to be covered in research projects includes manufacturers' organizations, consulting engineering firms, investment banking houses, and standard-setting bodies.

Not only is the inclusion of the meso level necessary if the societal effects of technology are to be analyzed; it is also required if we want to study the social shaping processes of technology. Sørensen and Levold (1992, p. 31) show that technical innovation depends on an "infrastructure of competence, skills, and knowledge," which they identify as a meso-level phenomenon. Elements of the infrastructure include universities, colleges, and research institutes as well as banks, venture capitalists, and networks of firms, not to mention the military, whose influence on science and technology has been significant (Smit 1995; MacKenzie 1990; Edwards 1996). The meso-level elements and units to which Sørensen and Levold refer are often used in institutionalist theorizing in political science and in economics to describe sectoral governance structures and innovation systems (Nelson and Rosenberg 1993; Hall and Taylor 1996). In fact the theoretical conclusion that it is necessary to pay attention to the meso level and to emphasize empirically *institutional aspects of technical innovation* is compelling; equally important, it opens up the prospect of integrating constructivist micro-level thinking into an institutional approach to explaining technological development.[3] Our attempt in this book to describe and explain standardization, one central aspect of technical development, is theoretically focused exactly on this integrative undertaking.

We do not aim in this book to develop a social theory of technical standardization, since it makes no sense to decorate every social phenomenon with its own substantive theory. Rather, we wish to gain an understanding of the process of standardization—of its technical peculiarities as well as its general features—by specifying and integrating existing theories in a frame of reference that combines an institutional perspective with an actor-related perspective. In what we call the *actor-centered institutionalist approach,* we relate meso-level elements (institutions) to micro-level elements (actors). Jointly, these elements constitute the social setting that shapes technology. Thus, our aim is not to abandon social constructivism but to link it to an approach that combines central theoretical components of sociology, political science, and economics.

We consider the argument that technology is socially constructed to be the starting point and not the result of social theorizing about technology. In this sense we will make use of constructivism, especially the SCOT framework, in combination with actor theory and institutional theory in order to find out how actors and institutions affect technical development. Only those who act are actors, individual or corporate. Technological choices can be explained as the outcomes of the interactions of intentional actors. Institutions do not act, intentionally or otherwise; neither do technical artifacts. Both artifacts and institutions channel, frame, and contextualize actions and interactions.

The term "actor-centered institutionalism" is new. Crucial concepts of this approach are institutions, actors, and actor constellations (figure 1.1). Actor-centered institutionalism is understood to integrate actor-theoretic individualist components with institution-theoretic structuralist components (Mayntz and Scharpf 1995). Many of these components, including elements of constructivism, have been around

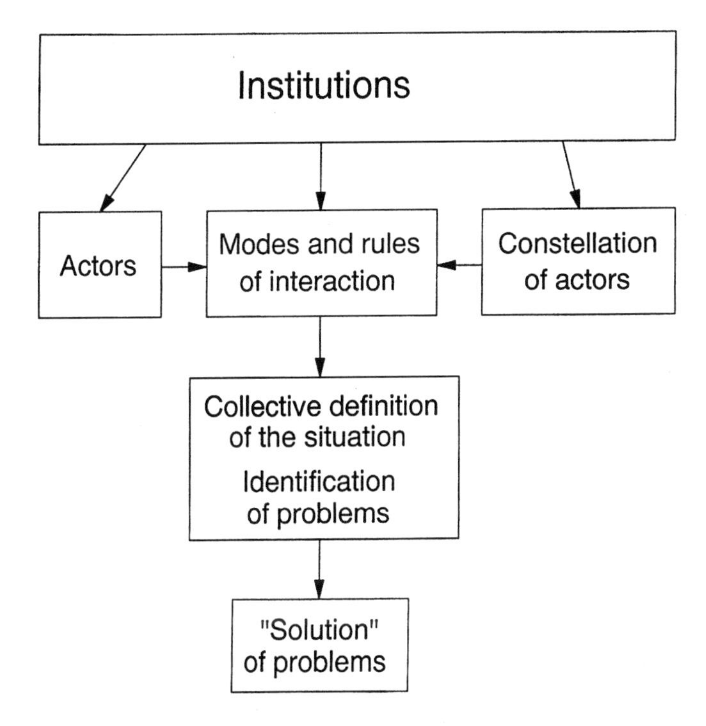

Figure 1.1
Actor-centered institutionalism.

for some time and so need not be reinvented. But they must be made to relate to one another. What is gained by this effort is a better fit between theoretical perspectives and observed processes of technical choice and technological development. Below we briefly outline the concepts of institutions and actors, drawing on a recently published book by Fritz Scharpf (1997).[4] We will add central concepts of constructivism to this approach in order to check whether they fit into or can be made compatible with an actor-centered institutionalist perspective. We draw these concepts from Bijker 1995a, the most thoroughly elaborated version of the SCOT approach. This does not mean that other variants of constructivism are less important. However, for the purpose of discussing the relationship between institutionalism and constructivism it appears appropriate to focus on one constructivist approach instead of mixing several.

In the SCOT approach, as in other constructivist theorizing on technological development, the concept of *institution* plays only a minor role. This comes as no surprise to those who know that constructivism concentrates on the micro level. Donald MacKenzie is one of the few authors who explicitly use this concept. He refers to institutions in connection with technological change. Institutionalization, in his view, "indicates the existence of a relatively stable organizational framework within which the technological change takes place," and institutions channel the resources to support this organizational framework (MacKenzie 1990, pp. 386, 387). In contrast to the constructivist tradition, the concept of institution, though often ill-defined, is frequently used in other theoretical approaches. Often the terms *organization* and *institution* are regarded as equivalent (see, e.g., March and Olsen 1989). In line with Scharpf (1997, p. 38), we "prefer to restrict the concept of institution to systems of rules that structure the courses of actions that a set of actors may choose. In this definition we would, however, include not only formal legal rules that are sanctioned by the court system and the machinery of the state, but also social norms that actors will generally respect and whose violation will be sanctioned by loss of reputation, social disapproval, or even ostracism." (See also Knight 1992.)

An organization, as distinct from an institution, can be conceived of as a specific selection, combination, operationalization, and implementation of these rules, with a higher normative obligation based on formal and informal sanctioning mechanisms. Similarity (isomorphism) of organizations in a field such as international standardization results from their common institutional basis (Zucker 1987; DiMaggio and Powell 1991).

Because actor-centered institutionalism is an approach or a framework rather than a theory, it is open for enrichment by theoretical components and modules including constructivist elements. As an analytical framework, it also helps to clarify theoretical concepts. Actor-centered institutionalism, for example, makes it possible to identify "relevant social groups" and to distinguish them from irrelevant groups. The concept of relevant social groups, originally introduced by Pinch and Bijker (1984), was criticized very early on because it traced groups only on the basis of ex post significance. Proceeding from the specific technical artifact, it declared groups that were more or less directly involved in "constructing" the artifact to be relevant. Without additional theoretical assumptions (a few of which are discussed in Bijker 1995a), this strategy for identifying relevant groups is almost tautological. It risks neglecting the struggle for relevance among groups as well as the meso-level and macro-level factors that, independent of specific construction processes, distribute opportunities and resources to shape technology (Russell 1986; Winner 1993; Rosen 1993).

Including institutional and organizational factors in the analysis helps us to deal with some of these problems. As we will show, membership and boundary rules of organizations, along with their institutional legitimacy and several other meso-level factors, influence the potential relevance of groups. They also constitute groups from individual actors who were not linked to one another before they joined a standardization organization.

Two other concepts central to the SCOT approach, "closure" and "stabilization," can at least be specified if we look at them from the institutional perspective. Bijker (1995a) uses the concept of closure to indicate that the meaning of a technical artifact has been consolidated in a common understanding. Closure, in this sense, means more than effective termination of a controversy; it "has come to mean the process by which facts or artifacts in a provisional state characterized by controversy are molded in a stable state characterized by consensus" (Misa 1992, p. 109). Standardization organizations provide arenas in which committees negotiate standards. The negotiations are governed by institutional rules that determine the general procedure, the decision process, sometimes the legitimacy of arguments, and the value of consensus. Depending on the specific rules, controversies are processed differently, and different forms of closure will prevail. Institutional rules, and not only the strategies, resources, and perceptions of actors, shape the process of closure, which from time to time may simply be a formal termination of a controversy (e.g., through majority voting) rather than a commonly accepted result. Thus, in an institutionalized context, closure can be achieved without the actors' converging on a common meaning of what would be the best or most appropriate standard.

From this point of view it appears not only to make sense but indeed to be necessary to distinguish closure from what Bijker calls "stabilization." In his earlier work Bijker treated "closure" and "stabilization" as synonymous, but he has since used the latter term to indicate that an artifact has reached a degree of obduracy at the semiotic level as well as at the material level (Bijker 1995a). Stabilized artifacts can channel change in a certain direction. In organized standardization, stabilization is a process that, as will be shown, can only partly be controlled by those who approve a standard. The chances of what we call implementation of a committee standard (i.e., of general coping with the technical rule) will often be enhanced by the formal legitimacy and broad membership of the issuing institution, but implementation is not at all guaranteed by these factors. Social and market processes outside the standardization organizations will be taken into consideration in our case studies, although the negotiation of standards in committees will remain our main topic. Stability of a standard emerges through the interplay of these forces. Standardization organizations often contribute to a continuous development of standards; in this sense they channel standardization in a certain direction, because they will hesitate to radically alter technical rules that have been approved in earlier periods of work.

In the concept of actor-centered institutionalism, institutions not only influence social processes by creating and constraining options; they also constitute actors and shape their perceptions and preferences (Scharpf 1997, chapter 2). Individual actors are assigned roles and competencies, and they are confronted with generalized expectations and rules. That institutions affect individuals is not such a new insight. Yet actor-centered institutionalism regards other units as actors too. Unlike individuals, these units are composite actors—mainly collective and corporate actors such as business firms, government ministries, and agencies or associations. In technology policy and in standardization we have to deal with these composite actors, represented by one or many individuals, rather than with individuals acting on their own account. "These composite actors," Scharpf (ibid., p. 39) writes, "are institutionally constituted—meaning that they were created according to pre-existing rules and that they depend on rules for their continuing existence and operation. In general, these rules not only define their membership and the material and legal action resources they [the individuals who act for them] can draw upon, . . . but also the purposes which they are to serve or the values that they are to consider in arriving at their choices."

Thus, the outcomes of choices will also be evaluated by the individual actors according to the principles and values of the corporate actor. Through constituting corporate and collective action and channeling individual action, institutions and systems of rules "in any historically given society not only organize and regulate social behavior but make it understandable—and in a limited conditional sense—predictable for those sharing rule knowledge" (Burns 1985, p. 256).

Even given this view of actors, we do not regard them as determined by institutions. Institutions often only define a scope of acceptable action, leaving room for diversity of strategy and choice. Individuals, on the other hand, whether acting as representatives, delegates, or agents on behalf of a corporate or collective actor or acting on their own account, have different social backgrounds, socialization histories, knowledge, experience, and worldviews. This does not allow them to be simply reduced to instruments of organizations and institutions. What from an institutionalist perspective may sometimes appear idiosyncratic constitutes a significant element of social interaction. Therefore, actors must be regarded as enjoying a considerable degree of autonomy. The dynamics of interaction processes result from the actors' heterogeneity and from the institutional opportunities and constraints that frame a situation.

Actor-centered institutionalism is focused on corporate action more than on individual action. Corporate actors are treated as having goals, orientations, perceptions, values, interests, and preferences, and as acting in an intentional and purposeful manner. Our interpretation of corporate action starts from an assumption of rationality. However, it is no general or comprehensive rationality that guides actors. Rather, the rationality assumption is reduced to the act of choosing. Where they have a choice, actors are assumed to decide on the basis of information easily available to them and to pick the option they prefer most (Werle 1995a). The actors' preferences are partially shaped by institutions and then modified by situational factors. The situation is defined by the actors, and the definition includes generating a rationality of action appropriate to the situation. Thus, the concept of rationality is soft and is grounded in social rather than economic theory (Scharpf 1997, chapter 1).

Actors approach technical artifacts with their specific rationalities. Because their perspectives are heterogeneous, the actors differ in how they conceive of artifacts, in functionality, in safety, in performance, in design, and in their status in a technical system (and in the resulting compatibility requirements). If we try to relate this premise to constructivist concepts, we can infer that heterogeneity of perspectives relates to "interpretative flexibility" (Pinch and Bijker 1984; Bijker 1995a). Different images of a technical artifact (or, in the case of standards, different versions or options) coexist and compete for some time. The institutional embeddedness of the standard-setting process, on the other hand, constrains flexibility: not just anything goes; rather, a limited set of feasible options are processed. Thus, interpretative flexibility is institutionally shaped.

With our emphasis on heterogeneity, we always have a *constellation of actors,* including their interests, perceptions, or preferences, in mind. The constellation can be more or less heterogeneous. Therefore, the constellation of actors is another crucial module of actor-centered institutionalism. As this approach aims predominantly at explaining the outcomes of interactions and of decision processes, the mode of interaction and the strategic interdependence of actors in a given constellation must be scrutinized. The actor-process dynamics that unfold in these constellations significantly affect the outcomes. Here, too, institutional variables have to be considered, because they shape actor constellations (even, we have argued, constituting relevant social groups) and because they frame interaction processes. Framing includes structuring the "rules of the game" in negotiations (Scharpf 1997).

The concepts of frames and constellations, the latter under the slightly different designation of configurations, are also encountered in constructivist theories. Bijker (1995a, p. 102) tries to combine them when he makes his "jump from the SCOT descriptive model to an explanatory scheme that can account for . . . [the] social construction" of technology. What Bijker (1995b, p. 252) calls a technological frame, in his understanding, "structures the interactions between the actors of a relevant social group. . . . It is built up when interaction 'around' a technology starts and continues." The technological frame attributes meaning to a technical artifact (see also Carlson 1992). This comes very close to our understanding of frames, which, as we elaborate in this book, entails both a knowledge facet (a facet similar to the notion of technological paradigms or architecture) and an institutional facet. With regard to the standardization committees we analyze, the institutional facet even encompasses informal rules specifying legitimate and illegitimate types of reasoning in controversial debates on standards. Constructivists seem to be less concerned than institutionalists with where the frames come from and with how they are stabilized and reinforced.

As with the concept of frame, the constructivist notion of configuration parallels our use of the concept of *constellation.* Bijker distinguishes three initial configurations according to the dominance of the technological frames that guide interaction. In fact this distinction is based on constellations of groups of actors. If exactly one group dominates a configuration, one dominant technological frame will guide the interaction. Without a dominant group, no frame comes to dominate. Most interesting from the institutionalist perspective is Bijker's example with more than one dominant or powerful group. In this case, technological frames compete, and "external criteria may become important as appeals are made to third parties" (Bijker 1995a, p. 277). Institutions can be regarded as a type of third party. They supply criteria by which to evaluate different positions, and they provide rules governing interaction which are implemented and operationalized in organizations. Institutions as third parties can level power differentials, channel controversies, and maybe even homogenize divergent technological frames prevailing in heterogeneous constellations of actors.

Thus, actor-centered institutionalism provides tools that are extremely useful in explaining technical development. It is open to constructivist theoretical elements, but it adds concepts and instruments that permit the systematic inclusion of meso-level structural and procedural variables—especially organizations and rule systems—in the analysis of processes that shape technology.

In chapters 2–4 we will develop and specify our approach step by step. Before proceeding to examine our three cases of standardization, we will introduce a general model of the standardization process that specifies actor-centered institutionalism in a way that makes it applicable to this peculiar object of research. Use of this template in the case studies helps direct our attention to negotiation and consensus-building processes in standardization committees as subunits of the larger organization. Here standards are dependent variables. How standards (which are a part of technology) interact with and shape their technical environment (i.e., artifacts, systems, other standards) will only be touched on. Hence, it is necessary to start by considering the status of standards in technology and the general relation between standards and technical development.

2

Standards and Technical Development

Research into technical standardization contributes to explaining technology's development by examining coordinative aspects of its formation. Insufficient attempts at coordination—to give an extreme example—can result in incompatibility of various devices and to failure of a technology, whereas successful coordination provides a basis for the accommodation of diverse gadgets. Large technical systems especially rely on coordination: not only with respect to construction and maintenance but also with respect to operation and use. Technical standards serve as a medium of coordination. They are technical rules that specify relational properties of artifacts. Compliance with these rules ensures compatibility and thus ensures the artifacts' smooth interoperation in a system. For the development of technology, however, not only the existence or nonexistence of standards but also their specific shape is important.

In the literature on technical and organizational development we find several examples and concepts of the relationship between technical rules and the formation of technology in general. By way of a brief illustration, we have selected three historical examples of network technologies. For such technologies, the achievement of compatibility has always been "an issue of recognized importance" (David 1987, p. 209). The railway, electric power, and telephone systems show that markets, business hierarchies, governments, and committees have all promoted compatibility, either through standards or through directly imposed or negotiated accords. In this portion of the literature, standards are identified as a crucial factor in the development of large integrated technological systems.

Another portion of the literature approaches technical development from a more general point of view. Here, typically, the significance of knowledge as a coordinative factor is stressed. We shall examine some of these contributions because they explicitly relate standards to technical knowledge. In this context, we will then analyze the role of standards in the process of technical development.

Standards and Large Technical Networks: Historical Examples

Alfred Chandler's (1977) historical analysis of the rise of modern business enterprise throws light on the replacement of market mechanisms with the internalization of coordinative efforts in large business units as a managerial strategy. The rail companies and the telegraph companies were the first enterprises in the United States to require a large number of managers "to coordinate, control, and evaluate the activities of a number of widely scattered operating units" (ibid., p. 79). In the middle of the nineteenth century, the railway celebrated its victory—based on successful technical advances—over the waterways and other modes of transportation of goods and passengers. The economic advantages of a nationwide integrated railway system, however, could be obtained only through the cooperation of the many railroad companies, and through a concomitant coordination of technical and operational rules (including safety regulations).[1] Gauges, couplers, brakes, and block signal systems had to be made compatible. In those years compatibility was often synonymous with uniformity (Salsbury 1988, pp. 56–60).

As a response to the need both for coordination and for a trained workforce, quasi-professional associations were set up and education in civil and mechanical engineering was strengthened. In 1867, the Society of American Civil Engineers, "which railroad men had attempted to found in the years before the Civil War" (Chandler 1977, p. 132), was revived. Professional groups set up many working committees, which defined standards based on a common stock of knowledge. In many cases, as with air brakes, couplers, and signals, "implementation" of the standards was secured by government railroad departments and not by the manufacturers. Most prominent in this respect was the American railroads' acquisition of a standard gauge and a standard time in the 1880s, when the railroads were integrated in a single national network.

Technical and organizational integration coincided. The large enterprises that were to operate the railroad network throughout the twentieth century were established in the 1880s. Their emergence, however, was not so much a result of technical coordination requirements as of competitive and general business concerns in an industry that was extremely capital intensive. Besides the few large enterprises, there have always existed a great many small firms operating short lines. In 1920 there were 1085 different firms (Salsbury 1988, p. 48). For most of them, the standard-gauge network turned out to be of vital interest when interstate and trunk traffic began to increase in the 1860s. In these years, railroads often had to unload and reload goods at transfer points because of breaks in gauge. In 1861, eight different gauges ranging from 4′3″ to 6′ were employed (Salsbury 1988, p. 50). The elimination of this obstacle required a single gauge, although some manufacturers offered specially equipped "compromise cars" that could travel on two-thirds of the country's rail network (ibid., p. 55). Eventual conversion to a relatively small gauge standard appears to have been the "logical" result of three factors (ibid., pp. 48–55):

- In 1861 the gauge later adopted already accounted for 53 percent of all the trackage in the United States.
- The federal government specified it as the gauge for the first transcontinental railroad completed in 1869.
- Most of the "nonstandard" gauges were broader than the standard, and it was comparatively easy for a company to convert to a smaller gauge because bridges, tunnels, and clearances were not directly affected.

The standard was officially adopted by the American Railway Association's Committee on Standard Wheel and Truck Gauges in October 1896, six years after practically all railroads had migrated to that standard (Salsbury 1988, p. 55). This step of ex post standardization is by no means unusual. Here a standard is supposed to manifest and reinforce a "state of the art" and, in doing so, stabilize a certain direction of technical development. With the gauge standard of 1896 a firm basis for an efficient, integrated trans-American railway network was achieved; at the same time, the development of cars capable of running on differently calibrated tracks was rendered obsolete.

In his study of the development of "networks of power," Thomas Hughes (1983) devotes his attention to another one of the great construction projects of the last century. The electric power system entered its formative period around 1880. In contrast to Chandler, who focuses on the ascent of large, hierarchically integrated enterprises,

Hughes concentrates on technical development. Comparing the United States, England, and Germany over a period of 50 years, he tries to discern patterns in the evolution of "large technological systems," which include both technical and social components, such as power plants, manufacturing firms, utility companies, engineers, managers, and financiers (see chapter 1 above). Like the railways, electric power systems display characteristics of interconnectedness. Thus changes in one component have an impact on other components of the system (ibid., p. 6).

In Hughes's conception, electric lighting systems were built up by "inventor-entrepreneurs," such as Thomas Edison, who directed their efforts at designing complete systems comprising incandescent lamps, generators, mains, feeders, and the parallel distribution system. Inventors who concentrated on single components of a larger system were ignored (ibid., p. 27). With proprietary systems protected by patent rights, and without "open" compatibility standards, there was little opportunity for isolated components to become integrated into comprehensive systems during their generative phase. Although a "variety of options characterized the early years of the electric utility industry" (Hirsh 1989, p. 16), Edison's system soon became the incumbent technical formation around which complex manufacturing and financial interests evolved. The system's core technology consisted of power generators producing direct-current (DC) electricity.

At the end of the 1880s, however, Edison met serious competition—especially from the Westinghouse and Thomson-Houston companies, which launched the more recently developed alternating-current (AC) system. A "battle of the currents" or "battle of the systems" started, with low-voltage DC competing against single-phase AC for the incandescent lighting market. "Professional societies held debates concerning the merits of each system and station managers and engineers filled the technical journals with articles proclaiming the technical and economic advantages of one system compared to the other." (Hughes 1983, p. 106) The invention and subsequent improvement of transformers and polyphase motors in the early 1890s permitted the AC system to match the capacity of the DC system to supply both power and light. Although the DC utilities in densely populated areas continued to expand for some time, AC's greater suitability for long-distance transmission made it the ultimate winner (Hirsh 1989, pp. 16–21; Hughes 1983, pp. 106–139).[2]

That the battle ended without a dramatic destruction of one system by the other owed much to the invention of the rotary converter, a prototype of a gateway technology that helps to overcome incompatibility between systems ex post (David and Bunn 1988, p. 181). The rotary converter combined an AC induction motor with a DC dynamo to allow high-voltage AC transmission lines to be connected to DC distribution networks. As a two-way converter, it enabled complementary use of already-installed AC and DC technology, as well as substitution of AC for DC motors and vice versa. Thus, the eventual victory of polyphase AC, which in the 1920s became the de facto standard for the large electricity supply systems (the "universal system") in the United States and in Europe, in no way resulted from the technical prerequisite of system integration. It was, rather, a consequence of economic pressures operating to achieve economies of scale, of a related series of mergers among the electric utility companies, and of tighter coordination between the utilities and the manufacturers (David and Bunn 1988, p. 191; Hughes 1983, pp. 126–128; David 1992).

In addition, negotiated standards facilitated this development as they reduced the variety of frequencies that had evolved in the early years of alternating current. By the time the United States had completed railroad gauge standardization, nine classes of cycles, ranging from 25 per second to $133\frac{1}{3}$ per second, were being used with AC. At times during the attempts to produce a frequency standard, railroad gauge standardization was cited as a paradigm for similar considerations. Although a common aversion to having too many frequency "standards" existed, it was still difficult to reach an agreement because, as Hughes (1983, p. 128) points out, no single solution stood out as being technically superior: "Designers and engineers had chosen the frequency that was optimum for the particular set of characteristics created by the coupling of incandescent lamps, transformers, arc lighting, induction motors, synchronous converters, or other apparatus."

Standardization was initiated by the engineers of the Westinghouse Company, who were constrained in their choice. Owing to internal technical commitments, they preferred a dual standard: 60 cycles for lighting and 30 cycles for power transmission. After negotiations with other manufacturers and utility companies, a compromise that was nobody's optimum was reached: both 60 cycles and 25 cycles would be used. Subsequent technical developments in the area of rotary converters and turbine generators led to a conversion of the US market to the 60-cycle standard (David and Bunn 1988, pp. 191–197; Hughes 1983, pp. 127–129). Thus a negotiated compatibility standard comprising two options, one of which was ultimately stabilized in the market, contributed to the victory of alternating current over direct current. The reduction of the initially great variety of AC frequencies through standardization helped this technology recover from dispersion and significantly improved its competitive position.

The early decades of the American telephone system give an impression of the role of standards in the formative period of this technology and also indicate the tendency in this industry to internalize technical coordination. In many countries the latter tendency resulted in territorial monopolies in the hands of public administrations, or in regulated private companies serving as single network operators.

After Alexander Graham Bell discovered how to transmit and receive the sound of the human voice over wire and (more important) managed to get his invention protected by patent laws, he could start developing the business free from competition. Lacking capital, the Bell Telephone Company franchised its monopoly service on the basis of temporary licensing agreements, taking in return fees and, later, stock equity from the local telephone companies (Schlesinger et al. 1987, pp. 1–24). The Bell Telephone Company's decision to lease rather than sell telephones, and to produce all the telephones itself, was aimed at preventing patent infringement. Interestingly, it also secured a uniform technical standard. The local licensees raised capital, built up the network, and supplied the rather expensive exchange technology, and this led to a decentralized system of local entrepreneurs. They initially also provided technological inputs, which resulted in considerable improvements in switching equipment (Galambos 1988).

Local heterogeneity of switchboards and exchange technology became an issue when the invention of "loading" by engineers of the American Bell Company made long-distance telephony possible (Wasserman 1985). Before 1880, the maximum transmission distance on telephone cables was 30 miles. The right to build long-distance telephone lines guided the company's development efforts and general business strategy on the eve of its patents' expiry. As early as 1879, Bell's renowned general manager Theodore Vail announced the need to integrate the telephone business into a national system. Later, the company's strategic goal was embodied in the phrase "one system, one policy, universal service." In 1885, the American Telephone and Telegraph Company was set up to be the long-distance (toll) service subsidiary of the Bell Company.[3] Three years earlier, Vail had already arranged the purchase of the important electrical equipment manufacturer Western Electric, which later became the major supplier of the phones and equipment of the Bell companies (Smith 1985).

The introduction of long-distance telephony revealed the technical obstacles impeding the construction of a network that interconnected the local systems of the licensed operating companies. Western Electric had constructed switchboards according to the specific needs of its customers, which varied greatly (ibid., pp. 126–127). As long as they did not violate licensing agreements and other contracts with American Bell, the local affiliates could not be forced to adopt any standard designed to integrate the whole system. Moreover, no such a standard existed at the time (Garnet 1985, pp. 74–89). The uncoordinated development of local standards turned out to be a serious constraint on the efficient operation of an integrated long-distance and local telephone network.

During the 1880s, two developments exerted pressure on the local operating companies to collaborate on standardization. On the one hand, AT&T was striving to realize its hopes of a highly profitable long-distance business through the attainment of compatibility. On the other hand, the growing complexity of the telephone networks, especially in the larger cities, required a complex switching technology that could be produced at reasonable costs only when a certain degree of standardization was achieved. In 1887 a 10,000-line switchboard planned for the New York City exchange was estimated to cost about $800,000—the price of an average ocean-going steamship (Smith 1985, p. 128). The early attempts of some of the local companies and of American Bell, Western Electric, and (later) AT&T to share knowledge about the technical problems of the switchboards can be regarded as a reaction to this pressure. These cooperative moves were formalized in the National Telephone Exchange Association (NTEA), which was set up in 1880.[4] American Bell also sponsored specialized conferences on switching and cable development. In the late 1880s, the NTEA was officially divided into two specialized subcommittees, one dealing with switchboard systems and one with cable transmission (ibid., pp. 126–130; Garnet 1985, pp. 97–99). The specifications these committees adopted "were arrived at by consensus and were frequently adjusted for local conditions"; compliance "continued to be of a voluntary nature" (Garnet 1985, p. 98). Compatibility standardization included quality standards, since interoperability often depended on relatively high levels of quality.[5] Metallic circuits made of the high-quality copper wire required for long-distance transmission, for example, could not be used in conjunction with switchboards designed for the iron wire often installed in local networks (Smith 1985, pp. 130–131).

The need to standardize and to orient standardization predominantly toward the requirements of long-distance transmission and switching increased with the rising competitive pressure on the Bell units from independent telephone companies—a great many of which were set up after 1894, when the original Bell patents expired (Fischer 1987; Fischer 1992, pp. 86–121; Barnett and Carroll 1987; Barnett 1990). The ability to offer long-distance calls and to exclude the independent competitors from the use of these lines yielded the Bell system a decisive competitive advantage over the independents. Around 1900 it became clear, however, that neither the Cable Committee nor the Switchboard Committee had succeeded in promoting standardization of local facilities. On the contrary, AT&T's engineers' "longstanding effort to establish universal standards was in danger of becoming a victim of competition" (Garnet 1985, p. 120). Because competition had driven down the price of the telephone service, the local Bell companies had to cut costs, and they proceeded to do this by abandoning construction standards advocated by the committees. This development, which hampered interoperation in a situation where it was urgently needed, exemplifies the typical problem of committee standardization: even when a standard is unanimously approved, its implementation remains uncertain. One can also see that, once several divergent standards (like those of the local Bell companies) have emerged, it is difficult to migrate to a single standard, even a "superior" one.

The step-by-step transformation of the Bell enterprise from a simple patent franchiser that also arranged for the production and distribution of telephone equipment into the vertically and horizontally integrated AT&T Corporation, with its national sphere of influence and its worldwide significance, cannot be adequately explained by sole reference to the technical standardization requirements involved in building up a truly integrated American telephone network. Hierarchical coordination, however, did safeguard compliance with standards.[6] In a period in which standardization was often synonymous with achieving uniformity, the comprehensive vertically and horizontally integrated telephone corporation achieved a still-unparalleled degree of integrity and homogeneity in its technical system.

These three examples illustrate the general interplay of standards and technology in a systems context, rather than the effects of a specific standard on the design of a specific technical artifact. Standards result from the intricate interactions of company business strategies, standards committee activities, government interventions, and processes of market diffusion, and they are rooted in the perceived technical requirements for developing, manufacturing, operating or using devices that are meant to interwork with others. This latter view of technical requirements might suggest a harmonious picture of standardization. In fact, as the examples show, negotiations in committees were often contentious, and thwarted efforts to agree on a standard were the order of the day. Occasionally, powerful companies obstructed negotiations on a standard in order to eliminate potential competitors and subsequently pushed their proprietary solutions in the market. Many of those solutions found widespread acceptance and turned into de facto market standards on which the interoperation of different components manufactured by a large number of companies relied.

The same combination of variables influencing standards also channels their effects on the development of a technology. When a standard is supported by powerful vendors, by governments, or by standards committees, and when it fits into an already existing system of standards, it is likely to shape the future development of a technology. Whereas some standards serve mainly to manifest the "state of the art," others are meant to influence new technical developments. Standards are embodied in the features and functions of technical artifacts, but they originate at a purely cognitive level. For that reason, we now turn to the relationship between standards and technical knowledge.

Standards and Technical Knowledge

In order to elaborate on the relationship between standards and technical development more thoroughly, it is useful to draw a distinction between the *product aspect* and the *knowledge aspect* of a standard (David 1987). Most coordinative standards need to become incorporated in a product to affect technical development. In the case of market standardization, standards exist only if they are integrated into products.

Nevertheless, standards remain rules according to which products must be assembled or used when they are designed to interoperate in a systems context. As rules, standards are constitutive elements of engineering knowledge.

For example, as we discussed above, the two rails of a track must be installed with a fixed (standardized) gauge that is equal to the fixed distance of the wheels on the axles of the engine and the other carriages of a train. When a certain gauge has diffused in a market as a rule, its approval as a standard by a committee serves mainly to officially declare the rule effective. This provides useful information for newcomers to the market, who can then design their products in compliance with the standard and offer them to the railway market. When different gauges coexist, a government may impose a standard or a committee may adopt a prescriptive standard in order to effect a future convergence of technical development with that standard. Information about a standard does not necessarily prompt compliance. As a part of a business strategy, or as a result of new experience and new technical knowledge, technical specifications that differ sharply from those proposed by a standard may be employed.

When we look at modern telecommunications systems or at other high-technology products that are "systemic in nature" (Rosenberg 1992, p. 80), we discover compatibility needs with respect to many features. In telecommunications, transmission frequencies, protocols, modulation procedures, codes, addressing, and signaling conventions are all cases in point. The development of such systems or of parts thereof requires coordinated efforts within and between firms. The origin of systems engineering as a discipline combining different areas of technical knowledge with a collection of methodologies is frequently attributed to the need for an extensive understanding of systems with high compatibility requirements, such as the railway or the telephone (Chandler 1977; Rosenberg 1992; Laudan 1984; Constant 1984).

The development of complex systems, however, relies as much on classical engineering knowledge as on what Henderson and Clark (1990) call "architectural knowledge" (in contrast with "component knowledge"). Architectural knowledge is the "knowledge about the ways in which the components are integrated and linked together into a coherent whole": "For example, a room fan's major components include the blade, the motor that drives it, the blade guard, the control system, and the mechanical housing. The overall architecture of the product lays out how the components will work together. Taken together, a fan's architecture and its components create a system for moving air. . . . " (ibid., p. 11; see also Abernathy and Clark 1988) Component knowledge, on the other hand, is "local and focused" knowledge about elements of a larger system. From an architectural perspective, it is not necessary to transfer the full range of knowledge of the internal working of components across component boundaries (Henderson 1992, p. 127). The ongoing substitution of digital for analog switching technology in public telecommunications networks, for example, does not require full knowledge of the internal working of the new "intelligent" exchanges, which are more similar to process-control computers than to the old analog electro-mechanical rotary or crossbar switches (Chapuis and Joel 1990). All that is needed is for the interface specifications of the switching part and the transmission part of the network to be left unaffected so that analog lines too can be switched by digital exchanges. Interface standards guiding analog-to-digital conversion secure the interoperability of old and new components in the network.[7]

Compatibility standards contain architectural knowledge by specifying relational properties of technical artifacts. At the same time, these rules "harden" the knowledge base of a technical system with their inherent compliance expectations.

The sociology of technology uses several partly equivalent knowledge-related concepts to explain technical development. These concepts are unanimous in suggesting that common knowledge triggers cognitive coordination of autonomous but interdependent actors. The most prominent version of such sets of "notional possibilities" (van den Belt and Rip 1987, p. 136) is the "technological paradigm." As an analogous extension of Thomas Kuhn's idea of the scientific paradigm, Dosi (1982, p. 152) defines the technological paradigm as "a 'model' and a 'pattern' of solution of selected technological problems, based on selected principles derived from natural sciences and on selected material technologies." Paradigms, according to this definition, do not determine developments, but provide a relatively coherent source of variation and mutation (Dosi and Orsenigo 1988, p. 17).

Rather similar to the notion of a paradigm is Nelson and Winter's (1977, p. 57) notion of a "technological regime" that gives "a broadly defined way of doing things," focuses the attention of engineers on a certain direction of activities, and relates "to the technicians' beliefs of what is feasible or at least worth attempting." Nelson and Winter use the term "natural trajectory" to designate the path within a technological regime in which technical development is likely to proceed. We have already cited MacKenzie's (1990) critique of this phrase, which does indeed suggest technological determinism. Nelson and Winter emphasize the role of professional standards and consensus-based standardization activities, which, they suggest, constitute a significant part of the "nonmarket selective environment" of technology. These standards and activities are said to shape development because engineers and technicians, who are part of the environment, generally prefer certain codified technical specifications to other, less formalized ones (Nelson and Winter 1977, 1982). One of Nelson and Winter's examples is the Douglas DC3, introduced in the 1930s, which defined metal skin, low wings, and piston engines as the technological regime for aircraft for the following two decades. This regime directed the technicians' efforts into a trajectory toward improving engines and toward making airplanes larger and more efficient within the given standardized framework.

Another related, but less abstract, concept is that of "dominant design," which suggests that the emergence of a new technology coincides with a great deal of experimentation and with little agreement, or even with confusion, about what the major subsystems or components of a system should be or how they should be assembled (Henderson and Clark 1990, p. 14). Intrafirm and interfirm uncertainty, search behavior, and learning prevail in the early stages of innovation, but later the innovation shifts from a "fluid" state to one that is highly "specific" and "rigid" (Clark 1985, p. 235). A dominant design is established that "incorporates a range of basic choices about the design that are not revisited in subsequent design" (Henderson and Clark 1990, p. 14). The evolution of the automobile provides an illustration. In the early days of the industry, cars were built with wooden or metal bodies, with steering wheels or tillers, and with gasoline, steam, or electric engines. However, once the gasoline engine had been accepted as a core component of the dominant design, which emerged as a de facto standard, it was not revisited in every subsequent design, and changes to this component were only incremental (Henderson and Clark 1990, p. 14; Anderson and Tushman 1990, p. 613).The role of standards is explicitly addressed in this concept. Dominant designs permit standardization of interchangeable parts, and the standards are crucial elements of a stabilized stock of architectural knowledge (Anderson and Tushman 1990, p. 614). The establishment of a dominant design is by no means solely an engineering issue. Every single standard in that design results from complicated interactions of "organizational and collective forces," including firms, strategic alliances, industry associations, and regulatory agencies (ibid., p. 628).[8]

The concepts delineated above have usually been applied to detecting sources of innovation and change rather than to explaining stability or coordinated development of technology.[9] However, in evolutionary models of technological change, the emergence of paradigms, regimes, and dominant designs indicates the beginning of a period of consolidation after an era of discontinuity and an era of ferment.[10] In Tushman and Rosenkopf's (1992) four-stage cycle of technical development, these two eras precede the era of dominant design, which is followed by a period of incremental change. Standards evolve in the era of dominant design. They reduce technical uncertainty and permit system-wide compatibility and integration for producers, at the same time lessening product-class confusion and promising decreases in prices for consumers (ibid., p. 321). As a result, technical development may even become locked into a path-dependent developmental process in which individual rationality "forces" incremental and continuous, rather than radical, changes to a technology (Arthur 1989).

In his theoretical deliberations on the development of large technological systems, Hughes (1987) defines these constrained dynamics of path-dependent developments as "momentum." Momentum is not only a technical category; it also applies to the social setting within which technology is embedded. The structures of technical and social systems are loosely coupled. They stabilize and reinforce each other, and in this context standards are generated and develop repercussions (Schmidt and Werle 1994; Schneider 1992). Tushman and Rosenkopf (1992, p. 323) even argue that, in the era of incremental change, social structures "arise" which reinforce this period of incremental, "order creating" technical change. Indeed, when large technical systems have reached "maturity," displaying a considerable degree of technical, social, and cognitive integration (Hughes 1987), they appear as the infrastructural backbone not only of an industry but also of a societal "functional subsystem" (Mayntz 1993c; Werle 1990, pp. 19–40). At this general level, however, the specific contribution of standards to the development and functioning of such large complexes can scarcely be isolated from other influences. Here technical knowledge is just one of a number of factors shaping the system.

Similar considerations apply to the generation of standards. Existing technical systems and the concomitant knowledge base often seem to force standard setting to merely execute a technical logic. An incremental variation of an established and moderately complex system may necessitate only a slight modification of an existing standard or the addition of an option to it. One solution will appear as by far the best. Pure technical considerations seem to dominate other considerations and perspectives. These enter the stage, however, when the objects to be designed are novel and/or highly complex. Telecommunications is an example of a highly complex open system whose coordination through standards does not follow a pure, internally consistent technical logic. Such systems are characterized, as Tushman and Rosenkopf (1992, p. 333) put it, by multiple, partly incompatible "dimensions of merit." Thus, not only will technical professionals disagree on the appropriate standards; in addition, interest groups (economic and political) will engage in a struggle over standards and other aspects of the system's design, which is only weakly constrained by technical boundaries.

Prefiguring Tushman and Rosenkopf's proposition concerning open systems, Aitken (1985, p. 22) points out that standardization takes place in an increasingly institutionalized "'grey area' where science, technology, business, and government meet, overlap, and interpenetrate, where resources and information flow between the systems." The incompatibility of different dimensions of merit can thus be regarded as a result of the different perspectives actors hold (Burns and Dietz 1992). Science, technology, politics, and the economy have been organized around different values, have rewarded different modes of action, and have provided different kinds of signals to guide decisions.[11] Actors holding these perspectives "literally use different vocabularies, speak in different languages, and respond to different cues" (Aitken 1985, p. 17). This heterogeneity can hamper standardization and thus hamper coordination of technical development. In the case of committee standardization, an organizational reaction to this problem is to privilege the technical perspective institutionally to the disadvantage of all others (Schmidt and Werle 1993). Even then the internal technical discourse cannot be regarded as homogeneous. High technologies such as telecommunications "do not emerge from a single technology such as the computer or the microchip but rather from a coupling of material, energy and information flows" (Rammert 1992, p. 197). Here different stocks of knowledge or even different technological paradigms may clash and render standardization a highly intricate affair.

Those trying to coordinate the development of large technical systems through standardization are therefore confronted with two obstacles. First, the technical complexity imposes a consistency requirement that, as a minimum, demands compatibility of standards for different but interdependent components of the system. Second, the social heterogeneity of the actors and the interests requires that some compatibility of perspectives be secured. The social mechanisms for securing compatibility of perspectives in standardization will be analyzed in chapter 3. As we focus on institutionalized committee standardization in this chapter, these mechanisms will become at least partly transparent.

The common analytical approach to the first problem of technical complexity is to decompose technical tasks, creating a coherent, typically hierarchical cognitive order (Simon 1962). Using a "layering" technique, complex tasks are partitioned into smaller, simpler tasks, each performing some well-defined function and using only a well-defined set of services provided by the layer immediately below. Layering provides a methodology for defining a system's architecture so that evolution and growth can take place in an incremental fashion. Rules of procedure (protocols) determine the interaction between two adjacent layers. Layering is a modular approach that minimizes interdependence among the components of an architecture (Konangi and Dhas 1985; Bowers and Connell 1985). In telecommunications (more specific, in computer communications), different, mostly proprietary architectural models have been developed in an attempt to handle the complexity of communication networks. Such models also help to systematize standardization needs. In the early 1980s, the International Organization for Standardization ("ISO") and the Comité Consultatif International Télégraphique et Téléphonique (CCITT) adopted the famous nonproprietary "Open System Interconnection" (OSI) Reference Model, which provides a seven-layer protocol architecture that is supposed to help locate and integrate standardization activities for technical communication networks and, through this mode of cognitive coordination, to combat the disintegrative effects of uncoordinated standardization activities (Day and Zimmermann 1983).[12]

The last point, to which we will return in our case studies, once again demonstrates the close relationship between technical knowledge and standards and stresses their coordinative function in the development of large and complex systems. Unlike knowledge, standards quite often "are established by the visible hand of a few powerful organizations," whereas, like knowledge, their coordinative influence often unfolds as an "invisible hand" constraining the technological competition of a multitude of organizations (Tushman and Rosenkopf 1992, p. 324).

Standardization, however, is not always an appropriate means of achieving coordination. Its relative significance with respect to other means of coordinating technical developments depends on the specific type of technical interdependence. In this respect, a useful distinction of three types of interdependence "stemming from technological requirements within organizations" (Thompson 1967, p. 64) is provided by James Thompson's seminal analysis of "organizations in action" from an organizational perspective (ibid.). For each type of interdependence, a specific means of coordination is most suitable as a means of achieving what would be termed compatibility in a technical context. In the case of "reciprocal" interdependence, where "the outputs of each become inputs for the others," coordination is achieved by "mutual adjustment." When interdependence takes a "serial" form, with the output of one unit making up the input of another, a "plan" most appropriately governs the actions. Only with "pooled" interdependence, where each unit renders a discrete contribution to the whole and each is supported by the whole, may "standardization" achieve coordination.[13] Yet this also requires "that the set of rules be internally consistent" (Thompson 1967, pp. 54–57). In technical systems this consistency requirement has been proven to multiply with the increasing complexity of the systems.

As the examples of electricity and railways showed, standards are not the only means of achieving compatibility. Gateways, adapters, converters, or similar technologies can overcome preexisting incompatibilities that might otherwise prevent some components from being combined into an integrated system (David and Bunn 1988; Farrell and Saloner 1992). They transform or convert information from one technical component to make it "intelligible" to another one. Thus they serve to manage existing technical diversity without replacing it. Standards, however, do not lead to complete uniformity either, because only the "connection points" of technical components must be "standard" if they are to interoperate. All other parts of the artifacts can be diversified.

Thompson tends to conceive of interdependence and complexity as inherent features of technology. From a similar point of view, Perrow (1984) describes organizations and technical systems as "loosely or tightly coupled," and as displaying high or low "interactive complexity," albeit more by nature than by design. A closer look at the role of standards in technical development, however, suggests that not only does the degree of coupling determine the need for standards, but the availability of standards affects coupling. Interface standardization, for example, permits the design of modular systems consisting of interacting components which are nevertheless loosely coupled. Thus, the degree of coupling would be higher, and more system-specific dedicated elements would have to be assembled, if no standardized modules were available. (See also G. Thompson 1954.) Yet, whether standards are adopted is a result of social rather than technical factors, as is the form of the standards.

The discussion of the relationship between technical knowledge and standards has shown that, in the literature, standards are often regarded as specific codified knowledge elements with high normative significance. Standards emerge when technical knowledge has entered the stage of consolidation. They serve to reinforce this state. This does not mean, however, that standards embody a natural, a logical, or even a consensual end stage of development. Rather, they result—precisely because of their presumed stabilizing effects—from contentious processes of standardization, involving actors from different social groups holding diverging values and interests and different knowledge bases.

In summarizing the role and the impact of standards in technical development, we stress the following points:

- The history of industrial organization and the history of technology provide many examples of the importance of technical standards. The availability of standards not only helps firms to achieve economies of scale; it also facilitates the building of comprehensive, integrated technical systems. In the latter instance, the coordinative function of standards becomes visible. The early decades of the classic infrastructural railway, telephone, and electric power systems bear witness to this.
- Standards are elements of technical knowledge. They make particular reference to the architecture of systems—that is, to how the components of a system are related to one another. As "hardened" or officially codified knowledge, they entail a mixture of normative and cognitive expectations to comply.
- Standards are established when innovative technology enters a phase of consolidation and incremental change. As cognitive guideposts, they secure consolidation by channelling developmental opportunities and at the same time eliminating or frustrating attempts at "radical" change. Thus, standards influence technology by coordinating its

production, its operation, and its use, and they are themselves shaped by technology.

- Standards issued by committees directly influence technology only when they are complied with. Here implementation can be distinguished analytically from standard setting. As parts of the stock of technical knowledge, however, committee standards can also affect technical developments when they are not implemented in products.

3

Institutionalized and Alternative Modes of International Standardization

In chapters 1 and 2 we sketched the theoretical baselines of our approach, introduced historical examples to illustrate the coordinating role of standards in the process of technical development, and identified a specific type of technical knowledge mixing cognitive and normative elements evolving as standards in the phase of consolidation of technology. We now turn to the modes of standardization and to the production of standards. In parallel to the varying motives, perspectives, and interests of actors to get involved in standardization, we find different modes of standardization coexisting. The historical examples we have discussed indicate that no specific mode ever prevailed. Rather, the modes were often linked, or they were "used" in a sequential order that depended on historical circumstances.

Analytically, three modes can be distinguished. *Governments* may impose mandatory standards hierarchically as binding solutions to coordination problems. In *markets,* standards can emerge as de facto or industry standards. Their diffusion is based on market leadership or on frequency-dependent bandwagon and imitation processes, in which the number of actors attracted by a standard increases with the number of those who have already adopted the standard. A growing number of standards are agreed on in *committees,* which, whether explicitly institutionalized or less formal in nature, are dedicated to the joint elaboration of standards. Typically, these standards are voluntary consensus standards (National Research Council 1995; David and Greenstein 1990; Farrell and Saloner 1988; Blankart and Knieps 1993a).

In telecommunications, on which our empirical research is concentrated, pure market processes of standardization have been of minor importance. For a long time, owing to the high infrastructural (including military) significance of this industry, national governments played a strong role in telecommunications. Worldwide, we find a rather homogeneous "organizational paradigm" (Schneider 1991, p. 25), which remained stable until recently (Noam 1992, pp. 3–65; Bickers 1991). A single network operator—either owned or tightly regulated by the state—provided all telecommunications services. Whether the operator was a public postal, telephone, and telegraph (PTT) administration (as in most countries) or a private corporation (such as AT&T in the United States) was of minor significance with respect to corporate behavior. The close affiliation of the network operator to the government resulted in a hierarchical quasi-imposition of technical standards as regulations, although most standards were supposed to accomplish coordinative functions (which generally do not require regulatory authority). The territorial confinement of these national standards resulted in either deliberate or evolutionary diversity in the development of technology in different countries. Thus, functional equivalence of national technical systems by no means implied uniform or highly similar configuration of these systems.

In telecommunications, it has never been possible to establish some kind of supranational hierarchy to coordinate technical development on a global scale. Considerable variance of national systems has always prevailed. Whenever international coordination was regarded as necessary, it was achieved on the basis of bilateral agreements between PTTs. The arrangements included technical and operational specifications for the network interconnection points as well as commercial regulations such as accounting principles, rate sharing, prohibition of bypass practices, and reciprocal monopoly protection (Aronson and Cowhey 1988). The International Telecommunication Union (ITU), founded in 1865 as one of the earliest international organizations, served as the institutional umbrella under which principles for bilateral and multilateral transborder telecommunications were established (Genschel and Werle 1993). Rather than provide the institutional basis for a transnational hierarchy coordinating international telecommunications, the ITU in effect reinforced national monopolies (Cowhey 1990; Rutkowski 1991).

The institutional structure of nationally segmented coordination complemented by bilateral agreements has also prevented the international coordination gap from being filled by market standards. Historically, national markets and the prevailing standards were either directly regulated by governments or indirectly controlled by the PTTs on the basis of their procurement power. In contrast with the unregulated computer field, where the technical specifications of one powerful vendor (IBM) became de facto standards in a global market, such a hegemonic structure has been missing in telecommunications. Nor has there ever been a highly competitive world market in which standards have evolved. Where neither industry leaders nor truly free markets nor hierarchies could achieve thorough transnational coordination, cooperative committee modes of standardization have provided a solution. Examples of this mode of coordination are found in many other sectors, such as consumer electronics and banking, where industry coalitions, consortia, forums, and interorganizational task forces are formed to prepare and agree on standards (Hemenway 1975; Rosen et al. 1988, pp. 133–138; Stalk and Hout 1990, pp. 133–148; Kubicek and Seeger 1992). Some of these groups are transformed into stable institutionalized standardization committees; others disappear once a particular task has been finished.

Since in international telecommunications neither markets nor hierarchies have provided the standards for transborder communication, the ITU's standardization branch, the Comité Consultatif International Télégraphique et Téléphonique (CCITT), has filled the gap. In standardization it operates in a collaborative way similar to, but more formalized than, the operations of many consortia and forums in other sectors. Albeit traditionally the most important one, the ITU is only one of several organizations on the international landscape of institutionalized committee standardization which are involved in setting standards relevant for international telecommunications. Another top-level unit is the International Organization for Standardization ("ISO"), with its sister organization, the International Electrotechnical Commission (IEC). The ISO's and the IEC's standardization activities are much broader in scope than those of the ITU, and under their roof we also find many committees contributing to the stock of technical rules which are directly relevant for telecommunications. Recently a number of more informal consortia and forums have been set up in telecommunications. Also, larger firms have tended to copy the traditional model of computer and software vendors, who used to try to bypass committee standardization by strategically introducing their own standards directly into the international market.

We shall now examine the significance of standardization at the international level and, in this context, discuss concepts and explanations that complement the empirical findings. Starting with a look at the landscape of organizations producing international technical rules, we shall observe processes of expansion and differentiation at work. We shall then turn to the production of standards, which is characterized by similar tendencies. Already the quantity of standards indicates their growing importance.

Standards Organizations

Which organizations provide standards? In a "reference manual" of international telecommunication standards organizations, Andrew Macpherson (1990) portrays about eighty internationally and regionally relevant units. Even though the main focus of his book is on telecommunications, Macpherson cannot ignore organizations issuing standards in the area of computing and office automation, because the boundaries between data processing and telecommunications have become blurred over the last few decades. Nevertheless, despite the variety of organizations that abound, Macpherson points to the continuing dominance of the ITU/CCITT and the ISO/IEC, whose positions as the main international organizations involved in technical standardization can hardly be contested.

The International Organization for Standardization was created in 1947 by representatives from 25 national standards bodies. Today it has about ninety members, each one the standards body that is most representative of standardization in its country. The ISO's work is carried out in some 2500 technical committees and subcommittees, working groups, and ad hoc study groups. The participants in these bodies are business firms, research organizations, consultants, and sometimes government officials or representatives of users. All must be members of, or accredited by, their national standardization organization. The object of the ISO is "to promote the development of standardization and related activities in the world with a view to facilitating international exchange of goods and services, and developing cooperation in intellectual, scientific, technological and economic activities" (Macpherson 1990, p. 96). This includes harmonizing standards worldwide, developing and issuing international standards, taking action for their worldwide implementation, arranging for the exchange of information between ISO member organizations and technical committees, and cooperating with other international organizations on all related matters.

In the context of our study, the ISO's most significant subunit is Technical Committee 97 (TC 97), which was created in 1960. Standardization of information-processing systems "from the perspective of free-standing computer systems, office machinery and data communication implementation" constitutes the scope of TC 97 (Wallenstein 1990, p. 87). Subcommittee 16 of TC 97 has taken responsibility for the Open System Interconnection (OSI) reference model. In 1984 this seven-layer protocol architecture was adopted by both the ISO and the CCITT. As a frame of reference, it has achieved considerable significance in the realms of telematics standards and computer communication (Schmidt and Werle 1994).

A little less overt in its activities than the ISO is the International Electrotechnical Commission, founded in 1904. The IEC's standardization work covers almost all spheres of electrotechnology, including power, electronics, telecommunications, and nuclear energy. As with the ISO, national representation is the constitutive basis of the IEC. Each of the forty or so participating countries has one vote on any decision. The national committees are explicitly required "to be as representative as possible of all electrical interests in the country concerned: manufacturers, users, governmental authorities, and teaching and professional bodies" (Macpherson 1990, p. 112). Technical work is charged to about eighty technical committees and numerous subcommittees.[1]

The ISO and the IEC have always had a close working relationship, and this has resulted in a perception of them as "twins." They provide world standards in all areas outside the jurisdiction of the ITU/CCITT. In a period of rapid technical change, in which it is difficult to clearly separate organizational domains without any overlap, jurisdictional conflicts between the ISO and the IEC have been prevented by institutionalizing a collaborative forum in the area of information technology. A concomitant arrangement between the two organizations was concluded in 1987. The ISO's Technical Committee 97 (Information Processing Systems) was merged with the IEC's Technical Committee 83 (Information Technology Equipment) and Subcommittee 47B (Microprocessor Systems) into the Joint ISO/IEC Technical Committee (JTC1 Information Technology), the central task of which was coordinating the definitions of basic and generic information technology standards. Both the ISO and the IEC have transferred to JTC1 tasks and competencies concerning standardization of communication processes in computer networks. The work of this "super committee" began in seventeen large subcommittees and more than sixty active working groups (Eicher 1990). JTC1's Subcommittee 6 (responsible for

Figure 3.1
The ITU's headquarters in Geneva. Courtesy of A. de Ferron, ITU.

telecommunications and information exchange between systems) and Subcommittee 21 (responsible for interconnection of open systems, data management, and open distributed processing) have issued many standards in the field of telecommunications.

Formally, the ISO and the IEC are nongovernmental organizations. In this respect they differ from the International Telecommunication Union, which is an intergovernmental United Nations treaty organization set up under a convention and based in Geneva (figure 3.1). In February 1996 the ITU reported a membership of 185 countries. The scope of its jurisdiction is substantially broader than that of a dedicated standardization organization. The ITU is responsible for the regulation and planning of telecommunications worldwide, for the establishment of equipment and systems operating standards, for the coordination and dissemination of information required for the planning and operation of telecommunications services, and, within the United Nations system, for promoting and contributing to the development of telecommunications and related infrastructures (Codding and Rutkowski 1982; Codding 1991b; Codding and Gallegos 1991).

Until 1992 the ITU's standards-making organs were the Comité Consultatif International des Radiocommunication (CCIR) and the

Comité Consultatif International Télégraphique et Téléphonique. The latter was set up in 1956, when international technical standards still played a minor role in the operation of telecommunications systems (Genschel and Werle 1993). The CCIR devoted only about one-third of its time to the making of standards, whereas the CCITT concentrated nine-tenths of its time on this activity (Codding and Gallegos 1991). According to the ITU's convention, the CCITT's duties were "to study technical, operating and tariff questions and to issue recommendations on them with a view to standardizing telecommunications on a worldwide basis" (Constitution, Art. 13 II, in ITU 1990: 16). As a result of the reorganization of the ITU in 1993, standardization is now concentrated in the Telecommunication Standardization Sector (ITU-T) as one of three sectors of the union, next to the Radiocommunication Sector and the Telecommunication Development Sector. Radiocommunication standardization activities were also transferred from the former CCIR to the new ITU-T at this time (Codding 1991a).

Standardization in the Telecommunication Standardization Sector (and, before that, in the CCITT, on which we focus, since our interest does not just lie with current standardization processes) is not significantly different from standardization in the ISO and the IEC. Rather similar to the ISO's General Assembly, a quadrennial Plenary Assembly with delegations from the member countries governs the CCITT's activities. To a great extent, both assemblies are concerned with overall policy decisions. Since the 1992 reform, one of the CCITT Plenary Assembly's main tasks—approving recommendations—has lost much of its relevance. However, the World Telecommunication Standardization Conference (WTSC), which has replaced the Plenary Assembly, has not changed the character of this arena of technical diplomacy very much. Traditionally, the PTTs or the dominant private operating companies have been appointed as delegates to the CCITT Plenary Assembly by the national government authorities, whereas in the ISO the national standardization bodies (e.g., the ANSI in the United States, the BSI in the United Kingdom, the DIN in Germany, or the AFNOR in France) directly represent their countries in the General Assembly. This indicates the more political character of the top level of the CCITT (part of an intergovernmental organization) in contrast to the top level of the ISO (an "ordinary" international organization). The CCITT's study groups, working parties, and rapporteur groups resemble the ISO's technical committees, subcommittees, and working groups. In the 1989–1992 study period, fifteen study groups carried out business related to standardization or to standards.

CCITT Study Group VII (responsible for data communications networks) and Study Group VIII (concerned with terminals for telematic services) played central roles in the standardization processes we focus on (X.400, facsimile, and interactive videotex). The domains of these and other CCITT study groups partly overlap with what the ISO and the IEC have been engaged in. The historic division of labor among the CCITT, the ISO, and the IEC has been eroded because a clear separation of technical domains has proved to be unfeasible as information processing and telecommunications rely to a considerable extent on the same basic technology. A grey area has evolved, triggering jurisdictional conflicts. As to their mutual relations, the ISO and the IEC were the first to relieve the tension by setting up Joint Committee JTC1 (Rankine 1990). In many other cases too, conflicts and competition have been avoided—initially through the largely informal cooperation characteristic of the early 1980s, which was increasingly formalized in subsequent years. It is especially in the areas of standardization relating to the OSI standardization architecture that collaboration of the CCITT/ITU-T and JTC1 has become a matter of routine (Genschel 1995, pp. 162–168).

The well-established top-level standardization organizations are complemented by some similarly designed units of regional significance. These were institutionalized within the three great blocs of the highly industrialized world (Western Europe, North America, and the Asia-Pacific region). Some observers interpret the appearance of the European Telecommunications Standards Institute (ETSI), the US Standards Committee for Telecommunications (ANSI T1), and the Japanese Telecommunications Technology Committee (TTC) in the area of telecommunications as a potentially serious rival to the CCITT and the ITU-T (Hawkins 1992).[2] Yet not one of them was founded with the express purpose of competing with the CCITT (Mazda 1992). In 1990 the three regional organizations and the CCITT decided at a conference in Fredericksburg, Virginia, to coordinate their activities and to find a work-sharing arrangement acceptable to all.

The international organizations, and some of the regional ones we have referred to thus far, typically rely on nationally based participation. Be it as parts of national delegations, as members of the national constituent standardization bodies, or on the basis of direct membership at the working level, the participants' influence is mediated by the principle of national representation. Thus, national interests as driving forces of *political* control of international standardization enjoy an institutionally legitimized status in this type of organization. That the procedures for voting on a draft standard presuppose the existence of a unitary national position appears chimerical on many an occasion.

The internationalization and globalization of telecommunications and data processing requires not so much national as organizational coordination of technical interdependencies and complementarities (Genschel 1995). The principle of national representation in international standardization is therefore increasingly regarded as an obstacle to the development of commercially appropriate committee standards. To overcome it, manufacturers, vendors, service providers and large business users especially have created—primarily out of *commercial* motives—"para-standardization" bodies of regional and global significance (OECD 1991, pp. 84–86). Some of these bodies are relatively old. The European Computer Manufacturers Association (ECMA) was established in 1961 with the objective of becoming active in standardization. The ECMA's membership is restricted to computer manufacturers engaged in Europe, though associate members from other parts of the world are also accepted when they have a general interest in the association's work: because they own manufacturing facilities in Europe, IBM, Xerox, Honeywell, and DEC were accepted as regular members.

The ECMA has a number of technical committees and task groups, in which all members—regardless of national basis—have full voting rights. Its standards are usually meant to complement the activities of the major international standardization organizations by filling perceived gaps, but it also works out proposals to be fed into the ongoing standardization work of these organizations. Owing to their broad backing and their high state of elaboration, these proposals often have significant impact (Macpherson 1990, p. 215). The ECMA has been particularly concerned with standardization in the computing realm, and therefore it has cooperated with the ISO rather than with the CCITT. Only with projects that include telecommunications in their remit (e.g., X.400) has this changed.

Although the European thrust in the ECMA was moderated very early on by opening the organization to global players with a base in Europe, the underlying commercial interest in neutralizing existing power differentials became apparent in other efforts to set up standardization committees. Examples are provided by several facets of the "open systems movement" (Rosario and Schmidt 1991). One is the European Workshop of Open Systems (EWOS), which was initiated by the Standards Promotion and Application Group (SPAG) and by others. SPAG, initially set up in 1983 by eight major manufacturers of information technology based in Europe, aims to facilitate the implementation of the base standards issued by the ISO and the CCITT through the specification of "functional profiles." In 1987 SPAG initiated the formation of EWOS, which was created by "the most representative European federations of technology suppliers and user organizations" (Macpherson 1990, p. 269; Genschel 1995), including CEN/CENELEC (the European counterpart to the ISO/IEC) and, as a member of the steering committee, DG IX (the General Directorate) of the Commission of the European Communities. Thus, commercial interests, especially those of the European vendors, were supported by political action at the European level.

Another rather prominent member of this movement has been the X/Open group, which was founded in 1984 by five European computer manufacturers (Gabel 1987).[3] Originally formed as a move to protect the European manufacturers' market shares against IBM, whose proprietary standards dominated most of the world, the group—which is similar to the ECMA—has grown to include all major international computer vendors. To give but a few examples: after DEC joined in 1986, Fujitsu and AT&T followed in 1987, IBM in 1988, and Hitachi and NEC in 1989 (Grindley 1990). Its general objective being to "level the playing field" by supporting or creating nonproprietary or at least nondiscriminatory "open standards," the group was dedicated to industry-wide cooperation. Hence, applications for admission by the large non-European computer manufacturers could not be refused. We need not speculate on all the reasons why the big manufacturers joined X/Open, which subsequently became vigorously involved in the promotion of an open computer operating system (UNIX) (Cool and Gabel 1992). Being involved in the committee work was certainly beneficial for them, as they could influence or at least monitor standardization.

Another source of committee standardization was the professional, scientific, and trade associations, as we discussed in the example of railway standardization in chapter 2. The most noteworthy association, active for more than 100 years, is the Institute of Electrical and Electronics Engineers (IEEE), whose core membership is in North America. It was formed in 1963 through a merger of the American Institute of Electrical Engineers (founded in 1884) and the Institute of Radio Engineers (Macpherson 1990, p. 161). By now it is a transnational society with about 300,000 individual members in more than 130 countries. The gathering, organizing, and disseminating of technical information is seen as pertinent to the IEEE's scientific, educational, and (above all) professional objectives. From this perspective, standards conflate technical information and present a consensus of opinion on a particular issue. Thus, the IEEE has also developed and published telecommunication-related standards through some of its technical committees. Prominent examples are the standards for local area networks, which were approved by Technical Committee 802 (see chapter 4 below).

In contrast to the IEEE, the International Federation for Information Processing (IFIP) is an organization not of individuals but of national technical and professional societies dealing with information processing. Founded in 1960, the IFIP has rather broad aims and is not explicitly focused on standardization. Its concerns range from the general promotion of information technology and processing to the advancement of international cooperation, research, and education, along with the dissemination and exchange of information (Macpherson 1990, p. 304). Some of its activities, however, are directed toward providing input into the formal standardization processes of other organizations. The IFIP's Technical Committee 6 (data communication), for example, conducted much pre-standardization work in the field of electronic mail and message-handling systems.

Nowadays, no discussion of professional organizations issuing standards for technical networks can ignore the organization that provides technical rules for the Internet. Since its foundation in 1992, the Internet Society ("ISOC") has taken responsibility for that network. At its inception, the central concerns of this rather new nongovernmental organization were "to facilitate and support the technical evolution of the Internet as a research and education infrastructure, and to stimulate the involvement of the scientific community, industry, government and others in the evolution of the Internet" (Articles of Incorporation of the Internet Society 3.A). Although the United States is the Internet's country of origin, the network of computer networks was never seen as being necessarily restricted to its home country (Malamud 1993). This has been mirrored in the ISOC's board of trustees. On the initial board, three out of fourteen trustees were from Europe and one from Australia. With the network's global expansion, Internet standards have gained international significance similar to, and in some cases higher than, those issued by international organizations. Since the ISOC's establishment, competencies in the area of standard setting and technical coordination that originally lay exclusively with committees that had evolved at different stages of the Internet's development have been shifted to the ISOC. Most of the committee work was done by volunteers from public and private research organizations. The majority of the input was related to research and development activities directly funded or indirectly supported by public sources like the (US) National Science Foundation or, in earlier days, by the US Department of Defense. Each of those agencies played a crucial role in the history of the Internet (Denning 1990; Abbate 1994a).

The central unit of standardization under the auspices of the ISOC is the Internet Engineering Task Force (IETF), which meets three times a year. The IETF is split into numerous working groups covering eight to ten functional areas. In May 1996 some 76 working groups were active in a total of nine areas. Working groups can be created easily, and most of them are dissolved after they have finished their task. The groups are managed by area directors. In contrast to most standardization organizations, participation in the IETF and its working groups is open to anyone; membership in the ISOC is not required. Much of the work is done on line via mailing lists. A steering body, the Internet Engineering Steering Group (IESG), consists of the IETF's chairperson and area directors. The IESG coordinates the activities of the working groups, assigns group chairpersons, and approves the results of the groups' work. Before standards are adopted, at least two independent implementations must have demonstrated that they really work. Moreover, when a standard is proposed, it is published electronically, and at some stage of the standards track it is introduced as a Request for Comments (RFC). Thus, broad and unrestricted discussion of a proposal via electronic discussion groups and mailing lists is possible. Although every standard is published as an RFC, not every RFC is a standard: other statements and comments relating to some (mainly technical) aspects of the Internet may be distributed as RFCs too. To be approved as a standard, a draft must be accepted by the IETF and the IESG on the basis of consensus. Until the standardization procedure was reorganized in 1992–93, the Internet Architecture Board (IAB) had to give its approval too. The IAB

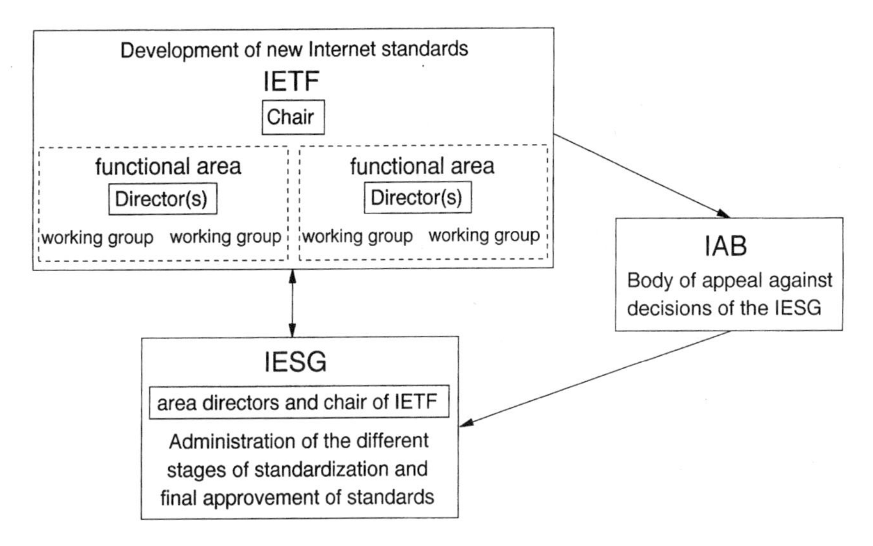

Figure 3.2
The organization of Internet standard setting. IETF: Internet Engineering Task Force. IESG: Internet Engineering Steering Board. IAB: Internet Archecture Board.

is the highest committee in the hierarchy below the ISOC Board of Trustees. Since the reform, it becomes involved in the standardization process only if conflicts cannot be resolved at the working level. Each member of the IAB is appointed for a two-year term by an IETF nominating committee. Figure 3.2 provides an overview of how Internet standardization is currently organized.

With the Internet's globalization, the number of individuals and organizations involved in standardization has increased enormously. Each of the latest IETF meetings was attended by 1000 or more people. The IAB and the IESG reacted to this trend by formalizing their working procedures. This is expressed, for example, by the length of a memo on the Internet standards process jointly issued by these two bodies. The document, published as RFC 1602 in March 1994, is 36 pages long, yet it constitutes only one-third of the volume of a comparable directive of the ISO/IEC from 1995 dealing with the procedures of technical work of JTC1 on information technology. Some observers regard the recent changes of the "Internet Standards-Making Model" (Rutkowski 1995) as an adaptation to "traditional" international standardization organizations such as the ISO (Lehr 1995). This brings us back to the starting point of our short and selective *tour d'horizon* through the universe of standardization organizations.

We can conclude that three constitutive coordination interests have left their imprint on the institutional fabric of international standardization: a country-based political (control) interest, an organizational or business-based commercial (profit) interest, and an individual or professional knowledge (consolidation) interest. Elements of these three aspects are present in all standardization organizations, but they differ in intensity. The interests are complementary rather than substitutive. This, in addition to other factors,[4] explains why the relationship of standardization organizations is characterized by cooperation and division of labor rather than by competition.

The standardization processes we will analyze have been organizationally centered on, but not exclusively confined to, the CCITT. Hence, working relations and patterns of coordination between the CCITT and other international, regional, and national standardization bodies concerned with facsimile, videotex, and X.400 will be scrutinized in the case studies.

Figure 3.3 gives an overall impression of the standardization world from the ITU's perspective at the end of the 1980s. It was during this time that standardization activities in the cases we analyze came—after some turbulent years—to be consolidated. These activities had been influenced by some of the organizations in the ITU's environment, which was already highly complex. However, not every organization in the field of institutionalized committee standardization was equally relevant for the ITU. This is indicated by the lines between the organizations or their subunits in figure 3.3. Bold lines mark strong interworking relations; missing lines indicate that no such relationship existed at the end of the 1980s. Interworking of the CCITT with JTC1 is very intensive, and the ties with ETSI and ANSI T1 are rather strong too. The TTC and the ECMA are also referred to as relevant partners of the CCITT, as is ETSI's forerunner, the CEPT (Conférence Européene des Administration des Postes et des Télécommunications; Conference of European Posts and Telecommunications Administrations), which has retained some administrative and regulatory functions in the traditional domain of the PTTs. The CCITT's relations to the other "rapidly growing regional and quasi-global standards communities for applications, network management, network operations, operations systems and conformance testing" (Irmer 1990, p. 23) re-

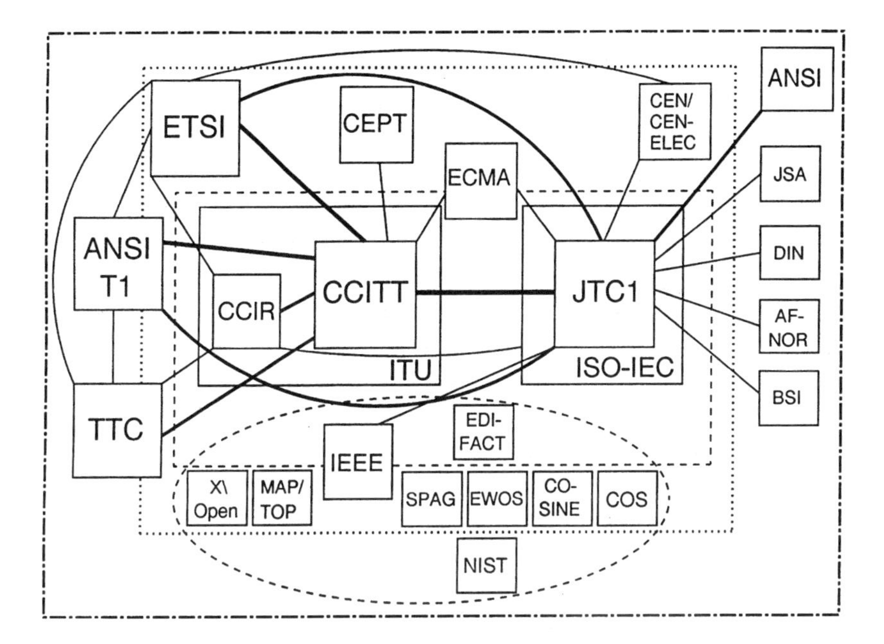

Figure 3.3
Information-telecommunication standards-making architecture at the end of the 1980s: the ITU perspective. Dashed rectangle: global. Dotted rectangle: regional. Dash-dotted rectangle: national. Dashed ellipse: rapidly growing regional and quasi-global standards communities for applications, network management, network operations, operations systems, and conformance testing that interact in complex liaisons with most other organizations. Connecting lines indicate the existence and strength of interworking relations between the organizations at the end of the 1980s. AFNOR: Association Française de Normalisation. ANSI: American National Standards Institute. ANSI T1: Standard Committee for Telecommunications. BSI: British Standards Institution. CCIR: International Radio Consultative Committee. CCITT: International Telegraph and Telephone Consultative Committee. CEN: European Committee for Standardization. CENELEC: European Committee for Electrotechnical Standardization. CEPT: Conference of European Posts and Telecommunications Administrations. COS: Cooperation for Open Systems. COSINE: Cooperation for Open Systems Interconnection Networking in Europe. DIN: Deutsches Institut für Normung. ECMA: European Computer Manufacturers Association. ETSI: European Telecommunications Standards Institute. EWOS: European Workshop for Open Systems. IEC: International Electrotechnical Committee. IEEE: Institute of Electrical and Electronics Engineers. ISO: International Organization for Standardization. ITU: International Telecommunication Union. JSA: Japanese Standards Association. JTC1: Joint Technical Committee. NIST: National Institute of Standards and Technology. SPAG: Standards Promotion and Application Group. TTC: Telecommunications Technology Committee. (Adapted from p. 23 of Irmer 1990.)

main unspecified; Irmer notes only that they "interact in complex liaisons with most other organizations" (ibid.).

Figure 3.3 also shows that the national standardization bodies maintain relations with JTC1 but not with the CCITT. However, this results from a formal viewpoint that hesitates to classify government agencies (including authorized PTTs or the national CCITT "mirror organizations," which in many countries coordinate the domestic position) as standardization organizations. Depending on the size and composition of the delegations, national meetings are held to coordinate input into the CCITT. Owing to the CCITT's lack of formal rules on how to coordinate national positions, this level is the least uniformly organized, as will become clear when we look at national preparation of input to the CCITT.

Institutionalized producers of standards display, at least at the international level, substantial organizational similarity. An "organizational field" has unfolded of standardization bodies which resemble one another with respect to their goals and working procedures.[5] The "business called 'standardization' seems to have its own rules and culture" (Irmer 1990, p. 22). Without ignoring the peculiarities of each specific standardization organization, we suggest on the basis of the CCITT that the following characteristics are typical for most organizations at the international level[6]:

Participation is voluntary and not remunerated.

Participation is, within certain membership rules, open to those who are "substantially interested."

The work is committee-based, cooperative, and consensus-oriented.

Organization and working procedures are impartial, unsponsored, and politically independent ("due process").

The work is based on technological knowledge and follows the principle of parsimony of standard options.

Standards are international, nonmandatory "public goods."

These characteristics have been comparatively constant. However, just what "consensus" or "due process" or some of the other concepts mean in a specific situation can change with time and circumstances. There is some interpretative flexibility when it comes to collectively determining the meaning of these characteristics. On the other hand, this flexibility, which excludes arbitrary redefinition of these characteristics, guarantees their relative stability.

To explain these characteristics from both an institutional and an actor-oriented perspective, we first must recall the crucial role of compatibility for technical systems. In telecommunications this is obvious. The telephone network has developed over many decades into a truly global network. Older parts, such as electromechanical exchanges and relays or analog transmission technology, were partly replaced by and partly combined with new digital components. New features and functions, such as data transmission, were added to the familiar voice communication. Software-based network components and intelligent terminals were integrated into or connected to the telephone network. New logical networks were implemented as subsets of physical ones. Dedicated networks were connected to universal service networks. The Internet—the network of networks—has evolved partly by using its own lines and partly by relying on telephone lines. Complexity and heterogeneity of telecommunications systems have generally increased through differentiation (Mansell 1993), and coping with the resulting problems of interconnectivity and compatibility has become the universal challenge. Business users demand more interconnectivity in order to reduce costs and enhance the performance of their telecom equipment. In addition, manufacturers have articulated a demand for coordination especially through compatibility standards (Hemenway 1975). Compatibility of complementary components in general enhances the value of network technologies from the user's and the vendor's perspective, and this can be achieved through standards. Where a standard is needed, welfare is highest when just one standard, or as few as possible, can be provided for each compatibility problem. Actors, in view of their technically mediated interdependence, presumably have an interest in coordination (Schmidt and Werle 1992).

The actors' interest in coordination motivate them to voluntarily engage in negotiations with others on an appropriate solution. They are willing to bear the costs when they are (over)compensated by the individual utility they anticipate. But the solution must be acceptable for all participants; otherwise, universal implementation would be jeopardized. Maximum protection of the participants' interests is theoretically ensured by the employment of the unanimity principle as the decision rule (Buchanan and Tullock 1971). Unanimity is understood to encourage participation in standardization and stimulate the implementation of standards. When all substantively interested actors are included in a committee that proceeds under the unanimity rule, the resulting solutions are beneficial for all participants, or at least cannot be reached at the expense of single participants. But whether a solution results, instead of disagreement or deadlock, is contingent on the attractiveness of the available options and the actors' preferences (Scharpf 1993a; chapter 4 below).

The idea that welfare increases when coordination problems are resolved by standards designates them as "public goods," to which (at least normatively) everyone should have access. If legally enforceable property or patent rights on a standard are claimed, these should not inhibit diffusion. Therefore, license fees should be low and nondiscriminatory. Indeed, most traditional standardization organizations have issued rules which are supposed to secure open or at least equal access to standards. However, the provision of most standards is not without cost. As is well known from the theory of collective action (Olson 1971), the production of public goods is confronted with the "free-rider problem": if nobody can be excluded from using a good, anyone may hesitate to help with its production and prefer to let others produce it. Consequently, goods of the "public" type tend to remain underprovided.[7] If, however, the groups interested in a standard are not too large, and if the standard's potential benefit is unequally distributed, those profiting most will acquiesce to bearing a disproportionate burden. In our case of telecommunications standardization with a good number of great and powerful actors, not only this possible "exploitation of the great by the small" (Olson 1971, p. 29) but also the specific situation of technically mediated interdependence can be expected to reduce negative effects of free riding.

The definition of committee standards must be kept separate from their implementation. In this case, broad adoption is typically welcomed rather than rejected as free riding. In fact, the implementation of a standard, especially if switching from an old to a new one is involved, may turn out to be another public-goods problem. Early adopters help to initiate widespread adoption, but they risk incurring losses if others do not follow and adoption does not become widespread (Dybvig and Spatt 1983). Here free riding means delaying adoption until many others have implemented the new standard. It is therefore not surprising that, on the one hand, committee standards are public goods and that, on the other hand, they entail a normative expectation to comply. This expectation aims directly, but not exclusively, at those who participated in the process of negotiating a standard because they had the procedural opportunity to influence the output. In this sense, procedural rules exert normative power well beyond the context they govern directly (Luhmann 1983a).

Another aspect of standardization is its technical content. As standards are designed to facilitate technical interoperation of components in a systems context, work in standardization organizations starts from a technically defined problem. In general, the work is supposed to require technical capabilities, at least as long as the set of feasible options has to be taken out of the pool of "possible" options when viewed from an engineering perspective. Technical problem-solving methods, often perceived as "scientific," are then applied (Cargill 1988). Although in international organizations politics always matter, the technical work is seen to be clearly separable from political concerns. This was expressed, for example, by a Belgian delegate when including the ITU in the United Nations was discussed at an ITU conference in Atlantic City in 1947: "Our Union is an essentially technical and administrative body and, as a result, international politics must continue to be excluded from its discussions. Belgium is favorable to our Union being connected with the UN, but under the formal stipulation that the complete independence of the ITU shall be maintained." (cited in Savage 1989, p. 39)

In standardization organizations, work is based on "technological paradigms, routines, heuristics, norms and standards" (Rip 1992, p. 244), which provide a cognitive infrastructure. Since the beginning of the twentieth century this infrastructural knowledge base has been gradually transformed from a multitude of local "technical models" into a set of abstract cosmopolitan and global ones (ibid., pp. 245–248). For the engineering profession, "unanimity" and "uniformity" of design protocols and technical models has become a central value that distinguishes this group from other occupations (ibid., p. 246). In the practical work of standardization, this value is transformed into an institutionalized collective attempt to achieve consensus by searching for the best solution in a collaborative professional effort. Thus, institutionalized international technical standardization relies in good part on a technical problem-solving orientation that stabilizes the prominent position of the engineering profession in the control of technology (Rip 1992; Weingart 1984; Noble 1977).

So far we have focused on cognitive aspects of technical functionality or on economic considerations to explain the institutional and organizational characteristics of international standardization. One crucial weakness of economic approaches to explaining the evolution and functioning of committee standardization is, of course, their tendency to neglect power. They do not regard prevailing resource and power differentials between countries, or between organizations or individuals, as crucial variables affecting the phenomena under consideration. Instead they emphasize deficiencies and failures of markets such as network externalities (Krasner 1991). Aspects that are not purely economic, especially political interests and motives, are dismissed as external.[8]

There is no doubt that, with respect to specific technologies and technical interdependencies, markets do not provide optimal, or sometimes even sufficient, coordination of economic (and indeed technical) action. Without political forces interfering in these cases, one would generally expect a shift from market to hierarchical coordination by means of vertical and/or horizontal integration of firms, as has been suggested by the economics of transaction costs (Coase 1937; Williamson 1975). The historic emergence of monopolistic telephone-network operators can be regarded as a case in point. However, in the past, economic incentives or even pressures to establish these "natural monopolies" (Baumol et al. 1982) were usually bound by national confines. Transborder expansion was not politically viable. Also, political antitrust motives and subsequent legislation often limit concentration (Hollingsworth 1991). Consequently, compatibility and the establishment of standardization organizations may be politically promoted in order to prevent vertical integration (Braunstein and White 1985). This too can provoke an argument over the appropriate representation of political, professional, and commercial interests in standardization (David and Greenstein 1990, pp. 29–32). Conflicts among political control, commercial profit, and professional domain-consolidation interests result during the establishment and the reform of standardization organizations, and these conflicts are by no means simply a repercussion of market failure.

The history of standardization activities in the United States demonstrates their "hybrid private-public character" almost from their inception (Olshan 1993, p. 322). In some sectors it was trade associations, in others professional associations, and in others state agencies that were involved in standardization (ibid., p. 322; Gabel 1991; Reddy 1990; Thompson 1954).[9] The question under whose jurisdiction standards would be made, though virtually hidden from the public, has been an issue from the early years of the twentieth century. From a theoretical point of view, it is related to neocorporatist considerations of corporatist democracy, private interests, government regulation, and associative regulation (Streeck and Schmitter 1985; Eichener et al. 1991; Egan 1991).

Olshan (1993, p. 330) argues that the national struggle for control "is now being iterated" at the level of international standardization. Political control at this level, however, cannot be regarded as some kind of subordination of standardization under a single authority. Where the principle of national representation forms a constitutive element, as in the CCITT/ITU-T and in the ISO/IEC, representation of political interests is declared legitimate but is eliminated from the technical working levels of organizations (Mansell and Hawkins 1992). Thus, politics remains institutionally confined to what one might call a negative control of standardization. By collectively shaping the organizations' membership rules, decision rules, working programs, resources, etc., governments can arrange for a general protection of national interests. Hence, some technical issues are never placed on the agenda, or they can be diluted or rejected before detailed technical negotiations start. The last resort would be the power of absolute veto for every government—the other side of the unanimity coin. To prevent too many deadlocks, some kind of (qualified) majority voting has been formally accepted for top-level decisions, but what governs the decision-making processes in international standardization is the consensus principle as an informal rule effective at all levels of the organizations. Whenever possible, explicit voting is avoided. The chairperson's assertion that no objection is discernible usually substitutes for formal voting.

In telecommunications especially, the coexistence of political and other interests has a long tradition. Cowhey (1990, p. 169) regards the ITU as a central unit of an "international telecommunications regime," which integrated political and commercial interests so successfully "that it eventually disappeared from sight."[10] Although the ITU has been facing internal conflicts between highly industrialized and developing countries for decades (Codding 1977), the pressure to change the rules due to a growing discontent with the distributive outcome of this regime has only increased in recent years (Solomon 1984). This indicates that the aversion to living in a world with completely uncoordinated telecommunications systems has been partly transformed into a desire for a coordinative solution and for rules that will facilitate such solutions. The United States, with only one vote in the CCITT/ITU-T but with a large number of carriers and manufacturers since the divestiture of AT&T, has challenged the organization's intricate coupling of regulatory and coordinative tasks such as service definition and standardization. The United States argues that this is likely to impede international competition, to the benefit of some European telecommunications monopolies (Drake 1988). Technological innovations have altered national power and policies in the telecommunications sector, including the question of how much coordination shall be achieved through committees and how much shall be left to the market. The issues involved have only partly been issues of market failure; not infrequently they have been concerned with the underlying regime structure of international telecommunications coordination (Krasner 1991, p. 353–357).

The reform of the ITU in the early 1990s was aimed at perfecting the separation of the technical work from political leverage and at increasing the efficiency of standardization (Genschel 1995, pp. 88–112; Genschel 1997). This included not only the creation of the ITU-T but also a modification of the membership rule, which now grants scientific and industrial organizations (SIOs) full membership status, a privilege reserved exclusively for PTTs and recognized private operating agencies (RPOAs) up to 1989. Before 1990, SIOs participated only in an advisory capacity. At the level of technical work, however, this minor status did not make much difference. Likewise, the reform of the membership rule has not changed the principle of national representation; there is still only one vote for each member country.

For political, professional, or business groups it may well appear more promising to establish new standardization organizations than to reform the traditional ones (Drake 1994). It may also be tempting to bypass time-consuming committee work and directly push a standard through. The ECMA and X/Open demonstrate that standardization organizations were successfully set up in the 1980s and before. A look at the ISOC acquainted us with a professional organization representing the domain of Internet standard setting, which for a long time has been almost completely ignored by the ISO/IEC and especially by the ITU—as can be seen in figure 3.3, where the ISOC (or, rather, the IETF) is missing. Only recently has the Internet been officially recognized by the ITU. At its TELECOM 95 Forum in Geneva, the ITU organized a special "Internet@TELECOM.95" conference with many companies representing the different facets of the Internet. During this conference Vinton Cerf, a co-inventor of the generic Internet protocol who is often called the father of the Internet, was awarded

Figure 3.4
Technical diplomacy: Vinton Cerf (right) is awarded the ITU Medal by Pekka Tarjanne of the ITU. (Courtesy of A. de Ferron, ITU.)

the ITU Medal by the ITU's Secretary General, Pekka Tarjanne (figure 3.4). Besides these largely symbolic gestures, the establishment of more formal links of cooperation between the IETF and other committees can be observed. One notable step is the recent creation of a formal liaison between the ISOC and the ISO/IEC JTC1 (Radack 1994).

The tendency to set up new standardization bodies instead of commissioning the traditional international organizations with new tasks seems to have gained momentum in the 1990s. Many new consortia and forums have been created, and others have extended their domains. In telecommunications the Asynchronous Transfer Mode (ATM) Forum and the Frame Relay Forum are two examples. Other examples are related to the Integrated Services Digital Network (ISDN). Many of these forums cover areas in which committees of the ITU-T have also been active. The consortia and forums are vendor-driven. A look at the area of information systems provides an even richer picture of these new units. Highly prominent are the X-Window Consortium and the Open Software Foundation (OSF), both set up in the late 1980s. These groups rely on large memberships, but the OSF in addition receives large contributions from a rather small circle of big influential members who created the organization (Dunphy 1991). However, X-Window and the OSF are just two out of an estimated 200 consortia and forums in the computer and related industries (Updegrove 1995). One more unit still to be mentioned is the W3 Consortium, which has been established recently with the aim of developing, supporting, testing, and disseminating the World Wide Web (WWW) protocols that are widely used in the Internet. The W3 Consortium, run by the MIT Laboratory for Computer Science, is open only to organizational membership. In March 1996 more than 130 member organizations contributed to the budget of W3, which is used to finance the activities of the consortium. That W3 and not the ISOC and the IETF provides the WWW standards shows that the Internet's committees are not going to absorb all network-related standardization. Unlike the traditional top-level organizations, however, the ISOC has never officially expressed imperialistic claims with respect to international standardization.

If one wanted to contrast the ecology of standardization organizations at the end of the 1990s with the situation at the end of the 1980s (figure 3.3), one would not only need two or more pages for an up-to-date chart but, first of all, some research funding in order to figure out the links and relations among all organizations. In future years the number of standardization organizations is very unlikely to decrease. Multimedia systems and national and global information initiatives may evoke the creation of even more standardization units (Drake 1995). Their appearance on the stage of international standardization puts pressure on the incumbent organizations to improve the efficiency and effectiveness of their working procedures, although cooperation and not competition is the prevailing pattern of the relationship of standardization organizations. In February of 1996, the OSF and X/Open consolidated their activities by creating the Open Group, which coordinates and partially merges the operations of the two member units. This illustrates, perhaps, that not only the traditional standardization organizations (such as the ISO, the IEC, or the ITU-T) but also consortia, forums, and less formal collectivities have to adapt to the rapidly changing environment. It clearly does not indicate that the need for standards will decrease or their future production will be reduced.

Standards Production in Organizations and Markets

How many standards for telecommunications technology are needed? Despite the transferability of microelectronics as a basic technology, even the largest corporations do not have the competence to produce complete systems for data processing and telecommunications. This shows that the hierarchical solution of coordination problems is declining as an option, especially when at the same time there is a growing tendency to design and treat single technical artifacts as parts of larger technical systems. Innovation and enlargement are "less the application of separate inventions than the integration of different new products and processes into new systems" (van Tulder and Junne 1988, p. 219). Keyboards, memories, processors, cables, fax machines, and other components are increasingly being manufactured by specialists. However, as one business leader puts it, they "are no longer seen in relation to a product but, on the contrary, in relation to a system" (Dekker 1984, p. xxxi; see also Tassey 1992, pp. 169–202). Thus, interdependencies and complementarities of large telecommunications systems imply a growing need for the coordinated production, operation, and use of technical components. This need has stimulated and accelerated the production of standards (including general basic standards and abstract standards architectures as well as specific functional standards and standard profiles) as a means of coordination.

Large volumes of detailed technical specifications and rules have been published in the last 20 years. The number of articles on computer standards, for example, has doubled every year since 1980 (Cargill 1989, p. 5). Research into international standardization provides evidence that, at this level too, the number of standards has grown considerably. In the same period of time, as we have seen, various international and regional organizations have evolved to issue many of these standards.

Still, the CCITT/ITU-T and the International Organization of Standardization (with its sister organization, the International Electrotechnical Commission) are the two most prominent producers of international standards relevant to the telecommunications industry. Since we will refer to more details of the CCITT's activities when we present our case studies, we will only emphasize in this context that the CCITT has traditionally been perceived as the top organization of

PTTs and of regulated private operating companies such as AT&T. Notwithstanding the recent growth in membership of manufacturers of telecommunications equipment, computer vendors and scientific organizations in the CCITT, most of the standardization activities of CCITT committees have been dedicated to the transmission and switching functions of telephone and data networks. On the other hand, standardization of information technology in general, and of terminals (with the exception of telephone sets) in particular, has hitherto been defined as the province of the ISO and the IEC.

In figures 3.5–3.7 we present an overview of the production of standards relevant to international telecommunications, comparing the output of the CCITT/ITU-T with that of the IEC and with the number of standards for the Internet issued by the Internet Engineering Task Force (IETF). Selecting the IEC instead of the ISO as one of two reference organizations appears appropriate because the IEC's work covers all spheres of electrotechnology (including electronics and the electrotechnical aspects of telecommunication); thus, data on that body tell us more about standardization of information technology and telecommunications than the general reports of the ISO might, because the latter condense a variety of activities (ranging from standards for information-technology applications, textiles, food processing, metals, and metrology to building standards). There are two reasons why we included Internet standards—although still rather few of them—in our comparison. One is that they exclusively specify rules for technical networks and thus compete with some standards of the CCITT/ITU-T and also the ISO/IEC. The other reason relates to the widespread tendency to regard the Internet standardization process as a promising model for the development of standards that, unlike the standards of the traditional international organizations, are very likely to be implemented in many networks (Branscomb and Kahin 1995).

The number of new CCITT/ITU-T "recommendations" (the official designation of the standards traditionally issued at the end of four-year study periods) has continually increased from barely more than 60 in the period 1969–1972 to more than 450 in the period 1989–1992 (figure 3.5). Although many recommendations were prepared and even partly finalized early on in a study period, most of them were not officially approved until the end of the period, and then published, only after some delay, in a series of volumes familiarly known as "colored books." Since the early 1990s, one element of the ITU-T's strategy for speeding up the standardization process has been to make

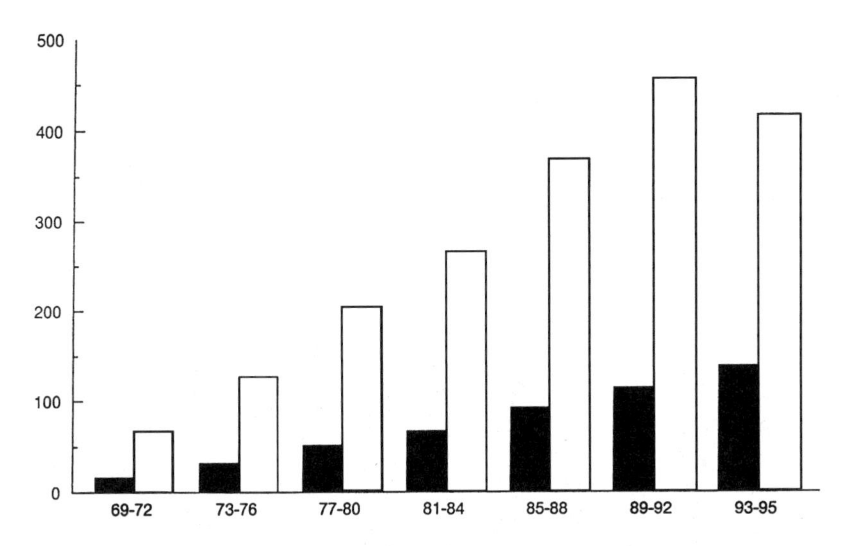

Figure 3.5
New recommendations issued by the CCITT/ITU-T. Black bars: average production per year. White bars: production per study period.

most recommendations available directly after their adoption. Furthermore, they will not be compiled and published in "colored books" anymore. Thus, since the end of the 1989–1992 study period it has been possible to determine the number of ITU-T recommendations for each year instead of a whole study period. The organizational changes, however, make it somewhat difficult to directly compare the pre-1992 output with production in subsequent years. These problems notwithstanding, we can observe a further increase in the number of new recommendations. In the years 1993–1995, an average of almost 140 standards per year were approved, whereas in all the previous study periods that production averaged less than 120. As the general four-year cycle of ITU-T activities has not been abandoned, we can observe a tendency to finalize and publish standards only shortly before the end of the study period (October 1996). Although statistical information is still incomplete, the number of new recommendations issued from January through October 1996 has already reached approximately 180. That means that the overall production of new recommendations in the 1993–1996 study period exceeds 500 and is thus significantly higher than in the previous period. In addition to the new recommendations, a comparable number of earlier ones are invariably updated and revised during any one study period. These are not included in figure 3.5. On the other hand, not all recommendations

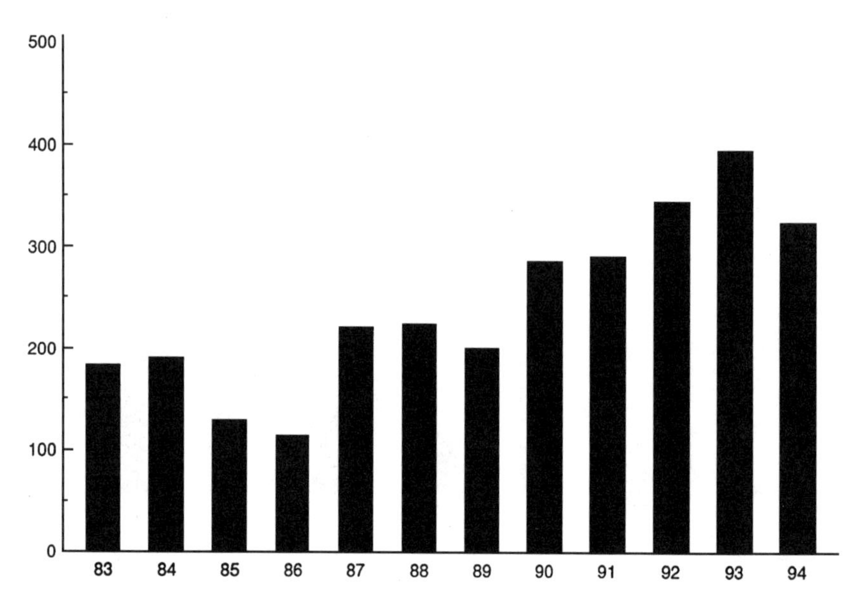

Figure 3.6
Standards issued by the IEC.

convey technical rules. A substantial number of them define such aspects as telecommunications services, tariff principles, and maintenance requirements for international transmission systems. The principles and rules governing the organization of work of the CCITT/ITU-T are also laid down in the form of recommendations.

The IEC's annual production of international standards between 1983 and 1994 shows a few fluctuations, but the trend toward more and more technical standards is obvious (figure 3.6). The accumulated four-year data show that in the period 1989–1992 there were 1127 standards produced—almost twice as many as in the preceding interval, when "only" 690 were produced. The number of standards issued in 1993 and 1994 already amounts to 731. If we add the recently published figures for 1995 (429) and 1996 (341), the IECs 1993–1996 record grows to a total of 1501 standards. This is in all periods considerably more than the CCITT and the ITU-T can report, but not all IEC standards are directly relevant to telecommunications. We can estimate their number by looking at the proportion of its technical work the IEC devotes to "Telecommunications, Electronic Systems and Equipment, and IT-Systems." Although the ratio dropped from 20 percent in 1989 to 13.8 percent in 1994, the absolute number of standards in this area rose from 40 to about 50 per year because the IEC's overall production of standards increased significantly, as figure 3.6 shows.

The declining proportion of telecommunication standards, and with it the rather moderate increase in their absolute number, is a direct effect of an organizational arrangement between the IEC and the ISO concerning the delegation of standardization tasks in the field of communication processes in computer networks to JTC1. At the end of 1989, two years after its creation, this organization had already published 61 standards, and by the end of 1993 the total number amounted to 458. The current annual production exceeds 50 (Gibson 1995). At the end of 1996, JTC1 declared 1170 standards, including draft standards, to be under its "direct responsibility." More than 30 percent of them are subject to the jurisdiction of Subcommittee 6 (telecommunications and information exchange between systems) and Subcommittee 21 (open systems interconnection, data management, and open distributed processing).

In 1991, the ISO and the IEC were responsible for 85 percent of all international standards. About 30 percent, most of them generated by JTC1, are devoted to information technology. A further 40 percent of the standards are in the electrotechnical sector, the domain of the IEC. In both areas we find technical rules for telecommunications and computer communication. The other 30 percent are standards in the nonelectrotechnical sector covered by the ISO (IEC Annual Report 1991, p. 4).

Figure 3.7 gives an overview of standards production in a less formalized context. The Internet standards are almost exclusively devoted to computer networking. Most prominent are the Transmission Control Protocol (TCP) and the Internet Protocol (IP), both published in 1981, which constitute the basic protocol suite of the "network of networks." TCP/IP has provided a platform for a multitude of standards which are functionally equivalent to OSI standards. Many of the TCP/IP-related standards have not been submitted to the official standardization procedure of the Internet Engineering Task Force (IETF) but have been adopted elsewhere—for example, in the W3 Consortium. This is one reason why the number of Internet standards is rather small. Remarkably, the majority of Internet standards have never been approved as international standards, although the specifications have gained a high international significance and reputation parallel to the global expansion of the Internet. It was not until 1994,

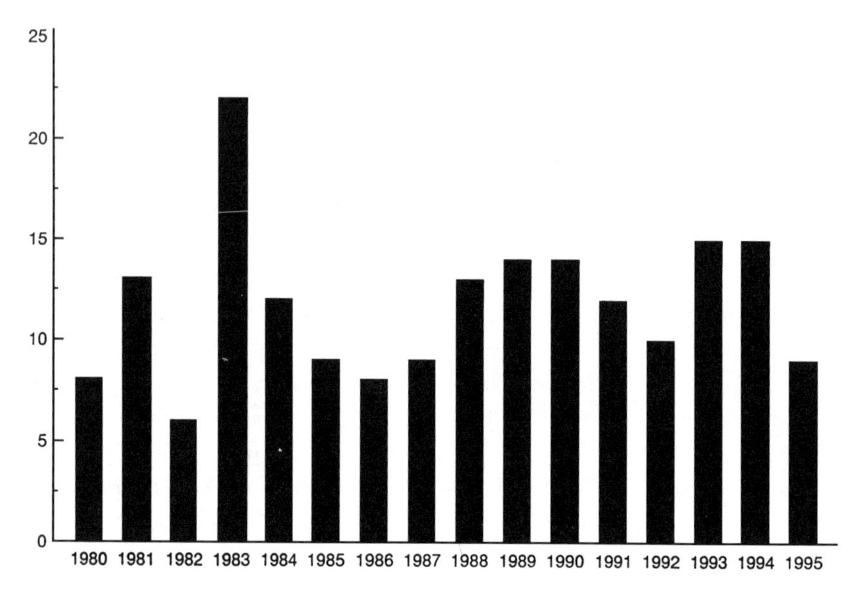

Figure 3.7
New and revised Internet standards issued by the IETF.

for instance, that a US Panel on Federal Internetworking Requirements established by the National Institute of Standards and Technology (NIST) recommended accepting standards developed by the IETF as open international standards (Radack 1994).

Figure 3.7, which shows the number of new, updated and revised Internet standards published per year since 1980, reveals that there was a peak in the early 1980s, when many of the basic protocols were defined. In the years following, the average annual production has been around 12. However, the IETF has announced a new generation of basic protocols (designated IPng) for the end of the decade, since the existing ones may not be able to cope with the rapid expansion of the Internet (Bradner and Mankin 1996). A new peak of standards production can be expected. Only since March 1992 (RFC 1311) has the Internet Architecture Board officially used the term *standard* for the technical specifications that have completed the full process of standardization in the Internet. By the end of 1996 the number of specifications designated as standards was 53. Since multiple RFCs can be included in one official standard, and since older RFCs in a standard can be replaced by updated or revised ones, the number of hits (RFCs) in figure 3.7 is almost four times the number of official standards. These hits, however, are a more reliable indicator of the amount of successful standardization work dealt with by the IETF each year.

Why have so many standards been produced? Did production result in too many standards? Some of the theoretical considerations that are relevant for an understanding of the institutional characteristics of standardization organizations can also be applied to their products. New institutional economics and industrial economics combined with the institutional approach in political science have contributed most to explaining the increasing quantitative and qualitative significance of standards. When it deals with standards, institutional analysis usually starts with a consideration of their economic aspects. Some commonly recognized benefits for economic welfare result from the fact that standards constitute markets by defining the relevant aspects of products (Voelzkow et al. 1987). Furthermore, they encourage market entry and enhance competition by clearly defining what is required to serve a market. The resulting transparency allows for competitive supply by firms, but might also serve as "cheap insurance" for customers, who can compare products on the basis of the standards (Cargill 1988, p. 64). For suppliers, standardization is a basis for scale economies: it simplifies training, maintenance, and inventory in manufacturing, as well as in operations of technical systems (Sirbu and Estrin 1989). Yet standardization may also provide increased and controlled variety for both users and suppliers, as when compatibility standards for audio components make it possible to combine them into tailor-made systems (Matutes and Regibeau 1988; Langlois and Robertson 1992).

Besides their capacity to facilitate economies of scale, the significance of standards in reducing transaction costs has been stressed (Reddy 1990; Voelzkow 1995). Neither of these effects will be achieved if there is an oversupply of standards. Although single segments of large markets may provide sufficient space for products to rely on different standards, the existence of too many standards hampers the development of markets (Arthur 1989; Berg 1989a). This general proposition has been substantiated for compatibility standards, which brings us back to the telecommunications domain.

The notion of compatibility is directly related to the concept of network technologies, which is widely used to characterize a modern form of "systemness" of technical configurations. Physical networks such as the telephone system obviously rely on the interconnectivity of their components. The growth of such networks generates what economists call *positive externalities*. The more users adopt the same service or compatible ones, the more valuable is the service for each of them; the demand is interdependent (Rohlfs 1974). In the most obvious example, a telephone user benefits from others' being connected to the same network. A customer of a telephone company will not be charged for any new subscriber to the telephone network, but the opportunity to call more and more friends, colleagues, businesses, etc. increases the attractivity of the network for incumbent and new users alike. Telecommunications networks typically combine network technologies, but the network character of a technical artifact is not confined to components of such physical networks. Computer hardware and software, operating systems and application programs, video recorders and video cassettes also indicate the characteristics of network technologies (Katz and Shapiro 1986). Whenever goods or services are complementary, their production or consumption can have positive (or negative) externalities, implying that individual actions affect the utility of other actors (Carlton and Clamer 1983; Tirole 1988, pp. 404–409; Gilbert 1992).

This rather narrow quasi-objective economic perspective can be extended using sociological categories. Whether technical artifacts are regarded as stand-alone devices or as components of a system largely depends on the concepts that designers, producers, regulators, users, and others associate with the artifacts. For example, these groups can look upon a personal computer as an intelligent typewriter and/or as a terminal of a computer network. As we have learned from what social constructivists call *interpretative flexibility* (Bijker 1995a), the perception may differ from one group to another, and it can change. Accordingly, the social and technical references of the personal computer are constituted differently. In the first instance, word-processing software, secretaries, typists, printers, and displays are relevant points of reference; in the second instance, e-mail facilities, transmission charges, rules of access to other computers, heads of computer centers, and PC-network interfaces move into the frame of reference. In both cases, coordination problems are apt to arise, and the attractiveness of a solution can depend on how frequently the solution is chosen. Thus, from an institutionalist point of view both cases can also be linked to the concept of network externalities, because these externalities are generated when the attractiveness of a particular type of behavior depends on frequency (Wärneryd 1990).

When externalities matter, markets perform very poorly, as "prices do not convey all relevant information" (Antonelli 1992, p. 9). This can result in a "critical mass" dilemma. Producers and users of components of a network technology tend to wait and see what others decide to provide or buy, because they do not want to end up with incompatible products (Allen 1988; Hohn and Schneider 1991; Rogers 1995; Granovetter 1978). When too many standards are available (i.e., when no consensus on the interface characteristics of the components has been established), it is likely that a technical system may never be built up. Thus, a situation with very many standards is akin to a situation with no standard at all.

De facto standards that evolve in markets may fill the gap. Such standards build on a solution originally devised by one or several firms cooperating to produce a product. However, often more than one satisfactory solution to a coordination problem is available, and different standards may be implemented in different market segments or different parts of the world (Ullmann-Margalit 1977, pp. 74–113; Majone 1992). This has tended to be the case with telecommunications. In addition, the more often autonomous producers are involved and the more often distinct components must be assembled to construct a system, the higher is the risk that incompatibility will result. Moreover, markets do not systematically choose the "best" standard. Rather, the first solution that comes along is likely to be stabilized and to "lock in" because network technologies have "positive feedbacks." In contrast to the conventional assumption of the economic theory of diminishing returns, which is considered appropriate for resource-based parts of the economy, network technologies and the knowledge-based sector in general are subject to increasing returns (Arthur 1989, 1990; Foray 1990). Among the reasons for these increasing returns are network externalities, which we have already discussed. Furthermore, initial investments in research, development, and tooling are invariably large; but once sales begin and use has reached a critical mass, incremental production becomes relatively cheap and increasing use does not depend on major additional investments. That makes it especially difficult to migrate from an "inferior" to a "superior" standard, and generally from one standard to another, if the actors decide upon adopting a standard individually, uncoordinated with others, and only from the perspective of their own utility (Leibenstein 1984). A well-known example here is the QWERTY keyboard, which is still in use although originally there was supposed to be a superior alternative (David 1985).[11] These few observations on the role of markets in standardization serve to exemplify some prevalent deficiencies of this mode of coordination—deficiencies that prepare the ground for the evolution of cooperative committee-based standardization and for the growing number of standards produced by committee-based standardization organizations (see also Church and Gandal 1992, pp. 99–100).

However, no committee in the world can prevent market standards from evolving or simply replace them with its own products. And nobody can force a firm to get involved in committee work or to devote its strategic potential exclusively to collective standardization. As the rest of this book concentrates on committee processes, we shall now present a few examples of successful and unsuccessful moves to establish de facto standards. Since market standards have hardly played any role in telecommunications, where technical coordination problems have typically been solved either semi-hierarchically under the guidance of a national monopolistic operator or in international committees, we have to turn to other technologies in order to present the main characteristics of market standardization processes.

Compatibility standards emerge in markets from the competition of at least two systems based on different technical specifications. Normally, the emergence of a market standard relies on the fact that technical systems generally exhibit positive network externalities. As a result, the market will converge on a single standard as one of two or more technical specifications with similar functionality gathers momentum and becomes dominant. Only if specifications offer a sufficiently differentiated functionality is it likely that different standards can be sustained, which could then be seen to form different, specialized submarkets.

The seemingly automatic convergence of markets on single standards relies on multiple social processes that provide positive feedback. Though the network externalities of a telephone network are apparently technical, relating to the size of the network, they are grounded in the interest of each user in being able to reach as many others as is possible. The social origin of these externalities is even more apparent when we turn to the well-known case of the QWERTY market standard. David (1985) explains the network externalities of this keyboard layout in terms of the complementary social skills linking the single typewriter to a virtual network. Before embarking on their training, typists consider which keyboard standard is likely to prevail; when selecting equipment, employers contemplate which standard will provide the best supply of trained typists. These and other individual choices such as those made by the vendors of typewriters lead to positive feedback that stabilizes the QWERTY design against competing standards.

Where alternative technical specifications compete in the market, actors often attempt to strategically influence the process in order to push their preferred outcomes. With a view to the opportunities and difficulties faced by actors when attempting to set a market standard, we will now present some case studies examining the characteristics of this type of process. We begin by looking back at the competition between Sony's Betamax and JVC's VHS over who should set the standard for the video cassette recorder (VCR). We then move on to discuss the standardization of the microcomputer. Our third case, the "Telepoint" mobile communications system, provides an unusual example in the field of telecommunications, because here the UK government, instead of imposing a standard, attempted to rely to some extent on market standardization processes.

A videotape recorder was first commercialized in the mid 1950s by the American Ampex Corporation (Gabel 1991). In the 1960s a number of incompatible products existed, and several Japanese and American companies as well as the European firm Philips were engaged in development. From a very early stage, manufacturers regarded standardization as necessary for the merchandising of mass-market products, and they repeatedly attempted to reach agreement on them. In 1969, all the Japanese manufacturers agreed on an industry standard for videotapes. In 1970, Sony, JVC, and Matsushita set another standard: U-Matic. In 1971 and again in 1972, the International Tape Association made an effort to solve the standards problem. But it was already too late: several different systems were about to be brought to market.

Sony offered Matsushita and JVC a license of its Betamax system in an attempt to convince them to drop their ongoing developments programs. After this strategic move failed, Sony introduced Betamax on the Japanese market in mid 1975. JVC, whose development of VHS was not yet completed, tried even more actively to forge a standards alliance for VHS. But only Sharp, Mitsubishi, and later Hitachi took out licenses; Matsushita, Sanyo, and Toshiba continued their own development programs. In early 1976 Sony and JVC approached the

Ministry for International Trade and Industry in an effort to bring about a compromise on standards. Only Sanyo and Toshiba, however, dropped their developments and switched to a Betamax license; JVC (still developing VHS) and Sony resisted any move toward a compromise (Gabel 1991, p. 66).

In late 1976, some 18 months after Sony introduced Betamax, JVC started selling VHS. JVC was soon joined by its parent company, Matsushita, which abandoned its own development program. As Betamax clearly had an important head start, the market situation for VHS was very difficult in view of the network characteristics of VCR technology. However, JVC, a relatively small company, had attempted from the beginning to pursue an open standards strategy allowing other firms to implement VHS. With the backing of Matsushita, JVC had gained access to a distribution network more than twice the size of Sony's in Japan. Sony had thought that it could capitalize on its technological leadership. But, whereas it had over 50 percent of the market in 1976 and 1977, its share dropped by the following year. One reason for this decline was the fact that VHS offered the advantage of longer recording time. Betamax was initially restricted to one hour. Moreover, VHS was easier and cheaper to manufacture, and this was complemented by Matsushita's initial low-price marketing strategy (intended to increase VHS's market share as soon as possible).[12]

In 1977 Betamax and VHS were launched on the US market, the former in an alliance with Zenith and the latter in an alliance with RCA. Zenith and RCA were at the time the biggest sellers of color television sets, each with a 20 percent market share. Despite the standards contest, the market developed rapidly, VHS slowly gaining ground as a result of longer playing time, a bigger marketing effort, and lower prices. With the increasing market penetration of VCRs, a complementary market for pre-recorded tapes developed in the early 1980s. Soon, use of these tapes amounted to 50 percent of the overall use of VCRs. Appearing at a time of relative decline for Betamax, the pre-recorded tapes accelerated its downfall.

In 1978 both systems were introduced to Europe, where a year later Philips even offered a third incompatible system (V-2000). The European market grew rapidly, with all systems participating. Yet VHS soon established itself as the market leader, again reinforced by the increasing use of pre-recorded tapes. In 1983 Philips abandoned its system and switched to VHS, having lost $55 million on V-2000 (Gabel 1991, p. 72). In 1988 Sony finally followed suit.

The VCR case clearly shows the importance of JVC's business strategy in the attempt to establish VHS as a market standard despite its late start. Not quite as apparent, but at least as crucial, was the early uncertainty as to precisely how VCRs would be used. Sony had originally focused on the market for "time-shift TV" (i.e., the recording of broadcast programs for later viewing), for which one hour of recording time was believed sufficient. Moreover, for a long time the market for pre-recorded tapes was considered to be, potentially, a separate market. In the late 1970s several companies were developing videodisc players, which could be used for playback only but which offered superior picture quality and lower software costs (Graham 1986). But in the end these dedicated devices could not establish themselves, because, with the increasing diffusion of the VHS standard, VCRs were being used for multiple purposes. What follows from this case is that, in a situation of uncertainty over the precise future uses of an innovation and its most crucial technical features, the corporate strategy may be decisive for the exploitation of market demand. It was essential to JVC's success that the technology had been licensed widely and that Matsushita brought in a very large distribution network. Moreover, JVC's relationship with RCA facilitated the supply of pre-recorded tapes that turned out to be an important positive feedback to the diffusion of VHS.

Similar uncertainties prevailed in another related conflict of standards. In the early 1980s the market for camcorders developed, after the increasing use of VCRs had devastated the market for 8-millimeter film equipment. Again it was not clear whether this market would form a part of the existing VCR market or whether the demand for home production would be separate from the demand for other uses. Because of the need for easy-to-carry equipment, it was difficult to integrate videography into the existing standards. In 1983 all the major VCR manufacturers signed an agreement for a new 8-mm standard which was suitable for videography and was thought to be the standard of the next generation of VCRs. Sony introduced a 8-mm VCR and a camcorder in 1985. Downward compatibility now could be achieved by attaching a cable from a camcorder to a VCR recording onto a VHS tape.

For those firms producing VHS machines, however, the implementation of the new standard was a clear threat to existing market shares, because competitors who had adhered to the Betamax standard now had the opportunity to gain an advantage with the next generation of

VCR equipment. Therefore, it is not surprising that, despite the agreement on the 8-mm standard, JVC introduced VHS-C, a new VHS specification for the camcorder, in 1986. VHS-C's smaller cassette could be played on existing VCRs with the help of an adapter. We will not present more details on this case; suffice it to say that this time Sony's product was quite successful on the market for at least some time. Sony managed to effect a reasonable tradeoff between a superior technology, optimized for a specific application, and downward compatibility with VHS machines, while JVC compromised much more on the "perfect" design for videography for the sake of compatibility (Grindley 1995, p. 97). The latter turned out to be the wrong strategy, because the market for camcorders tended to evolve separately from the market for general VCR use.

A look at the early stages of microcomputer standards suggests very similar conclusions (Gabel 1991, pp. 19–53; Grindley 1995, pp. 131–155). Microcomputers appeared on the market in the mid 1970s. The systems were proprietary and mainly dedicated to specific purposes, such as word processing, graphics, calculation, and computer games. As a result, software could not be transferred across a range of different machines. The Apple Corporation supplied the first ready-to-use machines perfectly suited to the mass market. Apple followed a strategy of open architecture, releasing technical information necessary for the supply of complementary products and software to independent manufacturers. Possessing the largest stock of software of the early 1980s, Apple managed to establish itself as a market leader, although it used a proprietary operating system and did not follow the loosely defined industry standard CP/M. However, the market of the popular Apple II computer was mainly restricted to educational and home applications.

IBM entered the microcomputer market relatively late, in 1981, when it introduced its PC (Hergert 1987). The PC had been developed outside of the established IBM organization, and it represented a break with IBM's tradition insofar as many of the components were purchased and insofar as the architecture was open. Microsoft held the copyright to the operating system PC/DOS and therefore lacked the incentive to withhold licenses from competing manufacturers. Only the ROMBIOS (read-only memory basic input/output system) was proprietary, though this was soon copied or reengineered.

IBM's entry immediately changed the microcomputer market radically. By 1983 IBM had already exceeded Apple's market share. In view of the existing uncertainty about the future applications available in the still-developing microcomputer market, IBM's success was very much based on its dominance of the mainframe market and on its reputation for preserving its customers' investments ("Nobody has ever been fired for buying an IBM"). The popularity of the IBM PC was not restricted to users; it extended to manufacturers, who offered a host of complementary hardware and software products as well as "clones" of IBM's computers. In 1986 the sales of clones exceeded those of IBM PCs. All in all, the case of microcomputers is a striking example of how a dominant firm can set a market standard because its commitment to a set of technical specifications creates a degree of certainty among users and other manufacturers that virtually guarantees success. Certainty becomes self-reinforcing.

That IBM compatibility emerged as an industry standard was greatly helped by the choice of an open architecture and by the fact that IBM bought most of the components of its PCs from outsides sources, which made cloning easy. Interestingly, IBM's open strategy was due less to a conscious attempt to set the standard than to time pressure and the endeavor to minimize risks by restricting in-house development (Grindley 1995, p. 141). Because of the fierce competition from clones, IBM soon ceased to profit from its leadership. This is why IBM attempted in 1987 to set a new standard—the PS/2 with the operating system OS/2—which it tried to keep proprietary (Grindley and McBryde 1990). Only the external production of software was encouraged. This time, however, IBM failed, and as a result its market share declined even faster. IBM had stopped producing PCs to stimulate the sales of the PS/2, but this had only helped the sale of clones. IBM had faced some problems—for instance, problems of timing—when introducing PS/2. Moreover, Microsoft announced that it would license the OS/2 operating system and not follow IBM's proprietary route. More significant than these obstacles was the fact that, at the time, the market had developed autonomously, and IBM could not seize control again. Users did not want to forfeit the investments they had made in their equipment; equally important, the amount of standardization achieved in the market provided significant advantages. Thus, it was possible to "mix and match" available components so as to tailor computers to individual demands (Matutes and Regibeau 1988). Compatibility allowed portability of software and complementary hardware, helping to augment and upgrade existing PCs and to avoid "locking in" to proprietary systems. Moreover, modular design made it possible to decentralize innovation, allowing manufacturers to respond better to customers' diverse needs (Langlois and Robertson 1992, p. 309). The PS/2 was not so radically superior to the PC as to allow IBM to establish a new network. In the end, IBM returned to the production of mass-market, low-margin PCs. Standardization of the PC market resulted in strong competition among manufacturers, bringing the largest gains to the suppliers of the complementary components of microprocessors, software, and operating systems: Intel and Microsoft.

The example of microcomputers is all the more interesting with respect to Apple, which managed to stay competitive in the market a remarkably long time despite its products' incompatibility with IBM products (Gabel 1991, p. 45; Grindley 1995, pp. 152–153). Apple achieved a stable market position relatively early because of the user friendliness of its computers. After several of its new products failed in the early 1980s, Apple was quite successful with the Macintosh, introduced in 1984. Like the Apple II, the Macintosh had an open architecture and attracted many complementary products, but the design of the computer was proprietary and not licensed by Apple. The Macintosh was particularly well suited to desktop publishing, and it was very popular because of its graphical user interface. It was in many ways sufficiently different from the IBM-compatible machines to constitute a separate market. Owing to the advent in the 1990s of Microsoft's Windows, Apple's niche has been steadily eroded. The Macintosh's advantages can now be successfully replicated by IBM-compatible computers.

We turn finally to an example of a market standardization process occurring in telecommunications: the introduction of "Telepoint" mobile services in the United Kingdom (Grindley and Toker 1994; Grindley 1995, pp. 235–268). Telepoint, which relied on new cordless telephone technology, was meant to develop into a cheap, mass-market mobile service and to target the sector below the more sophisticated mobile services market (already dominated by two firms in the United Kingdom). Telepoint was originally designed for outgoing calls only, which could be made from a number of base stations installed at public places. Only later were additional features, including a paging service, to be added. Thus, the service offered limited functionality from the outset, which corresponded to the idea of a "phone-box in your pocket."

The Department of Trade and Industry (DTI) awarded four licenses for Telepoint in 1989. No standard was specified at the time, but the licenses included descriptions of the service to be offered. The idea was that market processes should help to quicken standardization, to the advantage of UK companies that were attempting to establish the European standard. It was thought that having the suppliers initially compete on the basis of different technical specifications would increase the likelihood of the market's helping to select the appropriate ones. From the DTI's point of view, this procedure promised to avoid the common shortcomings of standardization bodies, which tend to give little recognition to users' needs in their work. After a year, the suppliers were requested to settle on a common specification, which would have to be implemented within another year to facilitate intersystem roaming.

This experiment at market standardization turned out to be a complete failure. Three services were launched in 1989 only to withdraw from the market after two years; the fourth service, which was started up in 1992, lasted only a year. The introduction of three incompatible services had confused potential consumers. Network buildup was slow, as was the availability of handsets for the service, and it was anything but clear what kind of equipment could be used where. None of the suppliers made a significant attempt to increase production and thereby augment its market share quickly. Instead of a strategy of low pricing to penetrate the market, all of them concentrated on differentiating their products from the others at comparatively high prices. This only confused the consumers further, making it difficult to converge on a single standard after a year. In addition to the Telepoint supplier' vying with one another, they faced competition from other existing services. Most important, shortly after Telepoint started, licenses were awarded for comparatively cheap new digital two-way mobile services.

The failure of Telepoint was partly caused by regulatory mistakes. Too many licenses were awarded to providers of a service that offered little functionality. At the same time, an attempt was made to quickly develop a mobile market in which several differentiated services were preferred to Telepoint's one-way only service. Aside from these (and other) mistakes on part of the DTI, the example highlights the typical difficulties of market processes of standardization in telecommunications (Grindley 1995, p. 242). Competition among several technical specifications meant that potential users ran a high risk of betting on the wrong service and ultimately being left high and dry. They therefore hesitated to subscribe to a Telepoint service.

In this respect, another look at the competition among different VCR standards is quite revealing. Compared to the truly interactive Telepoint system, network externalities in the case of VCR were much weaker. No matter what others chose, with each VCR standard the basic functionality of time-shift TV was guaranteed to the user. Hence the interdependence with other individual choices could not result in an absolute loss of functionality. All that could be impaired was the enjoyment of the additional benefits that pre-recorded tapes and videography brought if Betamax instead of VHS was chosen. For Telepoint, in contrast, having the "wrong" standard was synonymous with being incapable of communicating with other users through this channel. Thus, waiting until a standard was broadly adopted was the option with the lowest risk. The DTI's announcement that a joint standard would be implemented within two years explicitly reinforced this incentive to wait. After two years, however, the providers had been discouraged and their services had been withdrawn.

In summarizing our analysis of standards production, we suggest that the rising number of international standards mirrors the growing demand for global compatibility. Standards take the place of hierarchical coordination and also take the place of different modes of direct cooperation between firms in the joint production, installation, and operation of a technical system (Schmidt 1993; Lorange and Roos 1992; Hagedoorn 1993). The process of standardization, however, also requires cooperation—at least when it is a collective effort, as in the case of committee work. This mode of standard setting is widespread in international telecommunications; pure market processes have been exceptions. International organizations have been set up to provide an institutional framework for the cooperative enactment of standards, but they cannot guarantee that cooperation evolves and that it eventually results in the adoption of a standard on the basis of consensus. The process of standard setting is influenced not only by institutional and organizational factors but also by a range of other variables, to be discussed below.

4

Standard Setting

The number of standards, as well as the number of organizations producing them, has grown significantly. Market failure is only one explanation of this development. Political, economic, and professional interests in controlling technology are central variables of an alternative explanatory string that adds contractual elements and aspects of conscious design to the functionalist-evolutionary connotations of the market-failure explanation (Langlois 1986, pp. 247–251). Institutions, once they have emerged, gather momentum, and they are more likely to change incrementally than radically (North 1990, pp. 92–104; North 1993). It is especially through organizations based on them that institutions create opportunities and constraints which channel and contextualize, but do not determine, individual and collective action. The organizations provide an "action arena" that "can be utilized to analyze, predict and explain behavior" (Ostrom 1986, p. 460)—not only by researchers but also by potential participants.[1]

However, standardization organizations do not directly affect the interests and strategies of the actors involved in standard setting. To a considerable degree, interests and strategies are exogenous variables; that is, actors convey externally generated political goals, economic interests, or technical conceptions into the institutional arena. Within this context, the actors define and specify the problem to be processed (in our case a standard to be approved), enter into discussions and negotiations, and eventually achieve some result that terminates the procedure.

From an institutional perspective, participation in the standardization process results from a twofold choice. The actors—usually firms, PTT administrations, scientific organizations, and other corporate actors—decide to apply for membership in a standardization organization according to its rules; then, as members, they select the standardization issues in which they want to become actively involved.

Economic approaches conceive of both choices as being guided by the actors' self-interests. Actors do not invest time and money primarily in order to secure the viability of a standardization organization or to help increase general economic welfare regardless of their individual return. Membership status in such an organization offers an opportunity not only to initiate and influence, but also to monitor standardization activities and to keep abreast of technical developments. The public-good nature of committee-approved standards can also motivate actors to join the organization: once a standard has been officially approved by a committee, it may diffuse rapidly and exert soft pressure on non-adopters to accept it. For these actors, getting out of implementing a standard could be costly. Thus, for them the public good may turn into a "public evil."[2] If actors want to avoid this negative externality, they can join a standardization organization and try shaping its output for their convenience.

If, however, a large firm has at its disposal the technical capabilities and the market power required to design, construct, and operate a technical system, and to set the interface standards when interconnections are optional, the corporation may even do so without taking into consideration membership in an international standardization organization. In such a case, the dominant firm's standards may evolve into market standards (Gabel 1991; Besen and Johnson 1986; see also our examples above). IBM was such a dominant seller of integrated systems in the market for general-purpose computers (Adams and Brock 1982). The positions of monopolistic PTTs in their national domains were even stronger; as a result, formally organized national committee standardization for interactive telecommunications systems has been almost completely absent in many countries.

Dominant firms, however, have displayed an increasing tendency to apply for membership in the traditional as well as the newly established international standardization organizations. Participation relates to their interest in being present both in the market and in this arena, where potential competitors may agree on "hostile" standards that threaten the positions of the dominant firms. Technological leadership, moreover, does not in itself enable any firm to design and assemble a large technical system independent of other firms, although each firm may try supplying as many components as possible and bundling them into a systemic product (Teece 1988; Matutes and Regibeau

1992; Robertson and Langlois 1992). In general, the definition of the system's boundaries is vague and is apt to lie outside the discretion of a single organization. Hence membership in standardization organizations is seen as a normal and often necessary element of organizational activity, to be financed as part of the firm's overhead or out of the R&D budget. Besides these rather instrumental aims, reputational considerations or tradition may lead corporate actors to participate in standardization organizations.

The official reasons given by corporate actors for being present in committee standardization can be distinguished from the motives of individual actors for engaging in these processes. A recent survey of individuals involved in standardization activities, mainly in some technical committees of the US accredited standards committee X3 (information technology), revealed "strikingly personal" motivations for participation among the 54 respondents (Spring et al. 1995, p. 228). This is especially remarkable because X3 was set up by a professional trade association, the Computer and Business Equipment Manufacturers Association, and thus belongs to the category of standardization organizations which are regarded as "openly responsive to commercial market concerns" rather than as having technical, political, or merely idiosyncratic motives (National Research Council 1995, p. 37). Asked in this survey to pick a statement that best described their motivation in contributing to the standards process, two-thirds of the respondents stated prestige, curiosity, or a desire to positively influence future events. At the same time they stressed technical expertise, participation in meetings, and negotiation skills as important requirements for individuals who work in committees (Spring et al. 1995). In figure 4.1 we list the various motives of and skills required for participation in standardization according to the relative frequency of the responses.

In the area of telecommunications, the organizations participating as members in standardization organizations can be classified into several groups. Network operators, carriers, and service providers are represented either as specialized units or as functionally and organizationally incorporated parts of the traditional PTTs. On the side of the manufacturers, we have large corporations producing many components of telecommunications systems and other firms which specialize in dedicated parts (e.g., terminals, switching or transmission equipment, satellites, cables, branch equipment). Vendors of terminals and vendors of switching devices in particular have their origins in the computer and electronics industries and not just in the

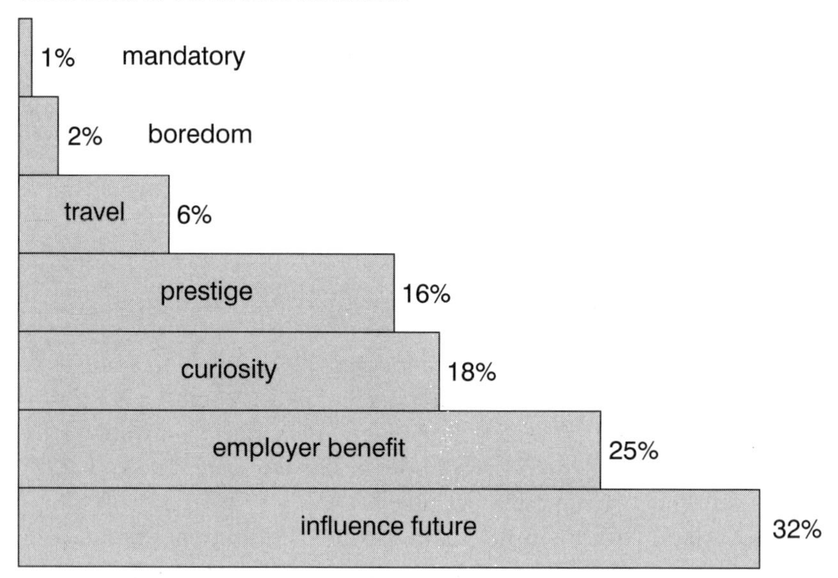

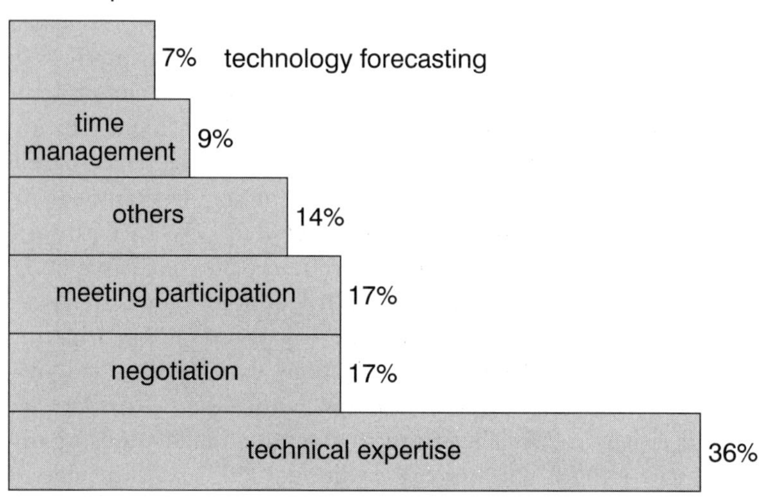

Figure 4.1
Motivations of, and skills required of, members of standardization committees. (Adapted from Spring et al. 1995.)

telecommunications industry. Specialized R&D bodies, consultants, user groups, professional and trade associations, regulatory agencies, public administrations, and ministries all send representatives as members of national delegations or in their own right to the international standardization organizations. Officials of other international, regional, or national standardization bodies can participate in certain organizations too.

The diversity of the members of standardization organizations indicates a considerable degree of heterogeneity and a broad range of interests being carried into every standardization matter, although users (especially small businesses and private users) are absent or inadequately represented at the international level.[3] Nevertheless, the interests of a number of members are mainly directed toward the possibility of exchanging information or of acquiring knowledge of ongoing technical developments and evolving company strategies, rather than toward directly influencing standardization.

Research on Standard Setting

The few empirical studies of standardization and the great number of "modeling exercises [which] have run well ahead of the solidly established fact base" have almost exclusively had an economic or an econometric foundation (David and Greenstein 1990, p. 35). Modeling especially tends to conceive of standardization as a market process involving monopolistic and oligopolistic constellations. Here the introduction, the adoption, and the diffusion of a standard are contingent on the (sometimes strategic) choice of the actors on the supply side and/or the demand side. Where a standard is seen as the partly controllable result of conscious acts, the analysis of standard setting assumes the prevalence of imperfect markets and power differentials.

Models of dynamic competition between firms that diverge with respect to vested interests, investments in an "installed base," market leadership, or patent-right titles analyze the conditions under which these firms choose to offer either compatible or incompatible technologies on the market. Pre-announcements, preemption, leapfrogging, predation, strategic pricing, and second-source production are among the elements of business strategies that are scrutinized.[4] These relatively stylized models concentrate on behavior in markets, although they are concerned with situations for which the neoclassical assumptions of atomistic markets are explicitly declared obsolete, since choices are interrelated. The models still assume a high degree of company independence and scope for voluntaristic choice, thus implicitly presupposing that the firms are capable of supplying "complete" systems. In this view, the firms do not decide on standards as rules; they decide on properties to be built into products, which happen to exhibit network externalities, positive feedbacks, etc. In most of these models, firms see neither a need nor an opportunity to directly coordinate on product properties in order to safeguard compatibility. They act and react in isolation from one another. This includes the decision to overcome technical incompatibilities by introducing a gateway technology (David and Bunn 1988, p. 170). Such reference to a market-compatible option represents a solution that can be easily integrated into the models. In point of fact, introducing gateways and similar technologies only shifts the analytical problem. Gateways, adapters, and emulators are real prototypes of technical components which are useless as stand-alone appliances, and, like market standards, they may be undersupplied or—less frequently—oversupplied (David 1987, p. 233; Blankart and Knieps 1993b, p. 47).

Very few models take into consideration that voluntary participation in standards committees is an often-used alternative for managing coordination problems of groups of interdependent actors. What has proved particularly valuable has been a comparison of markets and committees with regard to their ability to achieve coordination. With special and explicit reference to the choice of compatibility standards as a means of coordination, Farrell and Saloner (1988, p. 235) contrast committees and markets, whose mechanisms they define as follows: "The first [mechanism] involves explicit communication and negotiation before irrevocable choices are made: it represents what standardization committees do. The second mechanism, by contrast, involves no explicit communication and depends on unilateral irrevocable choices: it succeeds if one agent chooses first and the other(s) follow(s). This is a simple version of 'market leadership.'" Using game-theoretic methods, Farrell and Saloner (ibid., p. 238) arrive at the (welfare) conclusion that, when compared directly, committees, which are negotiation systems, outperform market processes with respect to their "relative speed, efficacy and payoffs." A further refinement of the model combines market and committee elements into a "hybrid game," with the result that unilateral actions improve committee performance. Such hybrid situations, in which firms do not confine their activities to a committee but seek advantages through the market too, appear quite realistic. In committees, for instance, one dominant supplier can precipitate compromise by threatening to otherwise use the "outside option" of advancing its proprietary specifications as a de facto standard on the market (Weiss and Sirbu 1990, p. 128). Farrell and Saloner's central tenet that from a welfare perspective "committees are always desirable" (1988, p. 251) is interesting in this respect. Thus, a switch of attention from modeling "pure" market processes of technology adoption toward analyzing "real," institutionally entrenched, committee-based standardization processes, which include potential acts of blocking, delaying, bypassing, or preempting committee work, appears highly desirable.

The number of empirical studies of the standardization process in international committees, in comparison, is very small. The well-documented battle for worldwide adoption of a single standard for color TV in the 1960s, for instance, was more a matter of high international politics than a subject of controversy in a standardization committee, although the Comité Consultatif International des Radiocommunication was partly involved (Crane 1979, 1992). Spectacular events of this kind bias the public's perception of standards-related activities. They represent an authentic but very small segment of international standardization.

Most relevant in our context is an examination of the development of X.25, a standard for packet-switched data communication between computers issued by the Comité Consultatif International Télégraphique et Téléphonique (Sirbu and Zwimpfer 1985). In 1972 the CCITT's newly formed Study Group VII (responsible for "new networks for data transmission") started considering the question of international data communication over public data networks and, within this framework, also paid attention to packet switching. This mode of operation was rather new for the operators of public networks, who were concerned with circuit-switched data networks. Sirbu and Zwimpfer show how a small group of highly committed telecommunication carriers, in a two-stage move, first managed to reach an internal consensus on divergent positions they were holding with respect to several variants of packet switching and then succeeded—by submitting and distributing a draft proposal well before the meeting of the study group—in eventually terminating the formal process through unanimous adoption of the proposal by the CCITT Plenary Assembly.

Sirbu and Zwimpfer sketch the formal constraints a Canadian and a US carrier faced as members of national delegations whose official positions differed from those of the two carriers. The CCITT's membership rules and the principle of national representation forced the carriers to seek and find support for their proposal from the PTTs of France and the UK, who determined their countries' respective positions. In negotiations outside the official meetings, coalitions were forged and a draft proposal was finalized. A "rapporteur group" on packet switching, formally set up by Study Group VII and attended by 40–50 people, proved to have worked too slowly to attain acceptable results before the end of the four-year study period.

The resulting standard, approved in 1976 on the basis of a highly technical and detailed 90-page recommendation, was imperfect; however, as Sirbu and Zwimpfer (1985, p. 42) argue, "an imperfect standard is better than none." Without a committee standard in 1976, the networks that were to be installed in the following years would have "been allowed to evolve independently, [and] the chances of reaching agreement later would have been drastically reduced" (ibid.).

Sirbu and Zwimpfer emphasize that the need to develop standards for computer communications placed unprecedented demands on the CCITT, which was accustomed to handling problems associated with voice communications. They refer to the fact that the convergence of telecommunications and data processing in telecommunications networks has exerted pressure on the CCITT since the 1960s. Gerd Wallenstein (1990, p. xiii), for many years one of the leading experts in international standardization, experienced this very early:

> Thirty years ago, I attended a CCITT study group meeting for the first time. It had been called for a working party focused on a single question (number 43) in one study group. The question concerned CCITT's possible standardization of data modems "for transmission of accounting data over the telephone network." The meeting was attended by more than one hundred people, many representing companies in the data processing industry. . . . The meeting, CCITT's largest up to that time, turned into a somewhat theatrical clash of two cultures.

In the specific case of X.25, a result was achieved by partly circumventing the formal procedures through informal meetings among the key parties. The participants—top managers rather than technical specialists—were able to commit their organizations to an agreement on a technically "weak" standard because "economic priorities were allowed to override technical concerns" (Sirbu and Zwimpfer 1985, p. 44).

An implicit institutional argument put out by Sirbu and Zwimpfer (1985) points out the shortcoming of CCITT procedures, which concentrate on technical problems and which therefore provide no adequate platform for official economic reasoning. This gets pushed into the background at informal meetings of exclusive circles, which may enable "the dominant economic unit" to force its standard by creating sufficient momentum by building coalitions outside the official arenas (ibid., p. 43). Official technical work of the standardization committee is thus reduced to elaborating on questions arising within the informally pre-defined frame of reference. Although Sirbu and Zwimpfer touch on a crucial element of institutionalization, namely the provision of a specific rationality of action corresponding to processes of functional specialization, they seem to overstate the distinction between formal and informal spheres, suggesting a dominance of economic calculation in the latter. Details of their case study infer that the technical content of standardization has never been completely eliminated, either in the formal or in the informal sphere.[5] For instance, the decomposition of X.25 into three layers, each requiring separate standardization, was clearly a technical achievement. It allowed economic conflicts originating from the vested interests of those who had already implemented packet switching to be specified and isolated and, as a result, to be overcome more easily. Moreover, the decomposition into three layers substantially helped to improve the standard later.

A comparable effort by the Institute of Electrical and Electronics Engineers to agree on a single standard for local area networks (LANs) failed. Approached by several semiconductor merchants, the IEEE's Committee 802 began standardization in 1980 after a special subcommittee of the International Electrotechnical Commission known as PROWAY, which had been set up in 1975, had so far failed to make any progress. As a first step the IEEE committee agreed to develop a layered standard in conformity with the Open Systems Interconnection (OSI) Reference Model recently published by the ISO (Sirbu and Hughes 1986, pp. 9–10). Three subcommittees started work requiring the involvement of many individuals, including the former chairman of PROWAY, who redirected their efforts from the IEC committee to the IEEE.

Sirbu and Hughes provide a rather detailed analysis of the standardization process, which began with a good opportunity for "ethernet" to become the single LAN standard but which ended with the approval of three incompatible standards (an "ethernet" variant, a "token bus," and a "token ring" standard). In the past, the IEEE had usually succeeded in achieving consensus on a single standard. All the relevant American and (subsequently) European computer manufacturers, chip producers, and retailers, as well as academics, government officials, and several users, participated in the enterprise. Some proponents of the "ethernet" standard had also approached the more exclusive European Computer Manufacturers Association,[6] which had already (in June 1982) adopted a draft standard very much in line with the major points of the "ethernet" adherents in the IEEE. The draft was officially supported by 19 companies in order to give notice to users and governments that standards other than those pertaining to IBM networking were available (Sirbu and Hughes 1986, p. 17). Yet this move could not prevent the IEEE from approving three incompatible standards, including an IBM proposal. Subsequently, the ISO also decided to adopt the IEEE 802 documents in toto (Stallings 1985).

In interpreting their findings, Sirbu and Hughes (1986) undermine the first impression one might have gained that differences in the parties' business interests adequately explain their behavior and the outcome of the standardization process. They argue as follows (ibid., p. 20): "Firms which might have been expected to have similar [business] interests made different decisions as to product and standardization strategy. Xerox perceived a value in LAN standardization for its products; Datapoint, producing similar workstation products, did not." By no means did the actors have an equal comprehension of the subject matter, and "many of the participants in IEEE 802 conceded they were there to learn rather than support any particular position. It means that even when there is a will to reach a standard the process can be lengthy as the participants each struggle to understand the other's arguments."

Few engineers—and the majority of participants in standardization belong to the technical profession—can be expected to grasp all technical issues equally well. Engineers' understanding of business issues is certainly even worse (Sirbu 1989, pp. 7–10). Moreover, in many areas the discussion on standards begins well in advance of clear market experience or even market perspectives (Barke 1993). During the process, new technical innovations and a new understanding of market needs change a firm's calculations (ibid.). The uncertainty triggers the coping strategies that Cyert and March (1963) describe as suboptimal decision making, satisficing, and sequential attention to goals. As with other negotiations, however, firms that invest time and money in preparation for the process can influence the outcome in a way that is favorable to their interests; thus, participation in committees always makes sense (Sirbu and Hughes 1986, p. 1).

Sirbu and Hughes (1986, p. 21) emphasize that, partly as a result of the factors delineated, committees tend to produce lengthy standards and to incorporate several options "with the recognition that the market will ultimately select from among them." In the case of LANs, however, IEEE did not adopt one standard that embraced several options; it adopted three incompatible standards. That this could happen is explained by the system characteristics of local area networks. LANs, Sirbu and Hughes argue, are bought and operated by single companies, and within an organization it is possible to interconnect workstations and terminals using one of the standards as long as the appropriate network interface unit is consistently purchased. Thus, every single buyer can secure internal interworking by adhering to one of the available standards. But different standards can coexist in the marketplace, because intercommunication between LANs of different organizations was not seen as an essential requirement in the early 1980s. In fact, the primary importance of LAN standards was related to the high investments necessary for the development and production of chips that could handle the complexity of LAN technology. The semiconductor manufacturers, therefore, were most interested in a standard that guaranteed a large market volume. Once this was seen to be secured, even with three coexisting incompatible standards, a considerable amount of variety of LANs could be tolerated because they were used as stand-alone systems (Sirbu and Hughes 1986, pp. 22–25).

Marvin Sirbu also contributed to the third study to be presented here, which proceeds from the premise that the complexity and the subtlety of the standardization process make it almost impossible to observe consistencies. Weiss and Sirbu (1990, pp. 111–112) employed quantitative methods to test "the conventional wisdom among observers that the outcome of a standards process is random" (ibid.). Selecting eleven decisions on standards, they used questionnaires to collect information on the historical and technical features of the decisions, and on the number and types of contributions and the number and types of participants involved in each alternative standard proposal. These questionnaires were mailed to the key proponents of the alternatives. In addition, background information concerning the market share, installed base, financial status, etc. of each firm involved was used (Weiss and Sirbu 1990, p. 117). The decisions had been made in several standards committees in the area of computer communications hardware. For Weiss and Sirbu, technical standardization is part of the competitive product-development process. Indeed, general project descriptions, often the starting point of standardization, resemble the product proposals of a manufacturer, and argumentation and debate in standards committees refers to these technical proposals (ibid., p. 113).

In each of the eleven cases examined, two firms or two coalitions of firms were fighting to have their preferred standard adopted by the committee. But Weiss and Sirbu (ibid.) emphasize that most standards "are developed [cooperatively] in committees or are adopted without opposition," which makes conflicts exceptions to the rule. Market power, size of firms, installed base of products, promotional activities of the sponsors (e.g., technical contributions submitted), the perceived superiority of the standard, and the political skills of the coalitions were used as explanatory variables for the decision on the contending compatibility standards. Statistical analysis disclosed some significant correlations. In particular, the size of the firms in the coalition supporting a proposal and the extent to which a position is backed by written contributions determine the choice of a standard. Market share only turned out to be an important factor on the side of the buyers of compatible products; that is, monopsony power was significant. In addition, standards supported by firms that placed greater emphasis on marketing than on purely technical factors were more likely to be approved. Weiss and Sirbu (1990, p. 118) show that coalitions between firms were issue-specific and not based on general mutual agreement. In one case, IBM and AT&T were members of the same coalition opposing a proposal from DEC; in another, AT&T and DEC collaborated.[7]

What has regularly been overlooked by economists is the technical content of the standards and the autonomous role of technical arguments and technical criteria in the decision-making process. Weiss and Sirbu explicitly include these factors and arrive at some interesting conclusions. Whenever a decision was made in a contentious case, the supporters of both the adopted and the non-adopted standard generally still felt strongly that the standard they supported was the best. The supporters of the non-adopted standards "had a somewhat narrower view of what constituted 'better' [whereas] the supporters of the adopted technologies [were] more eclectic as a group" (Weiss and Sirbu 1990, p. 130). For the "eclectic" position, which could also be termed pragmatic, Weiss and Sirbu (ibid.) add, "such a viewpoint may lead to greater possibility of compromise, if available, than a more narrow viewpoint, because tradeoffs can be made across multiple, evenly weighted dimensions." The conclusion of Weiss and Sirbu is twofold. First, although their data did not reveal a significant influence of technical superiority over technical choice, this apparently plays a central role in the discussions; moreover, the data are consistent with the view that a standard generally believed to be technically inferior to others could not be pushed through the committees. Second, as the analyzed cases deliberately concentrated on decisions in which significant dissent existed among committee members, the intense debates might have occurred only because the standards were technically approximately equivalent (Weiss and Sirbu 1990, p. 130). Only in such a situation of technical deadlock may economic and business interests and arguments be invoked in order to reach a solution.

The most salient results of the research on standardization processes are these:

• Besides the market mode and the hierarchical mode of technical coordination, elaboration of and negotiation on standards in committees embodies a distinct practice of coordination.

• The institutionalized rules governing committee work have created opportunities and constraints, different from those in markets and hierarchies, that must be taken into account when the standardization process is analyzed.

• Besides business interests and professional interests, national political interests play a crucial role in international standardization.

• Highly specialized technical argumentation constitutes an autonomous discourse in the standardization process.

• The development of a standard that also contains a possible number of alternative options depends, ceteris paribus, on the collectively perceived requirement of compatibility. This perception embraces mainly technical and economic elements.

• Economic interests and technical perceptions are often diffuse or ambivalent. Accordingly, participation in committees is often seen as an opportunity for learning and collective problem solving rather than for pushing a specific standard.

The last point especially deserves more elaboration. Notwithstanding the insights provided by the available studies on standardization, these studies remain essentially unsatisfactory. It is their difficulties in explaining the outcome of the standardization process in committees in general that point to a lack of complexity in approach. The economic models, for example, completely "black box" the technical content of standards, they concentrate on choices being made by actors on the basis of a presumed unambiguous "payoff matrix," and they tend to ignore processes of learning and collective problem solving. On the other hand, the few empirical studies that are slightly more sensitive in this respect have missed the opportunity to focus attention on the process dynamics of standards committees,[8] the impact of institutional norms on the willingness to cooperate, or the specific problems of the delegating organizations in monitoring the behavior of their delegates in the committees. Wallenstein (1990, p. 1) alludes to this aspect: "Coming together from competing companies and diverse countries, these standards makers form a small, transnational subculture held together by technical expertise and a specialized language of their own making."

Conflict and Cooperation in Standardization Games

Every approach that seeks to model or explain the standardization process is necessarily selective. Our own model, which will be introduced below, is itself incapable of overcoming all the "structural" weaknesses we have just listed. Some of them, however, seem to result from an internal conceptual vagueness, which cannot be regarded as a problem intrinsic to the different approaches. Most striking is their equivocal understanding or use of the concept of compatibility. This is intriguing since the processes that are analyzed focus on the creation of standards drawn up to secure this nebulous compatibility.

A closer look at what is generally meant by "compatibility" shows two concepts prevailing. Using a distinction introduced by David and Bunn (1988, p. 171), we find technical artifacts based on the same standard being viewed either as "compatible complements" or as "compatible substitutes." In the first instance, two technical components Y and Z are said to be compatible because they can be used with each other. When a specific format of film (e.g., 35 mm) is compatible with specific cameras or a certain word-processing program runs on a certain personal computer, the components are complementary—"vertically compatible," as Liebowitz and Margolis (1990b, p. 2) call it. In many cases, single components are useless unless they interoperate or interwork with compatible complements. Producers as well as users of

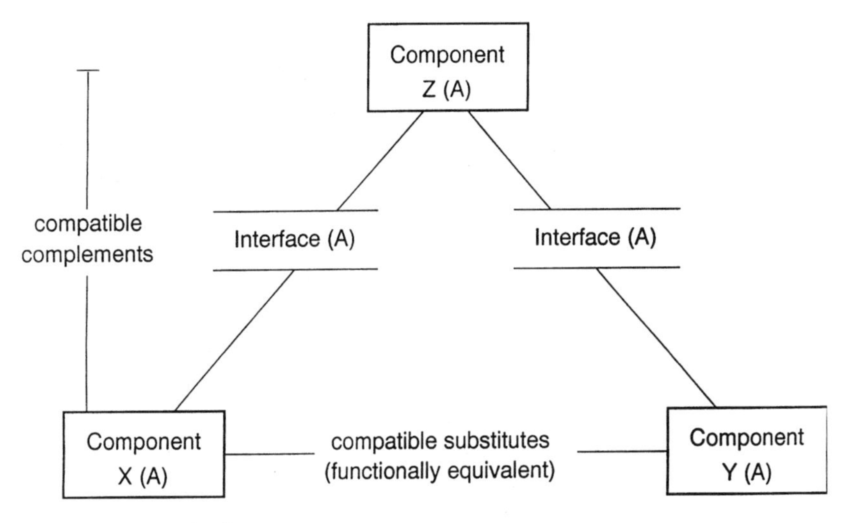

Figure 4.2
Two faces of compatibility.

these components have a vital interest in compatibility. However, if in the second instance not only component Y but also another functionally equivalent component X is compatible with component Z, then from the users' and the producers' point of view it makes no technical difference whether X or Y is employed. Thus, with respect to interoperability with Z, both components are compatible substitutes or "horizontally compatible" (ibid.).

Figure 4.2 illustrates the two faces of compatibility. Economists often have only the latter aspect of substitutive or horizontal compatibility in mind when they model the effects of compatibility standards (Pfeiffer 1989, pp. 17–25; Knorr 1993, pp. 27–32).[9] Substitutive compatibility is indeed a precondition for competition (Tirole 1988, pp. 298–301), and it is this aspect that is referred to when claiming that standards constitute markets. Substitutive components compete; complementary ones do not. In either case, a compatibility standard is the rule specifying the relational properties at the interface of the components.

To clarify the implications of the two aspects of compatibility, it is useful to draw on game theory. Here and in later stages of our analysis, we refer to game theory predominantly because its concepts can be used as a heuristic device to clarify crucial elements of typical standardization dilemmas, without needing to engage in the complexities and rigidities of the theory in greater depth.[10] Game theory offers a terminology and basic concepts for modeling and analyzing actors as "players" in games, where the individual "payoff" depends not only on individual choice but also on what the other players do. Their possible moves in the game have to be considered by the other players when determining their own moves. Thus, genuine interdependence is a constitutive element of a game, and the players take it into consideration. Schelling (1984, p. 214) notes that "there is no independently 'best' choice that one can make; it depends on what others do." For standardization, this underlying interdependence among the actors is a central characteristic (Hemenway 1975, pp. 101–102). However, interdependence should not be regarded as objectively given; it is rather a matter of perception by the actors involved and their collective definition of the situation, including the game's payoffs and rules. Technical as well as institutional factors certainly influence but do not determine the definition of the situation, and the definition changes over time. Thus, the game being played changes too.

Returning to the two faces of compatibility, we start with the case of *complementarity* (Economides 1989; Economides and Salop 1992). When only the components Y and Z have to be standardized, producers (and users) will generally be interested in securing compatibility by agreeing on a single standard, because in this case of complementary compatibility all will be better off with than without a standard. One may regard this instance as the ideal type of a pure-coordination game (figure 4.3), where there is no disjuncture between the collective rationality and the rationality of each of the two individual parties (Snidal 1985, p. 931). Here the main problem is the determination of a common solution, namely the concrete shape of a standard. When different standards can be chosen, choice is not conflictive if no player expects a special advantage of one standard over the other. Thus, standardization requires no more than communication, which in the institutionalized context of the standardization organizations will invariably suggest itself. In addition, once a solution to the coordination problem has been agreed on, it is in everyone's interest to comply, as is the case with driving on the left or the right side of the road (Tietzel 1990).

If the user/producer of Y, however, prefers solution (standard) A, whereas for the user/producer of Z solution B is clearly more attractive, then it may be difficult to achieve compatibility even though the parties

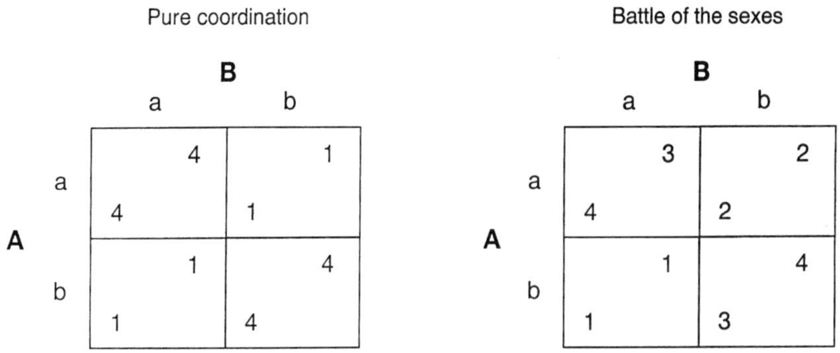

Pure coordination

		B: a	B: b
A	a	4 4	1 1
	b	1 1	4 4

Battle of the sexes

		B: a	B: b
A	a	3 4	2 2
	b	1 1	4 3

A, B = player (or consistent group of players) A, B
a, b = different options (in the battle: a is preferred by A and b is preferred by B)
Values at the top right = payoffs of player B
Values at the bottom left = payoffs of player A

Figure 4.3
Pure-coordination game and battle of the sexes.

do not dissent on the higher value of a common standard. In terms of game theory, this conflict is a "battle of the sexes."[11] In this game, the players seek coordination on a common solution which promises the highest total payoff. Yet each solution implies different relative gains for each party. One of the players has to settle on his or her second preference. The difficulties in overcoming this distributive issue of whose preferred standard is to be realized—against the mere second-best option—pose the danger that no common solution at all might be reached. Then both parties are worse off (figure 4.3).

In our view, possible gains and losses in such constellations need not be based only on economic considerations. Differing preferences can also be caused by political aspects, preferred technical orientations, scientific opinions, or reputational concerns. With the cluster of political, economic, and technical conditions affecting telecommunications, difficulties for successful technical coordination are naturally present. As the importance of standardization has grown with the internationalization and diversification of communications, the battle-of-the-sexes constellation has become more likely. The concerned actors have become more diverse, often with heterogeneous technical orientations involved in addition to competitive economic interests.

Pure-coordination games as well as battle-of-the-sexes games belong to the category of "positive-sum games," where a higher overall payoff can be achieved when an agreement on one solution is reached. Standardization organizations of the type we described in the preceding chapter are designed to enable the production of such surplus. However, they provide no guarantee that it will be produced, so coordination failure remains a realistic possibility.

Coordination failure becomes tangible when we look at the other face of compatibility: substitutivity. For this reason we reintroduce into the game the producer/user of component X, which is functionally equivalent to Y (figure 4.2). If both components realize their utility through interoperation with Z, the producer/user of either will be greatly interested in compatibility with Z. A single standard that is adopted and implemented will achieve compatibility, but at the same time it will lay the groundwork for competition between the substitutive components X and Y. If the producers/users of X and Y prefer different standards, they may fear they have a disadvantage if they accept the other party's proposal. For both of them it is certainly easier compromising with the producer/user of the complementary Z than reaching an agreement with each other. If the potential competitors prefer different standards (A and B respectively) and if the producer/user of Z is indifferent as to which standard to adopt as long as it is only one single standard, the differential advantages (relative gains) may become the center of collective attention and from either party's point of view become even more significant than the absolute size of the common advantage (gain). For the overall functioning of the technical system it makes no difference whether standard A or standard B is employed, but for the producer/user of component X and component Y the concrete choice may, in the most extreme instance, be synonymous with total success or total failure. From the perspective of these two parties in this extreme case, the game to be played transforms itself from a battle of the sexes into a zero-sum game: the one side's gain is equal to the other side's loss. In such a situation, no standard is likely to be agreed upon (figure 4.4). Examples of such zero-sum games could be found in our analysis of market standardization processes above. In committee standardization, such a drastic transformation appears less likely because a spirit of compromise has been institutionalized.

Game theory has illustrated that calculations of relative gains can change the payoff structure of a positive-sum game and thus modify the nature of the game. The illustrations rely either on considering the significance of individual relative gains compared to the joint

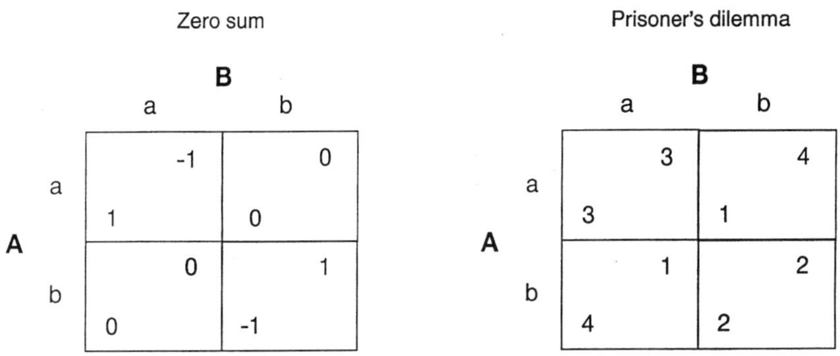

Zero sum

		B: a	B: b
A: a		-1	0
		1	0
A: b		0	1
		0	-1

Prisoner's dilemma

		B: a	B: b
A: a		3	4
		3	1
A: b		1	2
		4	2

A, B = player (or consistent group of players) A, B
a, b = different options (a is preferred by A and b is preferred by B; in the prisoner's dilemma: a = cooperation and b = defection)
Values at the top right = payoffs of player B
Values at the bottom left = payoffs of player A

Figure 4.4
Zero-sum game and prisoner's dilemma.

advantage of a consensual solution (Snidal 1991) or on introducing the concept of interaction orientations as variable individual properties which become relevant in the game situation (Scharpf 1989). If actors switch from concentrating on their own isolated advantage ("individualistic" orientations—typically presupposed in game theory) to looking at what the others get ("competitive" orientations), the game can change. The inclusion of possible changes in action orientations clearly "weakens" the objectivist basis of game theory, but this makes the approach more useful to social theory. The idea draws on Kelley and Thibaut's (1978, p. 17) distinction between a given payoff matrix depicting the outcomes as determined by both the environment and the actors' utility functions and an effective matrix representing the actors' perceptions. The perceptions depend on their evaluations, and this is what the actors base their choices on.

Thus, it is easy to imagine that, with changing evaluations and perceptions, the battle-of-the-sexes game can, for instance, be transformed into a "prisoner's dilemma," the type of game that has been most widely applied in the social sciences.[12] Its significance lies in the fact that it clearly demonstrates how rational individual action may lead to a suboptimal outcome, as each player is more affected by the actions of the other player than by his or her own moves (figure 4.4). In this sense the game is "the standard representation of externalities" of actions (Snidal 1985, p. 926). In the case of standardization, it is obvious that without external intervention in such a dilemma the actors would fail to agree on a standard that generates positive (e.g. network) externalities and positive individual returns. On the contrary, joint disadvantages would be likely to result.

A prisoner's dilemma can emerge when both the proponents of standard A and the proponents of standard B see an opportunity to compromise on a third standard, C, but find it rather expensive to draft C and later migrate to this solution. In this situation, both sides may hesitate to start preparing for this standard. Clearly they would both prefer to see their own solution realized, and each definitely rejects the other side's solution as the worst case from its point of view. Thus, as long as they are uncertain whether the other side will really cooperate, they will not cooperate either. In this dilemma, the actors end up with no standard at all—the worst of the possible options.

The typical situation in compatibility standardization, however, resembles the battle of the sexes sketched above. Stein (1982, p. 309) calls it a "dilemma of common aversions." The actors have a strong common aversion to a particular outcome (in our case, incompatibility), but they lack a common recipe for preventing it. They prefer different standards to overcome the situation. No single solution exists that is considered best by all actors. This is different from a "dilemma of common interests" (Stein 1982, p. 304), such as the prisoner's dilemma in our last example. Here a specific compatibility standard lies in the actors' common interest, but as they voraciously attempt to obtain their individually most preferred solution they arrive at no standard at all. Stein concludes that this dilemma of common interests can only be overcome by a neutral authority with obligatory rights to intervene or in a "collaborative regime" based on enforceable agreements, whereas the dilemma of common aversions can also be mastered by institutionalized self-coordination on the part of the actors in a "coordinative regime" (see also Stein 1990).

This leads us back to our starting point. Institutionalized standardization, especially at the international level, can be regarded as organized self-coordination combined with several elements that help to avoid or overcome the prisoner's dilemma and similar situations (Keck 1993). The stylized game-theoretic concepts introduced here overstate certain aspects of the standardization dilemma, which are mediated in the organizational reality of standard setting. The games being played are not completely noncooperative. Rather, in standardization organizations, the actors are institutionally expected to cooperate—they are "supposed to be able to discuss the situation and agree on a rational joint plan of action" (Nash 1953, p. 128). Reputational concerns and "the prospect of playing through time" can be regarded as additional sources of cooperation (Snidal 1985, p. 938). It would be most helpful if the spirit of cooperation were to lead to binding and enforceable agreements on standards. In contrast to the national level, where enforcement is possible in principle, at the international level this remains an open question (Siebe 1991).

A closer look at "real world" standardization processes points to a feature that has been neglected in the game-theoretic models of standardization. In many cases the technical solution to a compatibility problem and/or the economic implications of different alternatives are uncertain. In consequence, individual preferences may stay indeterminate and ambivalent during considerable parts of the standardization process. What Schelling (1960, p. 21) calls "the efficiency aspect" of negotiating—finding an adequate solution to a problem—often clearly overshadows its "distributional aspect." First, value must be created through joint action before it can be claimed individually (Sebenius 1991). At the beginning of a standardization process, actors find themselves in a situation of receiving incomplete information about what the others or even they themselves (should) prefer.[13] Although the institutional embeddedness of standardization reduces informational uncertainty to some extent, in every concrete standardization case new problems will arise, especially with respect to the technical content of the matter. Knowledge concerning the technical ability of a standard to achieve compatibility is incomplete, and collective efforts are deemed necessary in order to improve the knowledge base (Barke 1993).

In their studies on "constitutional choice," Vanberg and Buchanan (1989, 1991) introduce a distinction that helps us to treat the technical content of standards and the related knowledge problems separate from the actors' interests. In contrast to the widespread interpretation of choice in terms of preferences and constraints, where the notion of preferences generally blends evaluative and cognitive components, Vanberg and Buchanan (1989, p. 51) emphasize the need for a clear distinction between these elements: "How a person chooses among potential alternatives is not only a matter of 'what he wants' but also of 'what he believes,' and for some kinds of choices an actor's beliefs or theories may play a most crucial role. We suggest that the second element is particularly important for constitutional choices, that is, for choices among rules."

Rules are instrumental in nature and not objects valued in themselves. Although Vanberg and Buchanan (1991, p. 61) have in mind constitutional rules that are supposed, first and foremost, to govern behavior, constitutional choice can also be directed at more specific rules in a multi-layer system of rules. Thus, in our view, standards designed to achieve compatibility can be regarded as a family of specific technical norms or rules, whose creation is normally beset with knowledge problems.

As long as the knowledge base in a standardization process is incomplete, disagreements can be seen as resulting from this deficiency. Its elimination is a common concern (see also Morrow 1994). Agreement can be achieved through exploring and establishing what is true and what is false. From this perspective, controversies and competition among experts in standardization may even appear as a "discovery procedure" for finding desirable solutions and, in this sense, form a "constitutional discourse" designed to include all factors that help to alleviate the knowledge problem by eliminating disagreement (Vanberg and Buchanan 1991, p. 64).

Before we introduce our general model of the committee standardization process, we need to expand on the specific problem of this mode of standard setting, namely the compliance with or the implementation of its standards. As we have demonstrated, standardization organizations in general are designed to achieve agreement on technical rules, and these rules are supposed to coordinate technology. There is no guarantee, however, that the cognitive constructs will ever be implemented in a technical artifact. Rather, the achievement of coordination through compatibility standards issued by committees entails a public-goods problem. Especially in cases where—after a "war of attrition" (Farrell and Saloner 1988) involving heavy controversies—a weak consensus has been reached, early implementers will risk loss of functionality and money because adoption may not become widespread. Only rarely may there be first-mover advantages for early adopters, especially when the agreement on a standard has been difficult to accomplish (Farrell and Saloner 1985). Standardization organizations generally have no policing or sanctioning power and no other means to directly control or monitor the implementation of their "products." Thus, the actors who decide upon the adoption of a compatibility standard may find themselves in a dilemma: common

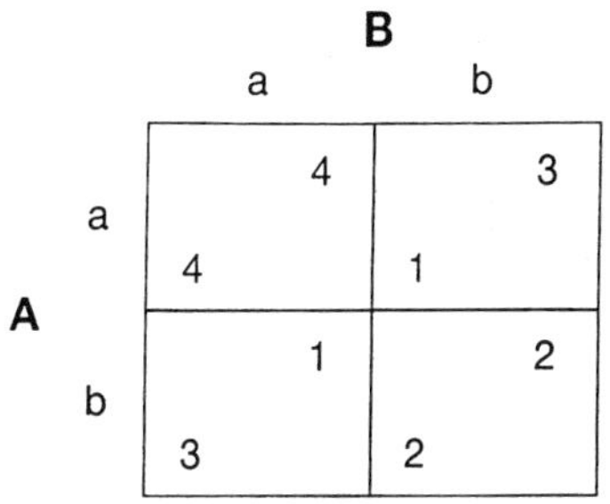

		B: a	B: b
A	a	A: 4, B: 4	A: 1, B: 3
	b	A: 3, B: 1	A: 2, B: 2

A, B = player (or consistent group of players) A, B
b = defection ; a = cooperation
Values at the top right = payoffs of player B
Values at the bottom left = payoffs of player A

Figure 4.5
Assurance game.

implementation gives the highest total and individual benefit, but isolated implementation is highly disadvantageous.

With network technologies, the implementation of a standard can be characterized as an "assurance game." Assurance depicts a situation where cooperation (i.e., joint implementation of a standard) gives the highest possible payoff to each player and at the same time the highest common payoff. This, in principle, provides a great incentive to cooperate, in contrast to the much more problematic prisoners' dilemma. The quandary of assurance lies in the danger that if the other player does not cooperate, the cooperating player will achieve the worst possible payoff, and he or she would be better off not cooperating also (figure 4.5). Thus, the players may be willing to cooperate only if there is enough assurance of mutual cooperation. Problems therefore arise if the players want to avoid any risk. In this case, neither player will implement the standard, and coordination will invariably fail (Colman 1982, p. 25).

Generally, risks are connected with decisions to invest in the implementation of a compatibility standard. If one producer implements a standard but others do not, and if there is mutual technical interdependence, as is often the case in telecommunications, then the investment will be lost. Assurance that a standard will be broadly implemented may result from the "signals" of strong firms preparing to implement (Schneider 1993). Other firms may quickly follow. This resembles the case of pure (sequential) market standardization; the only difference—from our point of view a highly significant one—is that the standard is not an intrafirm decision but a result of committee work. Assurance may, of course, also be achieved through negotiations and binding agreements, especially if the circle of interested actors is comparatively small. This transforms the noncooperative game into a cooperative one.

The game-theoretic view of standardization has helped to elucidate the tensions that autonomous but interdependent actors encounter in their search for compatibility. The uno actu realization of complementarity and substitutivity through compatibility characterizes the standardization process in most cases as a battle of the sexes. The game's distributive aspect entails considerable obstacles on the way to a consensus, while its aspect of problem solving and common welfare provides an incentive to cooperate. The nature of standards as rules and their technical content tilt the balance in favor of cooperation. At the same time, they indicate that standardization cannot just be regarded as a process of negotiation on two or more clearly fixed options; it often takes the form of joint research and development, with common knowledge instead of competition representing the central means of coordination in pursuit of compatibility. Nevertheless, standards may include several options, or competing market standards may rival a committee standard's status as single rule of coordination. Standards generally underdetermine technical choice, with the result that the individual decision to adopt a standard invariably entails a risk.

A General Model of the Standardization Process

We have discussed standardization and technical development, market standardization, and institutional aspects of standardization, highlighting the structure of standardization processes and games. The empirical material and the theoretical considerations we have presented were inferred from different theoretical approaches. Though they come from diverging theoretical perspectives, all the approaches we selected have a common focus: the problem of coordination. Autonomous actors involved in the production, operation, and use of such a large technical system as telecommunications rely on a minimum amount of coordination. Compatibility is the key objective of technical coordination, and it can be achieved through standardization. Hence, technical standards serve as a medium of coordination. What they coordinate is human action concerning the design, production, combination, maintenance, or utilization of technical artifacts. As rules, standards specify relational properties of these artifacts. Compliance with the rules promises compatibility and thus promises the artifacts' smooth interoperation in a system. The rules, of course, are addressed to actors, and a selection of these actors sets the rules in standardization processes.

The process of standardization is by no means the execution of a linear techno-logic. It is, rather, contingent on institutional factors, actor constellations, actors' interests and perceptions, technical knowledge, and what one might call the artifactual reference of the specific coordination problem. Hence, scrutinizing standardization is particularly suited to highlighting the social shaping of one significant stage in the generation of network technologies. As the social process of standardization refers to technology, technical factors as well as social aspects must be integrated into the analysis. Although they are results of social action, many features of technical systems appear as exogenously given to the actors involved in any single standardization process. In the prevailing mode in which standardization is institutionalized today, technology is not created from scratch, but the process is typically concerned with only one or a few particular problems. Thus, undisputed features of existing technology constrain or shape the pool of feasible options available to the actors. In this sense, one might say that, once a standardization problem has been collectively defined or identified, physical laws as well as practical system requirements and the like (in our terms, elements of the technological foundation) almost mandate the selection of certain solutions (Vincenti 1990, pp. 200–207). Normally, however, multiple solutions exist and are regarded as feasible, and the preferences and criteria for the attractiveness of the different solutions diverge and can be contentious.

The empirical studies of standardization—some of which we discussed above—touch on many aspects of the process. Similar to exploratory investigations, however, they do not provide a solid basis for generalization.[14] Our three case studies are intended to augment the descriptive knowledge of committee standardization and to go some way toward explaining the output of standardization processes. During these processes, standards take shape slowly. Some elements are finalized early; others are submitted for further study and discussion. Proposed options are approved or rejected. Experts are mobilized and coalitions are forged in order to strengthen a specific position with respect to competing alternatives. At the end—typically, but not inevitably—a set of technical rules and subrules results which is called a standard. From this point of view, the standard is the dependent variable, whose gestalt or shape is influenced by social factors such as institutional and procedural rules and politically, economically, scientifically, or otherwise motivated interventions in "purely" technical considerations.

Figure 4.6 gives an overview of our model of the standardization process (a model that is more heuristic than explanatory). Only parts of this general model are considered in our case studies, because the cases do not provide for all listed variables the variance needed to examine their causal influence on the standards. The first block of the figure contains three groups of variables comprising *structural elements,* which frame the process of negotiation and decision making on standards. The empirical material and the theoretical considerations discussed above deal primarily with such structural variables. A second block of "intervening" variables relate directly to the *process aspects* of standardization. The last block comprises certain standards features that indicate the output of the standardization process.

A detailed analysis of the institutional framework of standardization—the first set of structural variables—provides an adequate starting point for our explanation. The institutional context (its participatory rules, working procedure, culture, finances, and so on) is not only a precondition for attaining the output; it also shapes and influences the result of the work in a distinct way. Although (as we showed in chapter 3) the international standardization organizations display substantial similarity, their specific differences account for some variation in the output. An ISO standard for a terminal is likely to be different from a comparable CCITT standard. Our decision to concentrate on three standards processed mainly within the CCITT precludes a systematic comparative assessment of the impact of institutional variables on standards. Thus, by keeping the organizational context constant, we deliberately eliminate this source of variation. Otherwise we would need a much broader empirical basis. A closer look at the CCITT, however, does show a considerable amount of internal variation, for example between different study groups and subgroups or different levels of decision making. In addition, other standardization organizations directly or indirectly participated in the processes we analyze. Accordingly, the three cases will provide a certain amount of insight into the impact of institutional variables on standardization.

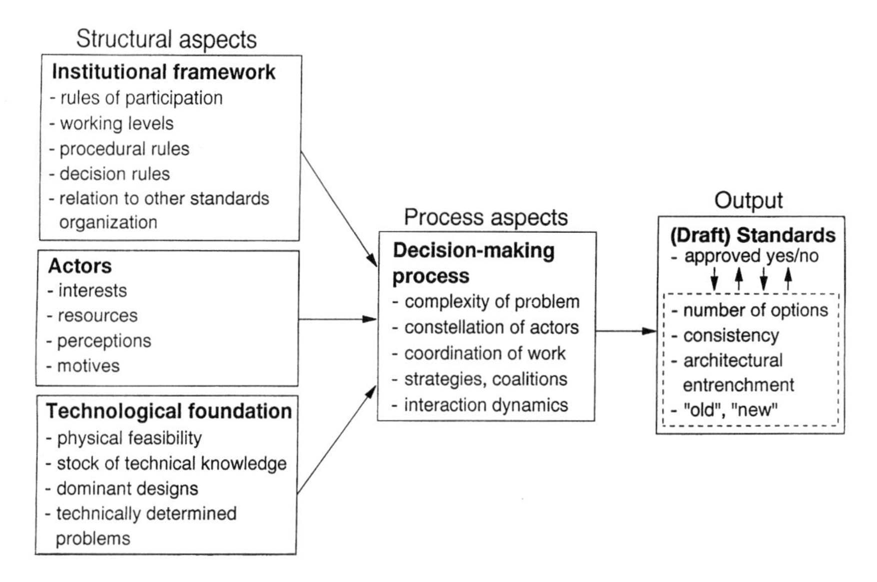

Figure 4.6
The standardization process.

The membership rules and participation rules of institutions channel the recruitment of actors who become involved in the standardization process. But actors will also at times choose the organization within which they prefer to push a standard, weighing the specific institutional opportunities and constraints each standardization organization provides from their point of view. Thus, the *properties of the actors* involved—their interests, resources, perceptions, and motives—constitute a second group of structural variables, which are independent of the institutional variables. Actors are, in principle, organizations which we regard as units capable of acting collectively.[15] The individual actors are usually delegates from these organizations, whose interests they are supposed to represent.

As standardization is charged to highly specialized experts, the delegating organization generally does not have the necessary knowledge to assess the relevance of participation, or to control and monitor the behavior of its own particular delegate (Irmer 1990, p. 22). Besides the rational grounds for engaging in the definition of standards that will secure compatibility in the organization's interest, involvement may be simply a matter of tradition; at least, it may be difficult to understand from a merely instrumental perspective.

For the individual members, participation might represent some sort of fringe benefit with at least potential usefulness for the organization. In a concrete situation it is often difficult to detect whether a person acts as an agent of an organization or from personal interest. In short, individual interests may be only loosely connected to recognized requirements for compatibility, to discernible corporate actors' interests in certain standards, or to technological developments in general. Not only the researchers but also the individuals directly involved are confronted with the problem of attributing motives—a problem which is aggravated when actions have to be recalled and reconstructed ex post.

In analyzing actors' interests, a loose distinction can be drawn between the motives for their involvement and the interests they carry intentionally into the standardization process. Although these two aspects obviously overlap, the motivation to engage in a standardization process can be understood as having been resolved by the time interests in standardization come into play. Not included in such interests are resources with which actors try to enforce their position in a standardization process, such as finances, market power, or standing in personal networks. A more distinctive resource is knowledge, especially technical and organization-specific procedural knowledge.

Interests are not necessarily economic. Although the desire for monetary gain in the area of technical standardization often leads to interested action, it is clear that political-ideological commitments and technical judgements and convictions, among other factors, also shape preferences in standardization (Marwell and Oliver 1993, p. 16; Sebenius 1991, p. 207).

The economic interests that actors carry into the standardization process will relate, for instance, to aspects of an already-installed technical base. In particular, major "investments in transaction-specific assets" (Williamson 1985, pp. 30, 52–54) can induce economic disadvantages in the form of high adaptation costs if new standards fit imperfectly into the established network context or into existing product ranges. However, not only past investments but also expectations of future market expansion and competitive advantages may guide standardization strategies. More narrowly, economic interests also pertain to pre-investment undertaken in the development of standards and to the resulting commitment to a specific solution.

Political interests are explicitly recognized in the CCITT and in other international standardization organizations, although they can be legitimately referred to only in specific situations. The international acceptance of nationally developed standards, however, generally means more to the actors than a mere competitive advantage for the domestic economy. To have such a solution recommended as the international standard by the CCITT is in many cases regarded as an important success for one's own country's technology policy and as a matter of national political reputation and prestige.

Technical interests come into play when, for instance, technical demands determined by the specific application concepts of a firm guide a choice between technical alternatives. Also, some technical options based on a specific standard may be preferred over others because they appear more elegant and not because they are economically better. Thus, concomitant strategies in standardization processes are rooted in the actors' technical interests. Comprehensive, more paradigmatic orientations of technological development are additional facets of such interests. An example of this is the Open Systems Interconnection framework, which is an orientation that excludes technological options violating its principles. Perhaps even more apparent are the differences in technological perceptions between the telecommunications and data processing fields, which have distinct views on how to solve a technical problem.

The third set of structural variables that influence standard setting encompasses the *technological foundation* or technological basis of the process. Standardization does not occur in a historical vacuum. Hence, the perceived need for standards and the choice of a solution are also shaped by technical factors that appear as objectively given. Many issues in standardization are, in this sense, technically imposed. Standardization builds on existing technical developments as well as on the stock of technical knowledge, including previous standards, and it enhances known functionalities. Which standards can be chosen or must be dismissed depends on the technological basis and the state of the art expressed in materialized dominant designs at a certain point in time. These designs channel the work in a specific direction and may constrain innovative standardization unless underlying concepts are deliberately altered. Like the institutional context, the technological foundation of standardization does not vary systematically in the empirical cases we have examined, and so it is impossible for us to determine its influence. However, as our analysis covers a period of up to 20 years, during which the technical foundation changed considerably and data processing entered telecommunications, it will be possible to find some indicators of the significance of the technological foundation.

The institutional context, the participating actors, and the technological foundation frame and shape what in figure 4.6 are termed the *process aspects* of standardization. Here, structural elements are invoked, activated, and combined to constitute the decision-making process. The internal dynamics generated during this process can be attributed chiefly to the specific configuration of the actors, to their heterogeneity or homogeneity, and to the strategies they pursue. The strategies and tactics are based on each actor's definition of the situation. The actor takes into consideration what the other participants may prefer, how they are going to act, and whether they are honest or dishonest. From one meeting to the next, the constellation of actors can change, and this can affect strategies and tactics. The individual cognitive comprehension of a complex technical problem is part of the definition of the situation, too. Also included in the dynamics of decision making are "process-generated stakes," which are produced by or pertain to negotiation behavior and the negotiation process (Underdal 1991, p. 106). In certain situations, for instance, an actor may be more concerned with saving face than with achieving a specific position. Apart from these situational dynamics resulting from the interaction within a specific constellation of actors, the decision-making process is mainly affected by the necessary coordination of activities and by the complexity of the problem.

Although the institutional framework does not vary with our focus on the CCITT, that organization's procedural rules are open enough to allow for different configurations of participating actors with quite disparate interests and for a general diversity in the possible organization of work. Our focus here is mainly on the study groups in which the relevant work leading to the recommendation of standards takes place. Coordination problems arise at three levels: within the study group, within the CCITT, and between different organizations dealing with similar or related standardization issues.

CCITT study groups have to accommodate hundreds of participants. Definitional work on standards is impossible under these conditions, which are rather similar to those of a small plenary assembly. The drafting of proposals for standards therefore takes place in working parties and their subgroups, or in ad hoc groups established for special issues. These meet on a more regular basis, sometimes being reduced to but a few experts working on a problem. The successful coordination of the division of labor between the groups depends much on the skill of the chairperson, who has to ensure that in the end all members of a study group agree on draft standards with which they are only partly familiar. Coordination problems also arise between the study groups. One group's output is often the input for another group, and complications already appear to arise when such input is delayed.

Because standardization often takes place simultaneously at the national, supranational, and international levels, and because standardization organizations with differing specializations (depending on the technology) are likely to be involved, additional problems of coordination will arise. In these cases, jurisdictional conflicts may occur because it is often unclear which organization should assume the main responsibility for the standardization of a technology.

The complexity of the standardization problem depends in large measure on the extent to which existing standards and mature technologies are available to build upon. Once the problem to be standardized is commonly regarded as fitting into an existing environment, and what might be called a core decision has been taken, the ensuing work is prescribed to a significant extent. With the integration of this work into the particular environment, a routine process begins. Certain technical options present themselves as direct consequences of the core decision; others are no longer available. Standardization, especially of a network technology such as telecommunications, necessarily takes place within a historically structured technical setting that has shaped future feasible options. Traditionally this has helped to routinize standardization and maintain a low conflict level. Once technologies from different contexts (e.g., telecommunications and data processing) are combined, however, complexity increases and the pool of feasible options invariably offers alternative solutions. The combination of different design traditions and architectural concepts, therefore, can lead to consequences which are difficult for the individual participants involved in standardization to anticipate.

What kind of standards emerge from a specific procedural context depends ultimately on the interaction of participating actors. Their constellation, the heterogeneity or homogeneity of their interests, and their general interaction orientations and strategies must be analyzed in this respect. The continuity of participation and the resources committed to contributions affect the output. Both vary across standardization processes in the CCITT and are only partly influenced by procedural rules. However, actors, especially those with long experience in committee work, may also try to exploit the procedural rules of the CCITT in a strategic way, either to secure compromises that conform to their particular interests or to block unfavorable solutions. Strategies also include trading concessions in different areas where simultaneous work is being undertaken. Also, asymmetries in influence and information between actors (as a result, say, of market power) come into play and can tilt the standardization process in a certain direction. In these cases, actors use different strategies to find support. Among the possible strategies are package deals or side payments, where support is rewarded in another area or subsequent support is promised. These strategies also provide a basis for long-term coalition formation.

We suggest further that the strength of the actors' economic, technical, or political interests (that is, the high or low commitment in these spheres), combined with the degree of their homogeneity, affects the intensity of conflicts generally. It can be said that standardization will be easiest in cases where commitment is high and the constellation of actors' interests is homogeneous. Low commitment in a homogeneous constellation might also succeed, but high or low commitment when coupled with a heterogeneous constellation is likely to lead to intense conflicts in the first instance and to an exit from the process in the second, with the likely result that no standard will be drafted or approved.

Actors' interaction orientations are also important (Scharpf 1994). As was seen above, depending on the orientation of actors, the underlying negotiation situation is transformed, facilitating or complicating standardization. Self-directed individualistic orientations—the standard assumption of neoclassical economics and game theory—can be distinguished from both competitive and cooperative interaction orientations. The predominance of certain orientations is not unconnected to the constellation of interests. If the latter is heterogeneously structured, competitive orientations are more likely, even among participants who in general follow a more cooperative orientation. In cases where cooperative orientations prevail, a smoother standardization process can be expected, with attempts being made to reconcile conflicting interests in a compromise. Changing interaction orientations change the nature of the game to be played. Cooperative orientations facilitate standardization, whereas with competitive orientations the approval of a standard becomes unlikely. In terms of game theory, switching from cooperative to competitive orientations transforms the game from a positive-sum into a zero-sum game.

The process aspects of standardization encompass the whole process of negotiating standards, including its normative and structural basis (actor constellation), its strategic elements, its learning components, and its patterns of decision making, as well as the internal division of labor and the different phases of the process. In our cases, we selectively refer to these aspects, well aware that in doing so we do not satisfy the standards of elaborated negotiation theory.[16]

Having sketched the "independent" variables in our model of the standardization process, we now turn to the "dependent" variable, suggesting that a causal relationship exists between the various structural and procedural elements and the properties of the *standards as the output of the process.* To simply present a list of the features of standards that are shaped by the independent variables is to expect too much of the comprehension of social scientists, since it would include many highly technical details. Notwithstanding the technical significance of such features for the functioning of technical systems, it does not seem necessary to refer to them as long as indicators are available which are easier for us to grasp as social scientists.

The most visible impact of the structural and the procedural variables does not concern any specific feature; it simply concerns the approval or non-approval of a standard. From the CCITT's perspective, a standardization attempt that results in nonagreement clearly embodies an organizational failure. Individual participants in the process, however, may register such a failure as a success when their actions have been aimed at blocking an emerging committee standard. Whether a standard submitted for approval will be accepted may depend on the features of the draft. In some cases, a draft with divergent features has a good chance of being approved. In other cases, the very divergence of features reduces the likelihood of a draft's being accepted.

A comfortable way out of a highly controversial situation lies in accepting differing proposals as equivalent options of a standard. For the sake of resolving the conflict without suspending the whole process of cooperation, participants may even approve several incompatible options, thus impairing the internal consistency of a standard. But too much diversity contradicts the institutional goal of parsimony of standard options (see chapter 3). In this case, implementations of different options of the same standard may render interoperation of different components of a system impossible.

The number of options included and the degree of the options' internal consistency is one of the standard's features. Another important general feature of a standard is its entrenchment or integration into a more comprehensive system of standards. System architectures, such as the abstract OSI frame of reference or a more concrete materialized technical network such as the telephone, constrain the pool of acceptable standards variants and thus the "zone of agreements" (Raiffa 1982, pp. 35–65). Adherents of OSI will prefer standards that match the OSI requirements over standards that violate them. The entrenchment of a standard includes not only architectural considerations but also—within and between architectures—its rather tangible links with other standards in the form of substitutability or complementarity. Within OSI, a standard may complement others upstream or downstream of the hierarchy, or may replace a former standard with less sophisticated features.

We can also draw a distinction between "old" and "new" standards. Old standards are those that have already gained standing in the technical landscape as de facto (market) standards or as recommendations from other international, regional, or (influential) national standards committees. Such standards are simply ratified ex post by the CCITT. This does not mean, though, that the standard to be approved is not subject to controversy. On the contrary, it may be easier to formulate a completely new specification (ex ante) than to reach an agreement on an already-existing version. When standardization affects artifacts to which actors have devoted significant resources by way of preparatory work, the sunk costs and the commitment of actors to a specific technical solution make it difficult for them to accept any other standard. Thus, a completely new standard may emerge from a process that set out to ratify an existing one. Efforts directed at formulating ex ante standards are confronted with problems of a different kind, especially in view of the uncertainty inherent in any anticipatory technical work. As the exact future use of the standardized technology is unclear and unforeseen developments may impose modifications on anticipatory standards, their general flexibility and their aptitude for refinement become crucial. Here, where standardization increasingly becomes an extension of innovation activities, participants have to develop a common perception of the crucial dimensions of a technology.

What is significant with regard to the question of old or new standards is the fact that, within the life cycle of a technology, it is not easy to define a point where standardization activities naturally begin. Often the focus of the work is on technologies already well known and established. Standardization thus indicates the beginning of a period of technical consolidation (Tushman and Rosenkopf 1992). However, exactly which factors determine the start of standardization of a technical artifact and how they bring it to the attention of potential participants must be substantiated in every specific case.

For global market or de facto standards, a certain degree of acceptance appears constitutive. Such standards express a common cognitive pattern of shaping the relational properties of technical artifacts in order to achieve compatibility with other artifacts. This practice is linked with a concomitant expectation to comply. A standard that is neither accepted nor implemented cannot be reasonably called a market standard. A committee standard, in contrast, need only be approved by a committee to become a standard. Such a standard is connected with an institutional expectation to conform, but not necessarily with widespread conformity.[17] In our cases, where committee standards are dependent variables, it makes sense to distinguish the "enactment" of a standard—resulting from the choices of committees—from its acceptance and implementation, which cannot be formally controlled and sanctioned by the respective standardization organization. In our analysis we clearly focus on the process of standard setting, which ends with a committee decision. Nevertheless, in order to give a more complete picture, we also present a certain amount of empirical evidence on the acceptance or implementation of the analyzed standards. At times, the market chances of standards may also become relevant as arguments in standardization debates. As we have shown, achieving acceptance and implementation is a central institutional rationale for the prevalence of the consensus principle in committee standardization ("compliance by consensus").

For the sake of conceptual clarity (which is missing in many essays on standardization), we emphasize at this point that the standards whose development we consider in our case studies belong to the class of *coordinative standards*. Thus, our model of the standardization process applies only to coordinative standards. Analytically, they can be

Table 4.1
Two types of standards.

	Coordinative	Regulative
Aim	Interoperation of components in technical systems, compatibility	Prevention of negative externalities of technology
Mode of generation	Negotiation of "interested" actors, self-governance, emergence in markets	Hierarchical political governance
Normative character	Convention	Legal rule, prohibition
Area of validity	Sectors, markets (techno-economic units)	States (political units)
Economic effects	Reduction of transaction costs	Internalization of externalities

contrasted with regulative standards by means of a distinction that was employed in more detailed fashion in earlier work (Werle 1993, 1995b). Table 4.1 gives an indication of some crucial distinctions between the two types of standards. It clearly shows that, unlike standards for, say, environmental protection, compatibility standards need not be externally imposed but can result from the activities of actors interested in coordinating with others. This interest is based, in our cases, on a collective acknowledgment of technological interdependence. In contrast with regulatory standards, coordinative standards tend to be self-enforcing—that is, they enjoy a considerable likelihood of compliance. They reduce transaction costs and cover economic sectors or markets linked to the respective technical components and systems, thus ignoring political frontiers.

Making Standards

5

Standardization in a Changing Environment

Beginning in the late 1960s, data occupied an increasingly significant proportion of the traffic on telecommunications networks in the major industrialized countries. The growing potential for new applications from the convergence of telecommunications and computing started to interest the Comité Consultatif International Télégraphique et Téléphonique in the 1970s. At that time, three internationally standardized, end-to-end-compatible telecommunications services existed: telegraphy as a service directly operated by the PTT administrations, telex for business communications, and telephony (the latter being the most ubiquitous). The technological potential for new applications and the developing demand for new services generated a need for coordination. The chairperson of the CCITT's Study Group I, the group most responsible for strategic considerations and management of services, from 1976 to 1984, expressed it this way:

> In view of the new systems appearing on the market, the CCITT must ensure that the developments are controlled and even directed. This calls for increasing efforts on the part of all concerned. Should the international standardization bodies no longer be in a position to keep track of events and developments or control and, as far as possible, direct them, the divergences between the various systems would widen, which would mean that compatibility could no longer be assured or even possible. Standardization procedures should not be carried out after the various applications are adopted, but well before. (Freiburghaus 1983, p. 564)

The CCITT responded to this challenge by creating a comprehensive concept of a new generation of "telematic services"[1] in order to integrate the potential of new applications into the traditional CCITT approach. The CCITT's standardized services provide complete end-to-end compatibility, including standardized terminals. Moreover, they comprise international directory listings as well as rules for charging and accounting. Targeted at a majority of users, the standardized services allow costs to be pruned through economies of scale in production (Hummel 1980; Cerni 1982, p. 52). It was on the basis of this approach that standardization on a set of new telematic services began in the late 1970s.

Originally, the new services included teletex (second-generation telex), telefax (a facsimile technology), and videotex (intended to be a telematic service for the mass market). With message handling, a fourth service was officially added in the early 1980s. The message-handling system (MHS), which is standardized in the X.400 series, concerns direct computer-to-computer communication (electronic mail). Building directly on the convergence of technologies, message handling is thus slightly different from the other three services. It has, moreover, been standardized for public networks as well as for private networks, so that it is not exclusively provided by postal, telephone, and telegraph administrations.

Figure 5.1 gives an overview of the CCITT's concept of telematic services. In our case studies, we describe and analyze three of these four standards for telematic services: videotex, facsimile, and X.400. The remaining service, teletex, is not considered separately, but our cases show that teletex was often influential in the standardization of the other services. The three cases we study show similarities and differences. The services offer capacities for the transmission of texts and include some graphics capability. However, standardization of videotex and, in particular, facsimile was mainly terminal-oriented, whereas X.400 comprised a whole series of standards (including routing and conversion functions). In line with the division of labor in the CCITT, message handling was allocated to Study Group VII (Data Networks), while telefax and videotex were entrusted to Study Group VIII (Terminal Equipment).

Our analysis takes account of the whole period of standards development for our three cases. In view of the long history of facsimile, however, our treatment of it is on occasion rather fragmentary. To contextualize the technology-centered activities, we will shortly allude to the historical background as a reminder of the significant political, economic, and technical changes that occurred during this period and had repercussions on standardization.

Significant measures taken with regard to market liberalization, differentiation, and growth have changed the telecommunications sector considerably, paralleling the history of our cases. As signified by

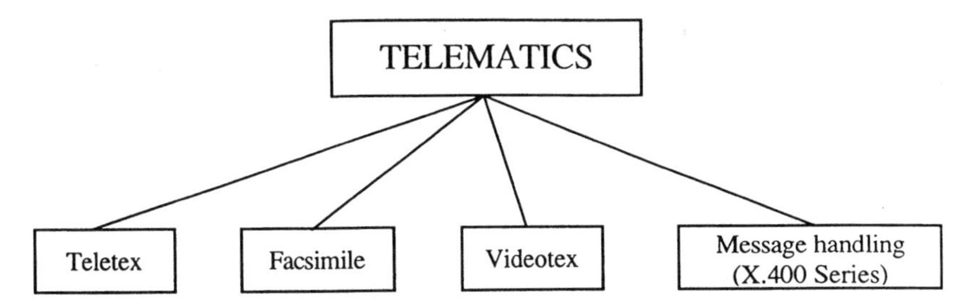

Figure 5.1
Telematic services defined by the CCITT. Telematic services: international telecommunication services (excluding telephone, telegraph, and data-transmission services) offered by administrations and defined by the CCITT for the purpose of exchanging information via telecommunication networks. The definition of service covers the full range of functions according to the Open System Interconnection model. Examples of telematic services: teletex, facsimile, videotex. Teletex service: an international telematic service offered by administrations and enabling subscribers to exchange correspondence via telecommunication networks. Public facsimile service: an international telematic service offered by administrations for the purpose of transmitting documents between facsimile terminals via telecommunication networks. Videotex service: an interactive service that, through appropriate access by standardized procedures, allows users of videotex terminals to communicate with databases via telecommunication networks. Videotex service has the following characteristics: information is generally in an alphanumeric and/or a pictorial form; information is stored in a database; information is transmitted between the database and users by telecommunication networks; displayable information is presented on a suitably modified television receiver or other visual display device; access is under the user's direct or indirect control; the service is user friendly; the service provides facilities for users to create and modify information in the databases; the service provides database management facilities that allow information providers to create, maintain, and manage databases and to manage closed user-group facilities. Message handling: services provided, by means of message-handling systems, through Administration Management Domains or Private Management Domains. Examples of message-handling services: Interpersonal Messaging Service, Message Transfer Service. (Adapted from Wallenstein 1990, pp. 12–15; Cerni 1982, pp. 61–66; CCITT Series F recommendations.)

the divestiture of AT&T, the licensing of Mercury as a competitive network operator in Britain, and the privatization of British Telecom between 1982 and 1984, liberalization became an issue in all major industrialized nations. Technologically, the digitization of telecommunications initiated a far-reaching transformation toward computerized multifunctional networks, services, and terminals.

Our cases are thus situated in a changing environment. Telefax, for example, began in the early 1960s almost as a monopoly of British and North American terminal manufacturers. In the 1990s, however, the fax market was overwhelmingly dominated by Japanese manufacturers, and the United States attempted to get back into the market, at least in the segment of fax modems and fax cards for PCs. This example indicates a radical change in the telecommunications industry. The 1970s and the early 1980s were turbulent years because the depth of the ongoing institutional and technical transformation of the telecommunications sector was exaggerated. It is no accident that on the alleged eve of the "information society" standardization work on the cases we have studied intensified. Videotex, in particular, was affected by these expectations and by an attendant desire to counter the growing importance of the United States in computing with a successful European telecommunications development. Although not to the same extent, our other two cases were also influenced by beliefs, visions, and forecasts that later had to be reassessed. The idea for a new generation of standardized end-to-end services emerged in this context. Embodied by it were specific concepts for the organization of telecommunications operation, for the growth of its use (the "paperless office"), and for the future pace of technical change.

All this has had an impact on the international standardization of telecommunications (Drake 1994). Although the CCITT itself changed little during the period covered by our case studies, the circle of participants became broader and more heterogeneous through the entry of computer manufacturers alongside new competing providers of networks and services. Meanwhile, the perceived demand for technical coordination has grown in parallel with the increased variety of networks and services and the accelerated technical change. As a result, the number of volumes of documents archived in the ITU has risen considerably. A large part of them were produced by the four study groups that are relevant in our context: SG I for service aspects, SG VII for data networks, SG VIII for terminal aspects, and SG XIV for facsimile equipment (until 1980).

6

The Institutional Framework

National, Regional, and International Levels of Standardization

The work of the Comité Consultatif International Télégraphique et Téléphonique, the central institutional focus of our case studies, is embedded in the activities of a variety of standardization organizations at the national, regional, and international levels (see chapter 3 above). Let us now consider the organizational interlinkages that bear directly on our cases in more detail.

The organizational interdependencies are chiefly of two kinds. There are, on the one hand, the direct links to the CCITT of organizations providing input to it or processing output from it. Here a sequential division of labor prevails, following the principle of nationally and regionally differentiated jurisdiction. On the other hand, there is a functional division of labor in international standardization (which is, however, becoming increasingly imprecise). The responsibilities of organizations have overlapped more and more as the initially distinct technological fields of telecommunications, computing, and broadcasting have tended to converge. This has led to some confusion and conflict as to which standardization organization is responsible for the solution of a particular technical coordination problem. Therefore, coordinative activities among the organizations precede or run parallel to technical coordination (Genschel 1995).

Within the sequential division of labor, the national level is the smallest relevant unit. CCITT membership is nationally based, and the countries involved are represented by national delegations. Depending on the size and the composition of the delegation, national meetings are held to coordinate input into the CCITT. This level is not uniformly organized[1]; it is subject to national discretion. Its organizational principles and practices have become more important recently with the weakening of the powerful postal, telephone, and telegraph administrations (PTTs) and the rising importance of international standards for manufacturers, service providers, and large users. The liberalization of telecommunications has induced legal requirements to reorganize national input to the CCITT. National coordination structures have been reorganized in response to demands for more pluralistic influence and as a result of the political motivation to maximize the aggregate national influence via coordination.

At the time of their extensive monopolies, the PTTs headed the national delegations to the CCITT. The separation of the regulatory authority from the operation of telecommunications that accompanied liberalization has transferred formal national membership to the responsible ministry in each country. At least in Europe, therefore, coordination of national inputs to the CCITT and to its organizational successor, the ITU-T (the telecommunication standardization sector of the International Telecommunication Union), is currently in a state of flux. The different ways in which those countries most important to our case studies achieve national coordination will now be described in brief. Because our interest dates back to the early 1960s, not only the changing structure of national coordination, but also the periods of unrestricted hierarchical PTT control have to be borne in mind.

In the Federal Republic of Germany, reorganization became necessary with the telecommunications reform of 1989 (Schmidt 1991). The Bundesministerium für Post und Telekommunikation was allocated the responsibility for standardization as part of its overall regulatory authority. In 1990 a federal agency, the Bundesamt für Post und Telekommunikation (BAPT), was founded to assist the Bundesministerium in its multiple tasks as neutral observer and regulator. In close consultation with the Bundesministerium, the BAPT now represents Germany in international standardization organizations, in particular at the CCITT. The BAPT coordinates the German telecommunications equipment manufacturers' and service providers' preferences concerning technical standards and regulations and prepares a common position to be taken up in international negotiations on standards. Notwithstanding this new body, changes can be expected to continue as long as the reform of the German telecommunications system remains incomplete. A second reform in 1993 prepared for the corporatization and privatization of Deutsche Telekom (Schmidt 1996), and the third reform package of 1996 arranged for full liberalization from 1998 on.

In the course of the reforms, previously existing groups, which served informally to coordinate the national position with participants from industry and major users, are being transformed into formally institutionalized input groups. These forums date back to the early 1970s, when, in the aftermath of a failed attempt on the part of the Deutsche Bundespost (DBP) to start a joint venture for data communications and processing, a committee (the Ausschuß für Datenfernverarbeitung) was formally institutionalized at the level of the PTT's central technical agency to discuss questions of remote data processing. The committee included relevant manufacturers and users, who met twice a year. Working parties were instituted, which met more frequently to deal with specific questions. The X.25 packet-switched data service and also the introductions of teletex, telefax, and videotex were coordinated here (Werle 1990, pp. 226–240).

In this structure of committees, small ad hoc groups existed to formulate the German position on certain CCITT topics, and international standardization was a natural part of several discussions on strategic issues, such as application profiles for the introduction of services. However, the coordination of positions did not have a formal character. The DBP had hierarchical control over the national system, and standardization was only one issue among others in its close, interdependent links with the relevant national "court suppliers" (of which Siemens was the most prominent).

Other European countries show similar patterns of reorganized national coordination. In the United Kingdom, since the demise of the British Post Office, standardization falls under the responsibility of the Department of Trade and Industry (DTI). However, this duty is often delegated to British Telecom, which will, therefore, still lead some British delegations in CCITT study groups, although now only in a commissionary sense. This change in British Telecom's position from principal to agent is not unproblematic. It makes it difficult to distinguish whether a British Telecom proposal represents a national position or merely a corporate interest. Thus, even in the UK, where European liberalization began, remnants of the former hierarchical relationships persist.

New structures intended to facilitate national coordination have been institutionalized. The DTI has tried to support the wider participation of firms in standardization by offering subsidies. National coordination groups for specific issues exist: the group for facsimile, for example, comprises two relevant manufacturers (Matsushita and

NEC), British Telecom, and Cable & Wireless. The manufacturers normally channel information from a larger industrial grouping, the British Facsimile Industry Consultative Committee (founded in the 1970s), which comprises all the relevant manufacturers (mainly Japanese subsidiaries) as well as British Telecom and Mercury. Most corporations are not present in the CCITT, but the operators are regarded as ready supporters of industry's position. A national coordination group also exists for X.400. In 1988, as an exceptional measure in national coordination, it was merged with the group within the British Standards Institution that works on message handling for the International Organization for Standardization.

Before liberalization, the United Kingdom, like Germany, had closed, informal national coordination structures. This was very apparent in the standardization of the videotex system, which the UK had pioneered. As will be seen later, international standardization attempts on the part of the Post Office could be agreed upon in this case only after consultation with the manufacturers.

A comparable situation could be found in France. As in all countries, the precise extent of consultation varied with the importance of specific standardization issues, particularly the core competencies of national manufacturers and the extent to which the national PTT regarded the realization of projects as depending upon close cooperation with manufacturers. In France, since the recent reforms, the government ministry is formally responsible. It has started to organize biannual meetings with the delegates to the CCITT for the purpose of coordinating the national position.

Outside Europe, national coordination follows a different pattern. Until the early 1960s, the CCITT was regarded as a "European Club," reflecting the greater need within Europe to achieve supranational technical coordination because of the relatively tight existing communications links and the more urgent need to harmonize equipment (Wallenstein 1990, p. 64). The United States, with its vast market, could develop its telecommunications system quite autonomously, as is still witnessed by the fact that some telephone network parameters differ fundamentally between the United States and Europe.

Until 1970, participation by the United States in the CCITT's work was low and virtually uncoordinated. International telecommunications hardly played a role in the US, and with its private organization of national telecommunications it did not square well with the intergovernmental character of the CCITT either. The US has always relied on private-sector participation. This has often been a target of criticism. Private-sector standardization was regarded as poorly coordinated, and the US government was understood to have no comprehensive national standards policy of its own, either in telecommunications or in most other industries (OTA 1992, pp. 3–36). In the mid 1970s, however, US interest in the CCITT grew rapidly as a result of national and international changes. International standards in telecommunications were seen to be particularly important in discussions on the worsening balance of trade. In this context, the national system of input to the CCITT was reorganized with the drawing up of a US CCITT Charter in 1977. Prior to this, frequent incoherent national positions had made it difficult for the United States to be influential in CCITT negotiations. Compared to the situation in European countries, the United States now has a more formally institutionalized and transparent national structure (Cerni 1982, pp. 83–96).

As can be seen in figure 6.1, the State Department is the United States' official representative to the CCITT. It heads the US National Committee for the CCITT, which comprises four subgroups (A, B, C, D), covering the whole area of the CCITT's work. Each group is responsible for particular study groups and their questions. Participants are actors either interested or involved in telecommunications, and meetings are generally open to the public. All US contributions to the CCITT have to be approved first by the US National Committee, and individual members' contributions have to be at least approved by the chairperson of the relevant subgroup (Cerni and Gray 1983, pp. 19–36). Formal CCITT conferences at which some issues are taken to a vote, especially the quadrennial Plenary Assembly, are defined as affairs of state. In these cases, a US delegation is constituted; it is headed by a State Department employee or a designated employee of an appropriate US government agency, such as the National Communications System (Wallenstein 1990, pp. 199–201).

Canada's Department of Communications fits into the structure of the CCITT as a treaty organization much better than its US counterpart. It holds the formal responsibility for Canadian participation, and it coordinates the input with the help of CCITT preparatory groups. However, as in the United States, transatlantic and transpacific communications were, for a long time, quite insignificant, making CCITT participation rather unimportant. Technically, Canada's telecommunications network is very similar to that of the United States, and traditionally the same technology has been employed (Cerni 1982).

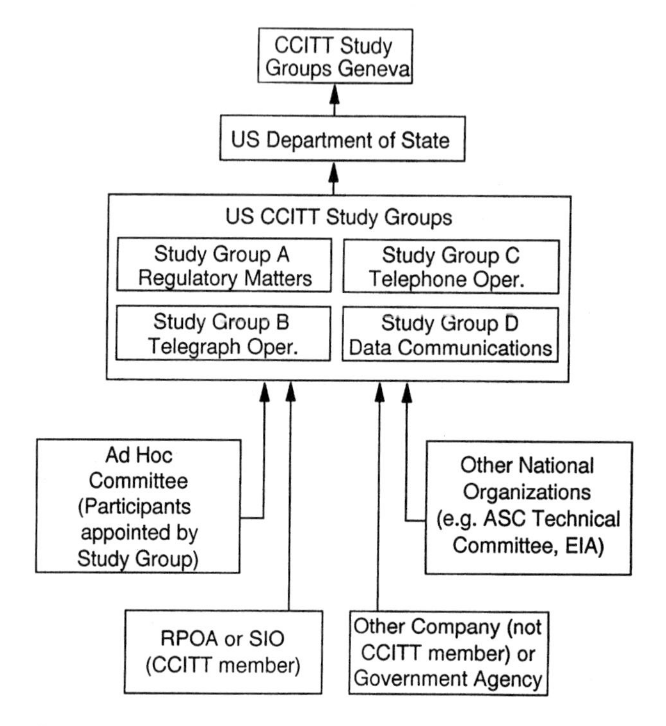

Figure 6.1
The formal "chain of approval" for US contributions to the CCITT in the 1980s. (Adapted from Cerni 1984, pp. 143–145.)

Japan, which also has a more clearly, formally institutionalized national coordination structure than the United States, has long stood on the fringe of CCITT activities. Japanese CCITT participation is controlled by the Ministry of Posts and Telecommunications. Coordination among the interested parties is achieved in the CCITT Committee, which is part of the Telecommunications Technology Council and has different subcommittees dedicated to ongoing CCITT work. The Telecommunications Technology Council, which belongs to the Ministry of Posts and Telecommunications, was founded in 1985, at a time of general restructuring of Japanese standardization (Müller and Kuhn 1988, pp. 40–42; Edelman 1988, pp. 415–427; Besen 1990a). However, the formal institutionalization notwithstanding, Nippon Telephone and Telegraph and the Kokusai Denshin Denwa Company still play a very dominant role in the Japanese delegation, which gives observers the impression that the Telecommunications Technology Council is a tightly, hierarchically coordinated group.

Not only the national but also the regional (supranational) level can be important for the coordination of positions. The success of a technical proposal to the CCITT depends on a broad consensus, so early supranational harmonization of positions facilitates influence. The recent importance of regional organizations in Western Europe, North America, and the Asia-Pacific region—already emphasized in chapter 3—has a precedent at the European level with the Conférence européenne des administrations des postes et des télécommunications (CEPT) (Labarrère 1985). The CEPT was founded in 1959, its expressed aim being "to harmonize and improve postal and telecommunication coordination and cooperation among European countries to form a homogeneous, coherent, and efficient unit on a continental scale" (Macpherson 1990, p. 192). With the foundation of the European Telecommunications Standards Institute in 1988, the majority of the CEPT's tasks were relinquished. In view of the changes brought about by liberalization, the CEPT had become increasingly unacceptable because of its closed "PTT club" character; it did not even admit manufacturers on an advisory basis (Besen 1990b; Temple 1991; Werle and Fuchs 1993; Genschel 1995). In comparison to the new regional organizations which are sometimes seen to rival the CCITT, the CEPT's relationship to the CCITT was never as ambiguous. With regard to standardization, it served as a platform for the loose coordination of European positions, which generally resulted only in common propositions shared by some of the CEPT countries represented in the CCITT. Moreover, the CEPT had a function with regard to the output of the CCITT. Since the CCITT's recommendations often included several options, the CEPT was the arena where greater harmonization among the European countries could be attempted.

Standardization was only a small part of the CEPT's activities in postal services and telecommunications, and it evolved slowly during the 1970s. It became more important after 1975, when the Comité de coordination de l'harmonisation was founded as a dedicated subcommittee for standardization. A series of working groups existed with a brief to examine the standardization of networks, terminals, and services. This activity became particularly well known through the search for a common European videotex standard. This CEPT agreement was replicated later as one option in the CCITT recommendation. It is predominantly in this context that the CEPT interests us.

The CEPT, like the coordination effective at the national level, fits into the pattern of a sequential division of labor where one level complements the others. For X.400 and videotex, further organizations were relevant because neither of these cases matched the established structure of functional specialization among the international standardization organizations. Because they combined different technical fields, X.400 and videotex had the potential to affect several organizations, and so new organizational interlinkages were created.

For X.400, the first organization of relevance is the International Federation for Information Processing (IFIP). The IFIP is not regionally but functionally oriented. Moreover, it aims explicitly to provide input solely into the formal standardization process (i.e., it is engaged only in pre-standardization), so its standardization work does not have an official status. Also, the European Computer Manufacturers Association (ECMA) has been important in the standardization of X.400. The International Organization for Standardization ("ISO") is the third organization that has visibly influenced the standardization of X.400. The ISO has been the organization where the few projects for the international standardization of computing technology through committees have taken place. These have mainly been directed at standardization of equipment or stand-alone devices, some of which were relevant in computer communications and X.400. Here the ISO and the CCITT have overlapping responsibilities, but the ISO's standard operating procedures and its membership and decision rules differ from those of the CCITT.

The formal members of the ISO are those national standards organizations that most appropriately represent their respective countries. Participation in the ISO is determined by these national member bodies and by their criteria for access, which normally include all interested parties. Technical work, as is typical of all standardization organizations, takes place in periodic meetings and through exchanges of documents. At the ISO, so-called Technical Committees are responsible for delineated areas of work, and they are divided into subcommittees and working groups.

In contrast to the CCITT, decision making on standards at the ISO is fairly decentralized. When a group has finished working on a standard, which is not predetermined by a comprehensive schedule, the organization will proceed to a postal ballot. First, a committee draft (CD) is prepared and circulated in the Technical Committee within a stipulated period. If consensus is reached, the CD is registered as a Draft International Standard (DIS) at the ISO's Central Secretariat. Then a ballot takes place among all participating members of the ISO (as distinguished from observing members, who do not enjoy voting rights). If the DIS is approved by at least two-thirds of the members on the committee and three-fourths of all members, it becomes an international standard. The CCITT usually reaches its formal decision at its Plenary Assembly. Moreover, differences exist in both organizations as to the informal legitimacy of dissent. In the CCITT, the right to veto may be exercised only as a last resort where the member has been actively involved in the standardization project. At the ISO, in contrast, such an informal rule of fairness is not institutionalized—a difference that is most likely attributable to the fact that the ISO employs postal ballots and the CCITT votes at an assembly. At the ISO, each member is free to submit counterproposals or reject the draft standard in the ballot, irrespective of earlier involvement (Cargill 1989, pp. 125–138; Macpherson 1990, pp. 95–104).

The overlapping organizational responsibilities for X.400 originally grew out of the aim to create a standard for exchanging messages between computers inside telecommunications networks. Whereas computer standards have traditionally been provided by the ISO, telecommunications networks have been subject to standardization by the CCITT. An even more complicated overlap of domains could be observed in the videotex standardization process. The combination of technical components from telecommunications, information technology, and broadcasting occupied more than 100 standardization organizations at the national, regional, and international levels (Clark 1981, p. 74). Though the ISO and (to a lesser extent) the ECMA were also active in videotex, particularly with regard to coding standards, the involvement of broadcasting standardization organizations had a greater impact on the course of events, because they were engaged in the standardization of the teletext service, which is the broadcasting equivalent of videotex.[2] The Comité Consultatif International des Radiocommunication and the European Broadcasting Union were particularly active in the standardization of teletext. Their work influenced videotex standardization mainly in an indirect way: constraints were imposed on the available technical options because teletext standards directly affected the display as a central technical part of both videotex and teletext standardization.

Institutional Structure of the CCITT

The CCITT and the ITU's Telecommunication Standardization Sector have been the central organizations for international telecommunications standards.[3] Though the CCITT was founded in 1956 as part of the ITU, its origins are much older. As early as 1865 a transnational organization was set up by a number of European countries to meet the need for a forum to deal with the coordinative issues of transborder telegraphy. From the inception, there was a pronounced connection between technical and political aspects. Although supranational technical coordination was the main incentive, the intergovernmental aspect played a significant role from the beginning. In continental Europe, telegraphy (and later telephony) was regarded in most countries as falling under the exclusive responsibility of the sovereign state, in view of its military implications (Thomas 1988). The early links between politics and technical issues were reinforced in 1947 when the ITU became part of the newly founded United Nations. In spite of this traditionally close relationship, a strong notion has always prevailed that work on technical issues can successfully be kept separate and independent from the formally institutionalized risk of politicization (Codding 1984).

In the ITU, the CCITT operates within the overall budget but under quasi-independent leadership (Codding and Gallegos 1991). According to the ITU's convention, which specifies the formal institutional structure, the CCITT's duties are to study technical, operating, and tariff questions and to issue recommendations on them with a view to standardizing telecommunications on a worldwide basis. The legal status of these recommendations is nonmandatory. Membership and voting rights are nationally organized (one country, one vote), and formally the majority principle is the valid decision rule. Equally relevant are the convention's detailed provisions regarding conferences. In principle these apply not only to the Plenary Assembly of the CCITT but also to the meetings of its subgroups. The chairperson is allocated the right to open and close the meetings, to set the agenda, and to control the course of discussions.

All in all, the operation of the CCITT is characterized by a hybrid hierarchical working structure (figure 6.2). The political mandate lies with the periodically convening (quadrennial) Plenary Assembly, which passes the recommendations resulting from the last study period. Plenary meetings are highly formalized and ceremonial (figure 6.3). The Plenary Assembly does not vote on various alternatives; it simply adopts or rejects a suggested standard. The Plenary Assembly also formally allocates the questions to be studied in the following period and decides on the number of chairpersons and on the structures and tasks of the study groups. Among these groups (there are about fifteen in all), Study Group I has the position of a primus inter pares. It is

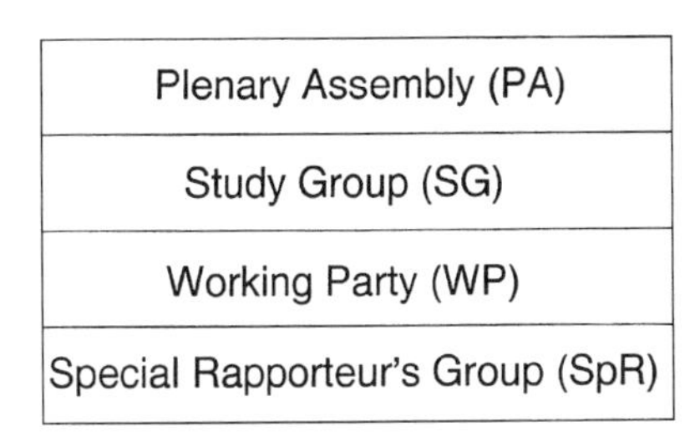

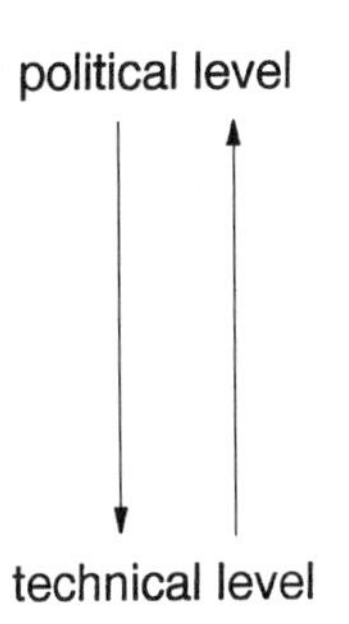

Figure 6.2
The working hierarchy of the CCITT.

Figure 6.3
The CCITT's eighth Plenary Assembly, Malaga-Torremolinos, 1984. (Courtesy of ITU.)

concerned with operating principles and service definitions, and in general the work of the other groups is meant to relate to it. This means in practice that the technical study groups may ask advice of SG I when there is uncertainty as to the direction in which work should proceed.

The technical work takes place in the study groups and their subdivisions, the working parties and the special rapporteur groups. This lowest level confines itself almost exclusively to the discussion and resolution of technical issues. Decisions taken here are then referred up to the level of the working party and then to the study group as a whole. If no opposition is mounted at these levels, the technical solutions are adopted and will be presented for final approval to the Plenary Assembly. Thus, the path is one of rising competence in political decision making and declining competence with regard to technical "content." The institutionalized separation of the political and technical spheres has increased in proportion to the increases in technical complexity, in the pace of technical change, and in number of participants. This has made it impracticable to refer decisions taken on the subject matter from the level of the study group to the Plenary Assembly. If dissent still arises, therefore, the issues can only be referred back to the appropriate special rapporteur groups, at the cost of a corresponding delay in producing the technical solution (Irmer 1990).

In 1989 it became possible to approve urgent recommendations in the middle of a study period under the procedure of "Resolution No. 2." This requires unanimous approval of the whole study group, and a subsequent majority of the administrations by means of a postal ballot. A similar provision had existed since 1968, but it was subject to the restriction that any recommendation be introduced as provisional until final approval by the next Plenary Assembly. Both provisions were meant to be exceptional procedures. Since the 1992 reform, the accelerated procedure relying on postal ballots has turned into the routine operation for adopting standards.

For a long time, the membership rules of the CCITT ensured largely homogeneous interests among participants—a situation that was reinforced and backed by comparable structures in the national organization of telecommunications. Traditionally, the PTTs have dominated the CCITT as formal members of the treaty organization. Because of their sovereign rights to telecommunications operation, they originally shared similar interests, and the absence of competition meant that they could try to realize the benefits accruing from their complementary relationships. Other interested parties needed to be formally admitted to the national delegation by the government representative responsible. It was mainly manufacturers who participated—until 1989 only in an advisory capacity—under the category of scientific and industrial organizations (SIOs). They still have to pay membership fees, which, added to the direct costs of unpaid participation, restrict from the start the circle of participants to comparatively large organizations. These costs point directly to the significant interests of firms in standardization. In recent years, as a consequence of liberalization, their number has increased considerably. According to the figures for early 1995, more than 320 private members—including major service providers, privatized network operators, firms from the telecommunications, computer and broadcasting industry, major user groups, financial and development institutions, and consultancies—are registered members of the ITU, whereas national membership (governments) stagnates at about 185. Because of their fairly exclusive technical knowledge and the intensity of their input at the level of the technical groups, SIOs often influence standardization considerably. Yet as an intergovernmental organization the CCITT faces fundamental difficulties in integrating SIOs on a nondiscriminatory basis. Owing to the nationally structured nature of decision making, the inclusion of a wider range of participants formally affects only the scale of internal coordination in delegations; thus, much depends on the way the national input system has been structured.

A comprehensive set of informal rules partly complements and partly replaces the formal rules. The *consensus principle,* guiding the CCITT's work despite the formal majority rule, is the most relevant. Its importance is bound up with the difficulty of sanctioning "deviance" in the international realm. A consensus decision adds weight to the nonbinding recommendation and makes widespread adoption more probable (Renaud 1990, p. 41). The implementation of standards is entirely separate from their definition, and the CCITT neither monitors nor keeps track informally of the degree of compliance. Although consensus is not identical with unanimity, certain dissenting opinions are most unlikely to be overruled (Hawkins 1995). A consensus would not be declared if, for example, one of the major industrialized countries, which are central nodes of the international communications network, rejected a draft recommendation. Hence, an informal ranking of countries can be observed.

Important in this respect is the question of which *institutionalized rules* exist to *facilitate consensus* once issues are contentious. In standardization, the predominant constellation is that all participants benefit from a standard, but that some can expect to get more benefits than others if a specific solution is adopted. Thus, a distributional conflict can arise during the standardization process. In game theory this constellation is depicted by the "battle of the sexes" game (see chapter 4). Theoretically, several means, which may be formally or informally institutionalized, exist to overcome the conflict and to reach a compromise in view of the distributive issues. A simple possibility is that one party always gives in, but this presupposes more critical asymmetries between the parties than are likely in the CCITT. Provisions for formal arbitration would also be conceivable, but they are not institutionalized. Another option is package deals, where participants swap consent ("I agree to your teletex, if you agree to my videotex"). This solution is not institutionally prohibited, but the implication is that there might be a difficulty in finding commensurable trading issues. Payments to compensate for relative disadvantages (side payments) are very unlikely in an organization in which not even economic reasoning is formally institutionalized. They are, moreover, only appropriate for resolving conflicts of a straightforward economic nature. Technically and politically motivated conflicts lack a comparable common denominator (Scharpf 1992, pp. 65–75).

Another possibility of overcoming conflicts of the battle-of-the-sexes type may be more pertinent. In the CCITT the issue is one of ongoing institutionalized work, where key actors participate on a long-term basis. Ongoing relations are favorable conditions for mutual trust and cooperation (Bradach and Eccles 1989; Chisholm 1989, pp. 114–120; Sabel 1993). In the CCITT this situation is supported because, whatever differences might arise, the professional homogeneity of participants remains significant. The technical purpose of the work leads to the sending of technical personnel, which generally feeds back a technical and often shared perspective on the task. Thus, where no "best solution" can be found, repeated interactions of a stable set of actors with cooperative orientations make for the emergence and stabilization of what is called "turn-taking": one party surrenders her or his turn knowing that the other party will grant the turn next time.

The CCITT's *informal rules* serve to balance conflicting individual interests with the common interest in a coordinated solution. They ensure that the decision process and its outcome are perceived as fair, by providing room for the participants to defend individual technical and economic interests alongside national ones. The need to reconcile divergent individual positions vis-à-vis the common interest in technical coordination has grown as a consequence of the greater heterogeneity of participants. But agreement is facilitated by several means, one example being the fact that it is impossible to veto the results of work in which one has not participated; otherwise, work could be obstructed at the level of the Plenary Assembly after years of effort to reach a compromise in the committees. Likewise, if single participants violate informal rules and obstruct work unnecessarily, there is the formal political level as the last resort: the leader of the delegation to which the disruptive person belongs can be asked to attend to his or her behavior. In the extreme case, members can face exclusion from the delegation and thus from the CCITT.

Because the rules of procedure do not leave much room to discuss anything other than technical aspects in formal meetings, coffee breaks are an important informal institutional pillar of committee work. Here, reasons other than technical ones behind opposition voiced in the meeting may be revealed. Once the legitimacy of these reasons is established, it is easier to overcome an impasse. Weaknesses in an earlier position may be exposed, and the search for a more encompassing solution that achieves consensus can again take place within the formal committee work (Schmidt and Werle 1993).

Several different types of documents form the output of the CCITT and the ITU respectively, and we refer to some of these in our case studies. The documents with the highest status are the Final Acts of the ITU Plenipotentiary Conferences, which often have constitutional significance for the CCITT. Other documents with official status include the standards which are published as recommendations after adoption by the CCITT Plenary Assembly, the minutes of the Assembly, and, possibly, adopted regulations that have a more binding character than recommendations. These documents were published on a regular basis as "colored books," each Plenary Assembly having a different color (the Red Book series, Blue Book series, etc.). As we have aleady mentioned, since the reform of the ITU, recommendations approved in the study periods are published as soon as possible.

The CCITT not only produces recommendations; it also publishes a number of documents with a more temporary, informal character. The day-to-day work of the study groups, the working parties, and the special rapporteurs' groups is accompanied by a variety of documents describing different technical options and comparing proposals and by the minutes of meetings. Although some of these documents remain temporary, many are included in the continuously numbered series of CCITT study group communications for each study period; these are designated "COM" and are collected and archived by the ITU in Geneva. Minutes of plenary assemblies of study groups receive a different designation: "AP." Also worthy of mention are the more formal opinions a study group may voice, which enable it to direct matters to the attention of other study groups.

"Bias" of Institutional Setting

The institutional setting, with its rules and procedures, structures decision making by imposing constraints and opportunities for actors' behaviors and strategies, thereby exerting a bias for certain decision outcomes over others. Since our description of the institutional structure of the CCITT has already touched on some of its possible effects on standardization, we shall now highlight a few attributes that typically influence the CCITT's work and its results. Some of these rules have been modified over the last few years. (As our cases were processed under the old CCITT regime, only its rules are relevant here.)

Regarding participation in the CCITT, the organization's reliance on voluntary and unpaid work implies constraints. Participants have to voluntarily commit resources to the work; this makes standardization especially vulnerable to attention cycles, as relevant actors might drop out in the midst of a study period or new actors might enter, thereby jeopardizing the continuity of the work. Widespread participation of all relevant actors is needed to secure support for a recommendation. Moreover, the CCITT, in contrast to the new European Telecommunications Standards Institute, has no provision for commissioning paid experts to elaborate workable compromises on contentious matters. Nevertheless, voluntary participation in combination with the comparatively high costs of membership does facilitate compromise; no organization sends delegates to the CCITT just to obstruct the work.

Another peculiar element is the principle of national representation to the CCITT. As a result, decision making is divided into two stages. In the first stage, the various interests and preferences of a nation's service providers, equipment manufacturers and network operators with respect to a standard have to be aggregated into a common position of a national delegation. In the second stage, the national delegations have to reach a compromise if their positions diverge. This contrasts with the alternative possibility of functional representation, where it is not nations but different functional groups (manufacturers and service providers) who settle on a common position first. Within national delegations, asymmetries tend to persist in favor of the national network operators. Whenever equipment manufacturers rely to a significant extent on a national operator's orders, they will hesitate to challenge its position (Schmidt 1996). This stabilizes national positions that have grown out of nationally specific technical and regulatory conditions. When these positions differ, the controversies in the CCITT are easily politicized, because they are (at least implicitly) related to fundamental differences in the national telecommunications systems. Increasingly, however, the interests of multinational corporations such as IBM cannot be framed in a national context. The principle of national representation can be undermined when multinationals gain a say in multiple national delegations, which they may use for their own purposes. Still, for the time being, as far as the major industrialized countries are concerned, the principle seems to restrict the influence of multinational corporations in the CCITT, because interests conforming to those of the national PTTs still enjoy priority.

Even with a dominant PTT, the need to arrive at a unified national position demands early coordination efforts at the national level, the results of which cannot be easily modified or sacrificed for the sake of an international compromise. This affects the process dynamics of standards negotiations. Yet it is not just from this perspective that standardization appears not as a single game but as a complicated ecology of games. Representatives of an aggregated interest position find themselves in a situation in which they have to act simultaneously with respect to different levels and reference groups. In game theory this has been conceptualized as two-level, connected, and nested games (Putnam 1988; Scharpf 1991; Tsebelis 1990). Two-level games capture the still rather simple two-stage decision-making process within the national delegation and in the international arena. If we also consider that no individual agent of an organization simply executes the will of the delegating corporate actor, a complicated picture arises in which each actor potentially has allegiances to different constituencies. There is the level of the individual player, the principal's (corporate) interests, possibly professional interests, then the aggregated national delegation's position, and finally potential allegiances to actors not directly taking part in the game. Here, the outcome of one game directly affects the connected games, and changes in positions may be due to changes in another arena. This results in unusual complexity, since the rationality of action must be assessed in the context of several games instead of only one (Putnam 1988, p. 434).

Other elements of the institutional setting also matter. The relative rigidity of the working schedule is especially striking. At the levels above the special rapporteur groups, the number of times meetings are convened is generally fixed and the dates are determined far in advance. The structuring effect of this schedule must be seen in the context of the CCITT's general four-year study period. Initially, much time can be given to defining positions because there is little pressure for agreement before the end of the study period. Contentious items can be periodically put on the agenda so that whether a compromise has become possible can be considered. But delaying agreements is without cost only up to a point. Controversies that do not seem amenable to solution risk being deferred until the next meeting. Toward the end of the study period, the threat of a possible delay for another four years, should there be still no agreement on the recommendation, can be expected to exert significant pressure for a compromise.

Other procedural rules have further implications. The division of labor among different groups may become important with regard to the number of groups involved in a decision and to the degree of mutual information and cooperation that exists. The chairperson with the responsibility for agenda setting may significantly influence the course of events by placing topics at the end of the agenda (thus leaving less time for their discussion), by not taking votes, by deferring controversial matters to the next meeting rather than formulating a possible compromise position, or by directing the way in which complex problems are divided and voted upon.[4]

The decision rule also must be considered. Though achieving consensus governs the work, the official majority rule is not entirely irrelevant. In imponderable situations, especially when it is not clear whose position will be the majority position, those actively involved in a standardization issue do not want the passive and indifferent participants to accidentally tip the scales in favor of one solution. Conversely, the avoidance of proceeding to a formal vote can further facilitate agreement, because open commitment to specific positions is thereby avoided.

In addition, a fundamental interdependence exists between the informal consensus principle and the nonbinding nature of the recommendations. The voluntary nature of standards facilitates consensus by allowing national delegations to agree to them without any further obligation as long as no specific contrary interests exist. There may be additional implications as to the shape of the output from the interplay between consensus and its nonbinding consequences. The difficulty of achieving consensus may be overcome by settling on multifaceted solutions, as we have already mentioned. Instead of forgoing broad agreement through a single selection among contentious choices, standards with multiple options may be the preferred compromise (Wallenstein 1977, p. 140). The bias toward multifaceted solutions is partly caused by the scarcity of institutionalized means for overcoming battle-of-the-sexes constellations. Another pragmatic option is circumventing conflict altogether by starting standardization before actors have developed definite opinions in favor of a specific solution (Sirbu and Estrin 1989). Early (ex ante) standardization tries to achieve technical coordination parallel to development work and is more likely to have the structure of a pure-coordination game. However, ex ante standardization is no panacea, since differences in existing networks and technical developments always imply a certain degree of path dependency and, moreover, different technical traditions may still give rise to conflict.

Finally, we need to take account of the fact that standardization may be discussed and evaluated from a plurality of angles. Technical, economic, political, or scientific concerns may lead to the preference of different standards options (Schmidt and Werle 1993). Not all of these different perspectives are equally legitimate, or indeed relevant, in the institutional framework of the CCITT. Its technical purpose is formally institutionalized, and, although the possibility of mobilizing politics as a last resort exists within the institutional structure of the CCITT, informally it tries to keep the technical work separate from politics. This institutionalization, together with the CCITT's consensus orientation and stable membership, characteristically restricts the appraisal of problems to a technical perspective. The system of formal and informal rules, meanwhile, does not provide equivalent room for arguments from other perspectives, particularly economic ones.

The CCITT constitutes the central institutional frame within which the three standardization processes we are now going to analyze have taken place. Such an analysis necessarily deals with many technical and historical details that might prove difficult for the reader to follow. In order to present the cases as lucidly as possible and to allow for easy comparison, we have chosen an identical structure for each. Each of the remaining chapters in this part of the book begins with an overview of the standard's history, which is followed by a description of the central technical and economic problems standardization has been confronted with. We then introduce the relevant actors and their interests before we turn to their minor and major conflicts concerning the shape and features of each standard. At the end of each chapter, we show how the conflicts have been resolved or terminated. Since the conflicts among the actors are rooted in their interests and in their perceptions of technical and economic problems, our manner of structuring the presentation can at times be repetitive. However, we regard this redundancy as the fair price we have to pay for rendering direct comparison and selective reading possible.

7

Interactive Videotex

The mid 1990s brought the sudden diffusion of the Internet, and, with it, explosive growth in a new and unexpected use of telecommunications. For subscribers and service providers, the Internet offers tremendous potential: all sorts of information can be obtained, new customer relations can be initiated, discussion groups can be established, and so on.

In the late 1970s and the early 1980s a comparable telecommunications service—videotex—was planned and heavily promoted by several European postal, telephone, and telegraph administrations. Videotex was meant to be a ubiquitous information service with the potential for professional applications. It built on the telephone network, using the TV set as a monitor with the aid of an adapter. This service, however, met a very different fate than the Internet. Prognoses about its diffusion were remarkable, but the slow adoption of the service in all countries except France frustrated expectations.

In this chapter we will analyze the standardization of videotex as an example of a standardization process that took place under extreme economic expectations, bringing aspects of market competition repeatedly into the game. For the actors involved, future market shares were tied to the success of their standards proposals. Subject to such pressures, they did not succeed in agreeing on a single standard. This failed attempt at technical coordination, in turn, made it much more difficult to realize the expected diffusion of videotex: markets stayed fragmented, so economies of scale in terminal production could not be achieved, and ongoing standardization meant that investment was put on hold in view of imminent technical changes. This, of course, is not to say that failed standardization alone was the kiss of death for the videotex service, although it allowed too much incompatible technical variety. An underdeveloped service concept and modest functionality would have also encumbered an internationally standardized videotex service. The short-sighted marketing notwithstanding, the standardization of videotex is a fine example of a process in which economic expectations and political interests were mingled with the technical work of standardization.

Overview of the History of Videotex Standardization

Essentially an interactive information-retrieval system, videotex, in its original design, employs the telephone system to access databases and a screen (often a TV set) to display information, graphics, and text. Videotex was intended to become the backbone of mass information services and to surmount the shortcomings of professional databases, which were too specialized, too expensive, and too complicated for the average consumer (Bouwman et al. 1992a). Among the services offered are telebanking, teleshopping, ticket reservation, advertising, bulletin boards, and electronic newspapers. As an "interlocked innovation," videotex combines an innovation in the telecommunications infrastructure, an innovation in the supply of new services, and a social innovation in the way users fulfill their specific communication and information needs (Bouwman et al. 1992b, p. 7). Standardization relates to the first of these three innovations and, as far as general service aspects are concerned, to the second.

In terms of international standardization, videotex is remarkable for the conflict it generated[1] and for the unusually high commercial expectations of all the participants.

The significance of the videotex project was closely related to the anticipation of an imminent societal transformation with "the coming of the information society" (Parkhill 1981, 1985). When computing first began to be more broadly diffused, videotex acquired significance in two respects: as a technology of information acquisition for the mass market and as a medium for the decentralized (remote) use of central computers. Prognoses for its growth were extremely high; indeed, it may have generated the strongest expectations in the history of telecommunications standardization.

Standardization of videotex in the CCITT began in 1978—quite unusually, in the middle of a study period. Normally, the Plenary Assembly allocates a study question to a study group; in this case, however, the CCIR initiated the study. Its Study Group 11 had adopted Draft Opinion 11/544, "inviting the CCITT to undertake urgent studies of a possible public-network-based data-bank service" (CCITT Circular No. 84/DIR. 10.3.1978). This proposal was based on the CCIR's ongoing studies of broadcast videotex systems ("teletext")—studies in which it had become apparent that several PTTs were examining the interactive network-based version of the service. In the CCIR, standardization had been introduced under a question (which had originated with the Japanese delegation) of "still motion TV." The United Kingdom, the first country to already have a broadcast videotex service in the early 1970s, had been strongly opposed to the idea of standardization, because it saw TV services as a national undertaking requiring no international standardization. Work at the CCIR was thus initiated contrary to the UK's interests. Because the CCIR saw broadcast videotex and interactive videotex as entirely complementary, it urged that attention be given to "the problem of compatibility between the symbol sets and the two modes of operation of the receiving terminal" (ibid., annex).

The CCIR's initiative was followed by an official meeting of the CCITT in May 1978 at which it was decided that delegates from France, the United States, Sweden, Switzerland, and the United Kingdom would draft questions for videotex regarding the definition of an international technical system, the definition of international services, and the specification of a multitude of operational procedures.[2] Furthermore, it was proposed that CCITT's Study Groups I and VIII study the questions. Meetings of both groups followed in the same year.

There had been additional preparatory steps toward videotex standardization outside the ITU. At the 1977 meeting of the Data-Communication Working Group of the Conference of European Posts and Telecommunications Administration's Telecommunications Commission, "Viewdata" had already been mentioned as a topic for future study by the subgroup CD/SE. Viewdata,[3] a British system announced in 1975, was in fact the first conceptualization of a videotex service (Fedida 1977).

Loosely connected with the activities of the Conference of European Posts and Telecommunications Administrations (CEPT), an important meeting of the Deutsche Bundespost, the British Post Office, and the French Centre Commun d'Études de Télédiffusion et de Télécommunication[4] took place in Rennes, France, in January 1978 (Marti 1979). Differences and commonalities among Viewdata, the CCETT-developed Antiope, and Bildschirmtext (a service undergoing early consideration in Germany) were discussed.[5] Despite existing differences between the national developments, an international standard was preferred by the participants, and the meeting closed successfully with a provisional compromise subject to domestic agreement in the three participating countries.[6] However, the agreement failed to materialize because the standard failed to get the support of British industry, which favored a harmonized national approach between broadcast and interactive videotex over a European approach restricted to the latter. Had the agreement been reached, international standardization would have been much easier.

Meanwhile, in the CCITT, once the decision to embark on videotex standardization had been taken (in May 1978), a new videotex working party within Study Group I, responsible for service definition, began work in earnest. Several meetings were convened, but on the whole the work of SG I was not very successful. Differences in service conception among the participants were fairly great and largely irreconcilable, so the resulting recommendation F.300 remained very rudimentary, with an explicitly provisional character. Study Group VIII, entrusted with the more technical aspects of the study, also set up a working party exclusively for videotex. The UK and France both contributed. This led to several meetings in search of a compromise, as the differences between both proposals were already apparent (Com VIII (58-E, annex), 1977/80).

It was not only the differences between the British and the French proposals that generated difficulties. Canada had also announced, in October 1978, that it was testing its own system (Madden 1979). The Canadian development, called Telidon, was then submitted to the CCITT as a further proposal in May 1979 (TEEGA 1985, p. 30). This meant that three systems were now being promoted, and each country wanted to see its own development succeed. Different degrees of sunk investment (the UK was already starting to implement the service) added economic reasons to the possible objections to seeking a middle ground. Thus, national prestige and economic commitment rendered a technical compromise among the three systems virtually impossible.

In view of ongoing domestic systems developments, the United States was reluctant to make a definitive decision. Japan, which had difficulty displaying its alphabet with the existing systems, had already announced a possible national system in April 1978. Design-system considerations of CAPTAIN,[7] as the Japanese development was called, were also submitted to the CCITT in this study period. As a result, it was impossible to agree on a technical specification. Hence, SG VIII submitted to the Plenary Assembly a recommendation that included as options the British, French, and Canadian systems and the Japanese development. This recommendation, S.100, was endorsed at the Plenary Assembly in Montreal in October 1980. The hope at the time was that S.100 could be a basis for future compromise.

In May 1981, as a further addition to the existing systems, AT&T put forward its Presentation Level Protocol (PLP) at the Videotex '81 conference in Toronto.[8] This system combined elements of the existing syntaxes contained in S.100. AT&T tried to promote its proposal as a synthesis, claiming that compatibility with Antiope and Telidon had been achieved. However, the other parties vigorously rejected the notion that AT&T's development was an integration that overcame existing differences.

Parallel to the CCITT, the CEPT continued work on a European standard for videotex (Marti and Schwartz 1980). Until 1980, this work was pursued less intensively than the work going on in the CCITT. However, with the emergence of system developments at AT&T and in Japan, the European countries put more emphasis on the work of the CEPT. Strong attempts were made to overcome the differences between the UK and France. With the Federal Republic of Germany acting as an arbitrator while also pushing its own development (von Vignau 1983), it became possible to present a first version of the European Unified Videotex Standard at Toronto in 1981, at the same meeting at which PLP was announced.

The CEPT standard integrated the divergent videotex standards of France and the UK. However, it did not constitute a hitherto-unforeseen common denominator between the different standards, nor did it directly replace the existing systems. Rather, it consisted of a more sophisticated system approach, which was considered to represent the future path of development. Alongside this common framework, called Profile 1 (the standard which Btx, the German system, uses), the British system, the French system, and a Swedish system were integrated as Profiles 2, 3, and 4.[9] By including different incompatible options, the CEPT standard was thus similar to the CCITT solution, but it managed to a greater extent to lay down a joint future plan for migration.

The 1981 version of the European Unified Videotex Standard was still incomplete, however, and so work continued in the CEPT (predominantly on more sophisticated extensions). The revised and enhanced version of the CEPT standard was finally recommended in June 1983 at Cannes. This European activity was part of a concentrated effort to bring the standardization of videotex to a conclusion and to see each country's or region's domestic developments succeed.

The CEPT countries had presented their standard to the CCITT in April 1982. With it, the common European approach superseded the divergent British and French national positions at the international level. This allowed a coalition to form behind one CCITT proposal. Meanwhile, the North American countries were trying to strengthen their position. Canada, having realized early that it would need the support of the United States for its proposal, soon made attempts to integrate its Telidon system with the AT&T's PLP. In June 1982 the American National Standards Institute and the Canadian Standards Association announced their agreement on a joint North American videotex standard (O'Brien and Bown 1983). Soon afterward, the United States submitted a preliminary version of this joint standard, called the North American Presentation Level Protocol System (NAPLPS), to the CCITT.[10]

Thus, in this second phase, two regionally based compromises, incorporating the different national system concepts, could be agreed upon. Japan, with its quite different system, was left in a somewhat marginal position, but it had attempted nevertheless to realize some commonalities with the North American compromise. Both regional developments at this time saw—or at least advertised—themselves as the appropriate means by which to merge the different national systems.

In 1982, a new attempt was made at international concord with the launch of the Worldwide Unified Videotex Standards (WWUVS) initiative under the Italian CEPT chairmanship. It was inspired by the hope of achieving broader support for the CEPT Profile 1. Shortly after the United States had submitted the North American PLP proposal to the CCITT, the CEPT countries proposed the WWUVS initiative in the CCITT's SG VIII. In November a general agreement based on the European videotex proposal was signed by the European countries, the United States, Canada, and Japan.[11]

Under general CCITT approval, the activities of SG VIII now took place within this WWUVS initiative for the rest of the study period. Meetings were held every 3 months, alternating in the three regions, each time under the respective national chairmanship.[12] Nevertheless, it was not possible in 1984 to recommend a unified videotex standard.

Instead, a new recommendation, T.101, was agreed upon, which included as annexes the three different data syntaxes. Annex B corresponded to the Japanese CAPTAIN system, Annex C to the CEPT standard, and Annex D to the North American standard. Annex A was a rudimentary neutral approach. This considerable variety was supplemented by the retention of the previous S.100 (now T.100). France insisted in maintaining T.100 as a way of having its system directly recommended as a standard by the CCITT, whereas the CEPT standard only indirectly represented the French solution as one of the different European profiles. Figure 7.1 gives an overview of the national, regional, and international videotex standards.

Despite these repeated failures to arrive at a unified international standard, work in the CCITT continued in the next study period until 1988, and in fact still continues today. Yet, after the unsuccessful WWUVS initiative, there were some changes in the standardization process. On the one hand, as time had passed, the major industrial countries especially had taken decisions toward the implementation of a particular system. It was thus increasingly impossible to abandon the technical specifications of the first generation, with the result that the T.100 recommendation was also maintained in 1988. On the other hand, implementation disclosed that the prognoses of the service uptake had been rather mistaken, as can be seen in figure 7.2. Videotex developed into a service for the mass market only in France, where significant government subsidies backed the service for many years (OECD 1993, pp. 68–74). The system's popularity exploded in France in the second half of the 1980s, and by 1993 more than 6.5 million terminals were connected to it.

A number of facts concerning use of videotex in the early 1990s that stand in marked contrast to the available forecasts propagated in the first half of the 1980s illustrate the low level of market diffusion of this service. Remarkably, only in France, where the forecast was comparatively low, could high diffusion be observed. In the UK and in Germany, where early forecasts had already been cut back to more "realistic" levels, diffusion was disappointing relative to the prognoses (figure 7.2). In the three European countries, the CCITT-standardized videotex services were provided by the national PTTs. In North America, the only CCITT-standardized service deserving consideration was the NAPLPS-based Prodigy, provided by a company that was owned by IBM and Sears. The regulatory situation in the United States made it difficult if not impossible for AT&T and the Bell

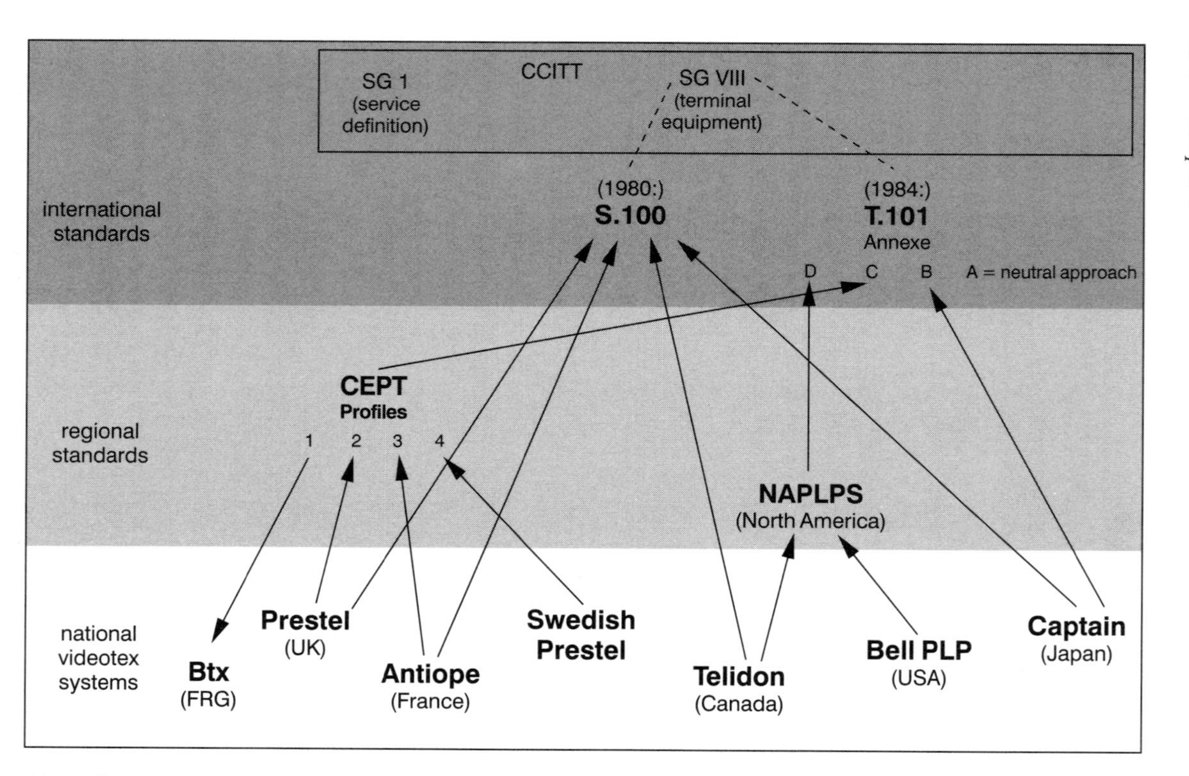

Figure 7.1
National videotex systems and the respective regional and international standards recommendations. (Arrows indicate temporal relationship between implementation and standardization.)

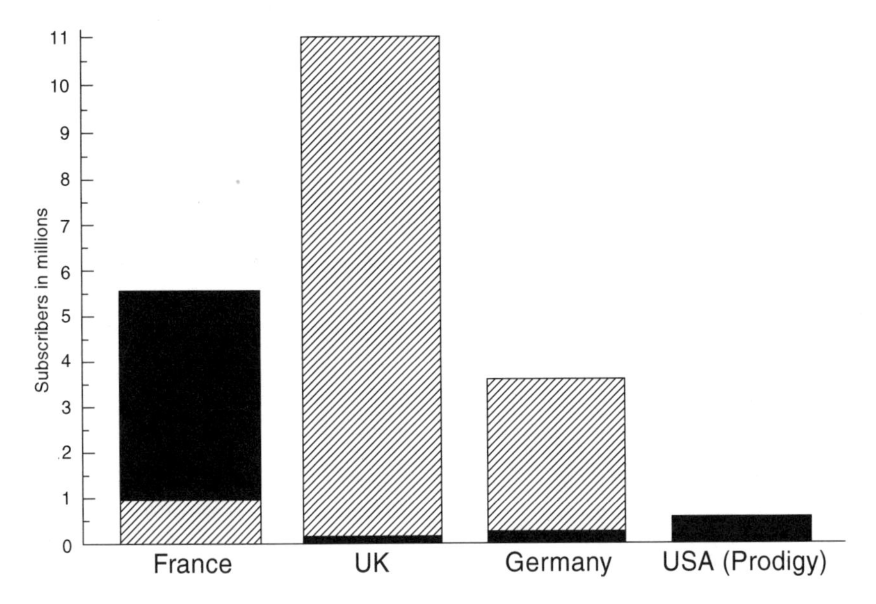

Figure 7.2
Shading: prognoses of diffusion of videotex services for 1990 (UK: 1989). Black: actual diffusion in 1990 (US: 1991). (Sources: OECD 1993, pp. 63–74; Schneider et al. 1991, p. 209.)

companies to enter the videotex market. A variety of other, non-videotex information services existed in Canada and the United States. Most of them were ASCII-based and had fewer than 100,000 users. By 1991, however, Compuserve, an international provider of online services, already had about 50 percent more subscribers than Prodigy.

While implementation increased the constraints significantly, the failure of the profit forecasts simultaneously reduced the stakes in standardization. It became possible in the study period up to 1988 to extend recommendation T.101 a good way toward "International Interworking for Videotex Services." This helped the CCITT videotex standard gain some common technical specifications beyond the mere enumeration of incompatible national ones.

Work was still pursued in the study period that concluded at the beginning of 1993. Because the original videotex design was by then generally regarded as a failure (Schneider et al. 1991), conflict over standardization was virtually absent. If there is a future for videotex, it is in the context of modern online services or the Internet.[13] But this happens, increasingly, outside of the ambit of the CCITT. Here, even though some parties are still interested in extensions, work continues

Figure 7.3
The use of the TV as a terminal, with a page demonstrating the capabilities of Antiope. (Courtesy of B. Marti, CCETT.)

to progress quite smoothly. In a way, it is as if, after 1988, the old conflicts were frozen and standardization started afresh in the (ultimately successful) attempt to achieve common technical specifications for all new extensions, such as those for sound and photographs. In the present analysis we will adhere to this assessment of the situation, focusing our attention on the pre-1984 study periods.

Central Technical and Economic Problems

Videotex consists essentially of a new combination of existing and well-known technical devices. The home TV set, upgraded with a decoder, is used as the terminal. Either a keyboard is attached to it or the remote control can be employed to enter information. The TV is connected to the telephone network with the aid of a modem (figure 7.3). The databases with the information services are held either on one or more centralized computers or on a series of decentralized

	Prestel	Antiope
Different character assignments in the open positions of the ASCII 7-bit code (ISO 646)		
2/3	£	#
5/11	←	[
5/12	½	\
5/13	→	]
5/15	#	–
6/0	–	'
7/11	¼	←
7/13	¾	→
Transmission of control codes in the alphamosaic graphical coding mode		
	serial	parallel
Page format		
	960 characters 24 rows/ 40 characters	1000 characters 25 rows/ 40 characters
Character resolution on display		
	6 x 10	8 x 10

Figure 7.4
Technical differences between Prestel and Antiope.

computers, often connected via a packet-switched (X.25) data network. This network can then be accessed through gateways from the telephone network.[14]

This basic conception was shared by all countries. The idea was to combine two widely diffused technologies to create an innovative mass-market service. Despite this common ground, differences in the precise technical design were large enough for a protracted debate during the standardization process, which had one central focus on coding procedures. Prestel and Antiope, the two systems submitted first, revealed comparatively few technical differences. Figure 7.4 lists the differences between Prestel and Antiope. Both use the international alphabet (ISO 646) or the ASCII seven-bit code, one of the best-known coding standards.[15] Only eight characters differ in the section of the coding table that had been designated by ISO for national use. For graphics, Prestel and Antiope use an alphamosaic method based on the principle that graphics have to be combined out of different elements, like generic pieces of a puzzle for which fixed codes exist.

There is a fundamental difference, however, in the transmission of control codes: Prestel uses serial coding, whereas Antiope employs parallel coding. Thus, in the Prestel system, the addition of attributes through control codes (blinking, color, etc.) always leads to a blank position (called "all-spacing"), so that it is best to change attributes between words. The parallel method of Antiope, in contrast, allows the user to add or take away attributes in a non-spacing way. Great significance was attached to this issue at standardization meetings because of the importance of diacritical marks in many European languages. The UK's low-key service concept was well served with the serial method. For France, however, serial coding implied that accented letters would always be accompanied by a blank, which of course was not acceptable.

To reduce the cost of the decoder, a seven-bit code was chosen instead of an eight-bit code. This provides for only 128 characters, including 32 control functions. By itself, this could not accommodate enough characters for a service. However, it is possible to use several different coding tables in parallel, indicating by way of control codes which one is being employed. Diacritical marks or entire accented characters can thus be placed on a different coding table.

Different methods exist for composing letters with diacritical marks. As the problem with accented characters was one of the main complications in the search for a common videotex standard, this was an important issue. Hence, the composition method, the multi-page method, and the direct method were all discussed in the attempt to find a way to overcome the differences between the serial and parallel modes.[16] Compared to the relatively simple means used to display the English language, diacritical marks in French demanded a more sophisticated technology. In the ensuing attempt to reconcile the display of diacritical marks with the aim of a simple and cheap technology, standardization was complicated still further because the requirements of other European languages were given equal consideration.

Compared to this problem of coding, the next aspect of Prestel-Antiope incompatibility to be discussed here is much simpler. Because videotex used a TV set as a terminal, the traditionally different standards existing for television resulted in different page formats. Prestel had a total of 960 characters in 24 rows of 40 each; Antiope had 25 rows with a total of 1000 characters. Although this one-row disparity is a slight difference in itself, problems emerged in the realization of so-called dynamically redefinable characters (DRCS), a sophisticated concept for the enlargement of the character repertoire. These are not contained in the terminal, but empty spaces are provided there which can be filled with DRCS from the videotex database. The two systems were based on different character resolutions: the number of dots for width and length was 6 × 10 for Prestel and 8 × 10 for Antiope. The resulting compatibility problems were thus jeopardizing the whole idea of DRCS (Marti 1982, p. 39). The compatibility problems experienced in North America as a result of the TV orientation were even more direct. Because of the different standard here, only 20 rows per page were standardized, which for the European countries was an unacceptable reduction of information content per page (Childs 1982). One of the most prominent proponents of videotex and the British Viewdata version sketched the consequences of this technical problem as follows: "The wide variations in television standards—PAL, SECAM, NTSC, etc.—and the resultant problems of compatibility have been reflected in Viewdata considerations. One result is that a format suitable for the 625-line standard used widely in Europe does not easily convert to the 525-line standard as used throughout North America. None of these problems are insoluble but they all add to the complexity of the standardization debate." (Ford 1979, p. 48)

The different TV standards imposed comparatively slight compatibility problems between Europe and North America relative to the problems resulting from the diversity of the approaches to graphical representation. As we have already mentioned, Prestel and Antiope use an alphamosaic method; graphics are treated like puzzles in which different generic pieces are combined. A completely different approach, involving alphageometrical coding, was taken by Canada. Here, the whole content of an image is mathematically described as a sort of drawing instruction which is transmitted to the terminal (Stukenbröker 1980, p. 26). Canada's Telidon consists of Picture Description Instructions (PDIs) containing the information necessary to draw a point, a line, a circle, and such. These elements can then be used to compose different graphics. This principle is very different from the alphamosaic method. In particular, it does not depend on the resolution of the display or on the transmission speed. Pictures will be sharper or not, depending on the display, with the network determining the speed of image transmission. But this is not yet the end of the story where diversity is concerned. Another approach to displaying graphics, called the alphaphotographical method, was used in the Japanese system. Here, all graphics are composed in the databases and transmitted as a whole, requiring much higher transmission capacities. Thus, Japan started with 2400-bit-per-second modems, whereas the rest of the world started with 1200-bps modems.

As to the pros and cons of these technical issues, various arguments were exchanged. Of the competing methods, the photographical was

different approaches to graphical coding	alphamosaic serial (Prestel)	parallel (Antiope)	alphageometrical (Telidon)	alphaphotographical (CAPTAIN)
page formats	Prestel 24 rows 625 lines	Antiope 25 rows 625 lines	North America 20 rows 525 lines	
major trade offs	alphamosaic quality of display		alphageometrical and alphaphotographical transmission time	cost
no common service definition	simple information retrieval (Prestel/ Antiope)		data processing for everybody (Btx)	medium for advertisement (Telidon/ NAPLPS)
interoperability with other services	Teletex			Teletext

Figure 7.5
Major problems in videotex standardization.

the one with the highest technical sophistication; however, the geometrical Telidon could produce better graphics than the alphamosaic method. The Europeans, however, argued against Telidon, whose graphics they regarded as "cartoon-like and stylized" (as one of our interview partners put it). Later though they extended their alphamosaic method toward a better quality of graphics ("picture Prestel"). Obviously the Canadian and Japanese systems were more demanding in terms of transmission quality and thus more expensive than the European system.

Discussions on these and similar tradeoffs abounded. In general simple and comparatively cheap service concepts competed against more sophisticated and expensive ones. In the considerations of the major problems, which are summarized in figure 7.5, the constraints imposed by each system's dependence on different technical environments had to be taken into account. The buildup on the display of a higher-quality image could impose an unacceptable transmission time for some committee members, whereas for others the inclusion of the diacritical marks for different languages was intolerably high a price to pay for marginal benefits. In the end similar tradeoffs were present throughout the deliberations over an international standard. By themselves these technical problems and arguments did not cause the contentious standardization debate. In particular, the Prestel-Antiope controversy, which marked the beginning of international stan-

Figure 7.6
One of the French developments of 1978: the telewriting geometric system Teleécriture. (Courtesy of B. Marti, CCETT.)

dardization, cannot be understood solely in terms of the technical differences in the two systems. In some respects, as has been illustrated by the specifications of graphical representation, the two systems were quite similar. This was even officially confirmed at a very early stage in a tripartite (United Kingdom, France, Germany) Memorandum on Compatibility that arose from the meeting held at Rennes on January 26 and 27 of 1978: "The main features of the systems are compatible. The differences lie mainly in secondary functions and in the interpretation of some escape sequences." Thus, the discussions and conflicts in SG VIII, in the CEPT, and in other bodies, were fueled not so much by the technical issues as by factors from a wider context, of which the late start of videotex standardization was but one. When the unusual step was taken of starting work in the middle of the study period in 1978, the intention had been to ensure an early standardization of this very promising service. Yet the British system was already at the field-trial stage, with plans afoot to have 1500 users by the end of 1978. In France, meanwhile, different national technical developments were being pursued, including "Teleécriture" (figure 7.6). Also, the decision on an experimental implementation was imminent (Mayntz and Schneider 1988).

From its inception, videotex was conceived as a mass-market service, based on the successful diffusion of telephones and TV sets. The projected growth rates promised a pace and breadth of diffusion more familiar in consumer electronics than in telecommunications. The notion of capturing markets influenced the talks in the CCITT and added significant political overtones to the issue. It became a matter of national pride and industrial policy not to back down from supporting a system developed in one's home country.

Moreover, it is mostly over the provision of international services that nationally based operators invariably have to cooperate for network access. Here committee standardization is essential. These international service aspects, however, were much less dominant in standardization, because language differences and the regional significance of most of the information envisaged for the service supported the national orientation of videotex. Only in those cases where neither the confinement of informational relevance nor linguistic variation was manifest did the incompatibilities imply heavy extra costs for information and database providers. Thus, standardization activities tended to remain focused on terminals. This facilitated the conflict, since terminals (i.e., TV sets) had traditionally been supplied competitively, and many standards for them had evolved in the market without the "help" of committees.

As a result, the effort to arrive at a common definition of the service failed to a large extent. Whereas the original British idea for Viewdata was to have a simple information-retrieval system, in Germany videotex was conceptualized as "data processing for everybody," which included much more sophisticated technical features. The combination of existing technical systems in videotex seemed to involve a large degree of contingency, since very different use structures were present or could be projected by the members of the standardization committees. These different elements produced functional ambiguity.

In Study Group I, which was responsible for service aspects, most of these difficulties could not be overcome. Accordingly, a recommendation (F.300) was agreed to that remained quite imprecise and programmatic. This imprecise outcome of the work of SG I implied that the range of necessary international standardization was never explicitly delineated for videotex. For example, the scope and the direction of interoperability of videotex with other services were not determined consistently. While most countries emphasized compatibility with teletext (the broadcast form of videotex), Germany in particular prioritized compatibility with teletex (the new-generation telex). Behind this controversy lay the decision on the technical embedding of the new service, and how far it was to be seen as part of the new telematic services environment. However, in the context of the development of the new OSI standards architecture by ISO and the CCITT, it was hoped for a while that general telematics standards could be agreed upon that would work for both videotex and teletex. More specifically, the failure of SG I's work reflected the low priority given to an international service and relieved pressure on the service-oriented PTTs to reach agreement.

Difficulties with the international standardization of videotex affected only a small part of the overall standardization requirements. In view of the conflicts encountered, aspects other than display standards were never examined closely. This does not mean, however, that other aspects were not contentious. In principle, international coordination for a new service such as videotex touches upon multiple issues. Apart from the whole area of display standards, including coding tables, attributes, and color and size of display, the networks and the databases used impose many compatibility problems. For a true international service that goes beyond basic interworking, the content structure of the databases must be transparent and homogeneous search procedures must be available. But neither these matters nor the development of principles for international tariffing and accounting (Broomfield 1981) have ever been on the agenda, although the CCITT was responsible for them. The general lack of interest in international cooperation and the continued absence of institutional pressure in the CCITT (where the efforts of SG I soon ended in a deadlock) created a vicious circle.

The most visible impediment to standardization was videotex's reliance on a recombination of existing technologies, which brought national differences to the fore. The reutilization of so many elements imposed several technical constraints. All technical elements had been optimized for different functional and national contexts. For example, nationally diverse TV standards became directly relevant to the new application. The choice of the TV as a terminal implied difficulties in regard to picture resolution and a definite constraint with regard to how many lines of information could be displayed. Similarly, the selection of the telephone network meant that transmission capacity was fixed at a low level.

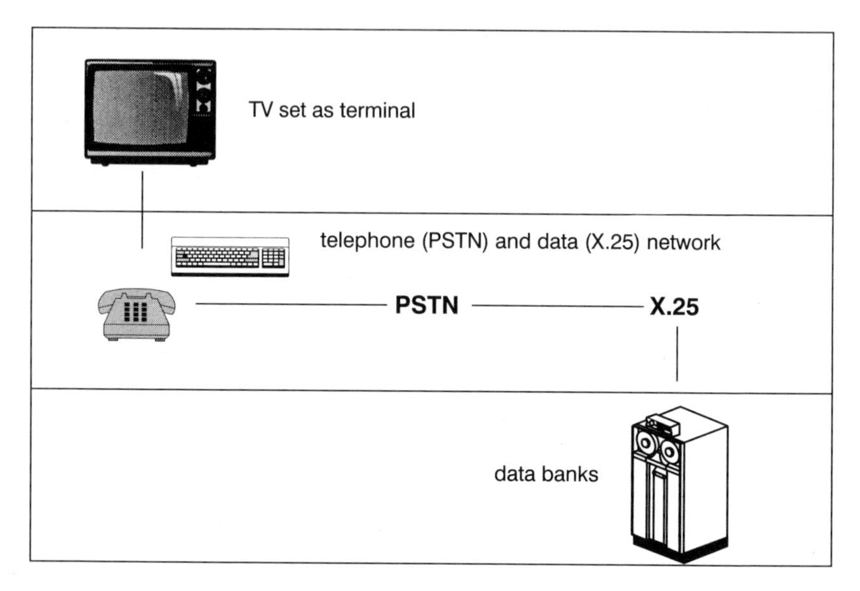

***Figure** 7.7*
Assembly of different technical environments of videotex.

As another side effect, the constraints imposed by the reconfiguration brought with them a tendency to compensate by making the greatest use of the different capabilities of the individual elements. In the case of videotex, this became most apparent in the attempts made to integrate as far as possible the different features of the (color) TV into the service, which led to "overkill" in the standard. Videotex standardization thus differed in many respects from the standardization of very new systems, such as cellular communications, because technical problems were linked to the novel use of existing technologies rather than to novel technologies. Figure 7.7 illustrates the assembly of the various technological environments of videotex.

With the CCITT's inability to mediate or solve the conflict institutionally, the standardization of videotex resulted in a deadlock. Additionally, videotex was the victim of a coincidental turning point in international standardization. Unlike teletex (which used data networks and was not aimed at a mass market), videotex was not reliant on a dedicated network. It marked the transition to early (ex ante) standardization, and it was a first realization of the convergence of computing and telecommunications.

Actors and Interests

Videotex (broadcast and interactive) was a highly contentious topic right from the start. Owing to the very high expectations of the system, early international dominance by any one national system promised immense economic gains. In the context of progressing computerization and the emergent theories of a coming information society, videotex was believed to be the generic information service for the public. Not only did it signify a telecommunications response to the growing relevance of computing; it was, moreover, a significant European challenge to US dominance in the area of high technology.

We will now try to specify the positions and strategies of the participating actors in greater detail. In this attempt we confront the fundamental difficulty that the principle of national representation in international standardization, as in other areas of international politics, transforms the process into a game at two levels because it requires accommodation of interests in the national and the international context (Putnam 1988). In the early, highly controversial years of videotex standardization, national positions in the CCITT were usually represented (as they were elsewhere in telecommunications) by the PTTs or by comparable private network operators. The positions were based on prior negotiations among domestic interest groups, with the PTTs usually regarded as the primus inter pares. Once achieved, such a domestic balance of interests constrains the international "bargaining set" in which agreements are possible (Sebenius 1991, p. 205). Therefore, we must sketch the domestic-interest constellations of the major international players before we turn to the conflicts at international level.

Videotex originated—under the name Viewdata—in the United Kingdom. The British Post Office had experimented with new services in the hope of generating new network traffic in the early 1970s. In 1975 this development was successfully merged with similar systems belonging to the broadcasting authorities. The BBC and the IBA (Independent Broadcasting Authority) had already begun in the mid 1960s, quite independently of each other, to elaborate the idea of a new TV service that would use the vertical blanking intervals of the TV broadcast signal for the transmission of "pages" of information services (Giraud 1991, p. 66). In 1972 these efforts produced the BBC's Ceefax system (Amos 1978) and the IBA's Oracle system.

In France, the development of videotex focused on the telephone network from the outset, although the possibility of transmitting data over the TV network had already been discovered in the early 1960s.[17] In the search for new services, the CNET experimented with a "service for calculation by telephone" (Marti et al. 1990, p. 3) in 1970. The first pocket calculators soon revealed the failure of these plans, which were then renamed a "service for consultation by telephone" (ibid.). The TV was included in the development only when a terminal display was needed.

The CNET continued work on the connection of the TV and the telephone network. In 1974 it demonstrated "TIC-TAC," its first system development. It was only then that the similar achievements in the UK were discovered. The British were already more advanced than the French, having profited from the early close collaboration with the broadcasting developers, who had opened up the knowledge base of the TV industry to the researchers of the British General Post Office (GPO).

Toward the end of 1974, work started at the CCETT (the joint development center of the Office de Radiodiffusion et Télévision Françaises and the CNET) on a French teletext system. The resulting development, Antiope, was then transplanted onto the telephone network as an interactive service.[18]

The UK's lead was significant in view of the fact that the first public specifications for Ceefax and Oracle had been announced a year before the CCETT had even started its development work. A pilot trial for Viewdata started in 1976. One year later the UK was already selling the system to the Bundespost.

The GPO's interest in videotex was mainly confined to the additional traffic it would generate. The precise design of the system was not a crucial issue at the GPO. Any new application could only make more use of the existing network. Thus, anyone who could directly or indirectly contribute to increasing traffic in the GPO's network was relevant to the speed of the British system's development. Close cooperation with the TV industry was crucial. The strong backing of the TV manufacturers resulted from a peculiarity of the British market, which involves a very high percentage of rented TVs. Those who rent a TV usually do not keep it until it does not work any more; they tend to replace it as soon as additional features and services become available on new sets (Marti et al. 1990, p. 2). Manufacturers can thus stimulate replacement and accelerate production by supporting new developments. Hence, there was a high motivation in the British TV industry to realize videotex and teletext. Soon after the emergence of Viewdata, videotex and teletext were made compatible so that the same decoder could be used (Schneider 1989, p. 76). This achieved some economies of scale in the production of decoders; it also helped to replace TV sets and to diffuse videotex and teletext, which was in the interest of all parties.

One consequence of this early success was a decrease in flexibility. Subsequent changes became difficult because of existing investment. Purely financial considerations, however, were only one impediment. The difficult diffusion pattern of network technologies also makes change difficult, since a service like videotex will rely heavily on expectations and on mutual confidence in its future success (Hohn and Schneider 1991). Moreover, a change in the design of the domestic system had to be approved by all members of the national coalition. The stringency of this situation was demonstrated after the tripartite Rennes meeting of January 1978, where a compromise on collaboration with the CCETT and the Bundespost in the creation of a common display standard for videotex had been achieved. The compromise failed because of the resistance of BREMA (the British TV manufacturers' organization) to the adoption of this standard. They were not willing to change their concept, particularly since it meant forgoing the common videotex-teletext decoder.

The UK therefore had little leeway in international standardization. Changes to its domestic system were not really an option. Apart from trying to sell the Prestel system through bilateral talks and agreements, the UK could only attempt to "market" its system in the standardization committees. The interests of the GPO were only indirectly tied to Prestel: it had devoted few resources to service development other than the investment it had made in the database software, from which profits could result through wider acceptance and sales (Heys 1981, p. 430). The GPO also had an interest in principle in any additional traffic generated by a more widely diffused standardized service. However, the GPO did feel an important responsibility toward the service as an advocate for those sections of British industry that had borne the brunt of the development costs and had been instrumental in the creation of the additional service revenues. Profitable services in an area of mixed technologies, in addition, required a high degree of domestic cooperation (Schneider et al. 1991). The manufacturers, lacking equal access to the international standardization committees, needed the "assistance" of the GPO in order to achieve a transnational market share (or, at least, to protect the national market by means of an adequate international standard—hopefully the British one). This mutual dependence between PTTs and manufacturers was also typical of the experience with videotex in other countries.

As a result of its domestic setup, the UK could not participate in international standardization with the aim of finding a true international compromise solution for videotex. Rather, committee standardization complemented the attempts at direct international marketing. Where the limited technical capabilities of Prestel prohibited direct international adoption of that system, it was important for Prestel that international standardization not proceed in a manner contrary to the UK's interests either. Any international agreement had to at least lie within the development scope of Prestel, and should not be a compromise isolating Prestel. The results of the CCITT and the CEPT work were satisfactory in this respect. Prestel was included as one option and was a valid starting point for a videotex system that could be enhanced later in the direction of the more sophisticated international standards.

As we will see below, the United Kingdom's involvement in international standardization ebbed away in the early 1980s, when a significant change of interest took place. The exaggerated prognoses of videotex's diffusion had to be reassessed in such a way as to remove some of the high expectations that characterized the early debates. But domestic political developments were more significant this time. In 1981 a first major step was taken toward telecommunications liberalization; with that, the backing for the unprofitable videotex was slowly withdrawn and British priorities were shifted to other areas.

In France, the initial developments of videotex and teletext were less well coordinated. In view of the technical inability of Prestel to meet the demands of the French language, mere adoption of the advanced British system was never seriously contemplated. France also seemed to have recognized early the international competitive aspects of the new developments. Unable to catch up with the UK, France introduced teletext as a topic for study at the CCIR in the context of a question, introduced by the Japanese, on "still motion TV." Standardization here gave France the opportunity to counter the UK's market lead with a different international standard. Not surprisingly, the British saw no need to standardize teletext internationally as a broadcasting system.

However, not only did the French see an international standard as a significant short-term strategic aim; they also saw advantage in an international technical agreement based on the collective experience and technical knowledge of several countries. That is why they did not hesitate to integrate the results of the Rennes meeting into the Antiope system, thereby profiting from the greater British experience with the technology. The resulting modifications were generally seen as improvements and not as an unwanted adaptation of British technology.

France was much more flexible than the United Kingdom at the time, and was willing to compromise for the sake of international standardization. This changed rapidly in 1979–80, when the decision was made to start a field trial with a view to national implementation in the near future. As in the UK, investment now rose sharply while international flexibility declined with the decision to implement Antiope. Consequently, France too became highly confrontational in the arena of international standardization.

France's implementation decision was precipitated by domestic factors. During the 1970s, France had managed to completely rebuild its national telephone network, which had ranked among the worst of the highly industrialized countries. After this project of national priority had been fulfilled, however, the country was faced with enormous surplus capacities in the telecommunications industry. Whereas it normally took 40 years to build a network, France had caught up in only 10. New telecommunications projects were needed if a politically unpopular contraction of the industry was to be avoided. Furthermore, the telecommunications sector was receiving great attention in the aftermath of Nora and Minc's (1978) famous report on the "informatization of society." In this context of "telematics," videotex was a highly suitable candidate for stimulating the growth of the telecommunications industry (Schneider 1989, p. 42). It went on to achieve the status of an industrial *grand projet.*

In contrast to the United Kingdom, where there was a coalition between public authorities and private industry, the emphasis in France was on industrial policy. This was particularly apparent from the French decision to couple the implementation of Antiope with the Minitel strategy. A dedicated simple videotex terminal called Minitel (figure 7.8) was offered free of charge as an alternative to the conventional TV set, and an electronic directory was offered to replace the printed telephone book. Because France's telephone network was expanding at four times the normal rate, phone books quickly became

Figure 7.8
One of the first Minitel terminals (1979). (Courtesy of B. Marti, CCETT.)

outdated, and an electronic directory had a lot to offer. Shortly before the François Mitterand was elected president (in May 1981), it was announced unofficially that DGT, the French PTT's telecommunication division, had ordered 300,000 Minitel terminals from CGE-Alcatel. This action contradicted the international efforts toward working out a standard for videotex that involved using TV sets as terminals. France's commitment, in terms of actual investment and general expectations, was now on another scale. In particular, the significant subsidies for the free Minitel terminals meant that there was a substantial political interest in seeing the development succeed. Consequently, France became very uncompromising in international standardization discussions. And, unlike the UK, France sustained its interest.

Germany did not want to miss out on the promising new videotex development (Schneider 1989). The Bundespost had purchased Viewdata very early, in 1977, for the purpose of conducting a pilot trial. Germany's broadcasting organizations had selected the British Ceefax system. Prestel could be adapted to meet the diacritical demands of the German language; however, it then lost its full compatibility with the original British system. This situation was unsatisfactory, and at the same time the apprehension was growing that a common European development was crucial to the profitability of videotex.

As we have mentioned, Germany, like France and the United Kingdom, had already participated in the supranational talks on videotex at Rennes before the beginning of the CCITT's standardization efforts. Germany's position was, however, very different from those of the two other countries. Instead of presenting a third, incompatible approach to videotex, Germany assumed the role of arbitrator from the start, stressing the need for a unified development, actively searching for common ground between Prestel and Antiope, and putting international standardization first. Only when failure loomed did Germany appear determined to specify its own national system. That Germany was not completely neutral is indicated by the priority it gave to videotex's compatibility with teletex over the preference of France and the UK for teletext. Intentionally or not, Germany's role perfectly combined the positively valued effort to pave the way for the production of an impartial standard as a collective good with the interest in promoting a completely new standard instead of adopting one of the competing solutions. This turned out to be important in the eventual success of European standardization (Schneider 1989, p. 113).

Instead of developing another national system to compete directly with Antiope and Prestel, Germany undertook to prepare simultaneously for international standardization and the introduction of a domestic videotex service. In 1979, technical working parties were instituted at the Bundespost to coordinate technical development, service implementation, and international standardization with firms (e.g. Dornier and Valvo) that were particularly active in international standardization.

One result of Germany's attempt to arbitrate videotex standards was the CEPT standard, which combined ideas and technical elements from several countries. Implementation of this common standard in the largest European market, Germany, would have been quite a coup for the participating German firms. In a way, then, Germany can be seen to have learned from the experiences of France and the UK: for those countries the export of the national developments was necessary, but the required adaptation of their systems to other domestic conditions had hampered international success.

If Germany's position can be interpreted as an attempt to bring about the reverse of the British and French attempts, Canada again was an underwriter of the national-system approach. Like other countries, Canada became aware of the new developments through a demonstration of Prestel. It was not long before Canada considered buying the Prestel system for a trial; however, development of a domestic system was pursued instead. As in the case of France, this was an outcome of a contingent process. When Prestel was demonstrated to the Canadians, several Canadian engineers who in the early 1970s had been involved in a military program involving the transmission of graphics over the telephone network had the idea for a development of their own. Thus Telidon started in a very experimental, unofficial way.[19]

Once some early problems had been overcome, Telidon received strong official backing.[20] The system was, clearly, seen in the context of the information revolution (Parkhill 1981). Telidon was introduced to the CCITT in October 1978. However, Canada, unlike France and the UK, did not aim to capture the whole international standardization process; it merely wanted Telidon to be considered as being on a par with the other existing systems. At the same time, Canada had great hopes of being able to dominate the US market. In this it shared a primary aspiration with the UK and France, since the US market provided the best opportunity for realizing the necessary economies of scale and profits for videotex (Savage 1989, p. 203).

Thus, despite some reservations in the CCITT, Telidon was in direct competition with Prestel and Antiope. It seemed to be more suited to the North American market. With its geometric coding, Telidon was a more sophisticated system. It was developed with a view to the rapidly decreasing costs of chips, reflecting North America's greater computer use. Its more advanced graphics were particularly suited to the demands of the advertising industry, which was seen as a major user of videotex and a major mover of its diffusion. The display-independent coding allowed Telidon to be used with TV sets, but its development had been less focused on the constraints of this medium than Prestel and Antiope. With a mean use time of 45 hours per week, the TV was not suited to further applications in North America. Nevertheless, Telidon could also be used as a broadcasting teletext system. Compatibility with teletex was not believed to be relevant.

The US proposal for a domestic system was not officially announced until 1981, at the beginning of the next study period of the CCITT (Wetherington 1983). This comparatively late active involvement was linked to a variety of factors. The conditions of domestic use in the United States were similar to those in Canada. A greater degree of computerization, as well as different network use and regulation, meant that AT&T was not motivated to realize new applications in order to increase network traffic to the same extent as the European PTTs. In North America the dominant flat-rate tariffing of the local loop had led to an average use of 3 percent instead of 1 percent as in Europe. With flat rates and a maximum switching capacity fixed to only 5 percent average use, there was less incentive to put more traffic on the network. Inasmuch as videotex was conceived as the European answer to American computing, the ambition to introduce it was, understandably, lacking in the United States. Unlike Canada, the United States lacked any government commitment to subsidizing development.

The regulatory situation in the United States had the most impact, since videotex development coincided with major regulatory turbulence. The Federal Communications Commission's 1970 "Computer I" decision had barred AT&T from any involvement in information services. Only with the 1980 "Computer II" decision did it become at all possible for AT&T to engage in the videotex market, under the condition that it establish a separate subsidiary. AT&T's divestiture came in 1984, at a time when international interests in videotex were again declining. In addition, the fragmented regulatory situation created significant problems for a service concept such as videotex. The federal principle, whereby (for example) the federal government was responsible for rates and entry but the states were responsible for regulating information content, made coordination extremely difficult in the United States, much as in Germany (Bruce 1981).

In the United States, therefore, Videotex appeared to be neither a priority nor a particularly feasible concept. At the same time, the US was regarded by the developers of the national systems of the UK, France, and Canada as the primary market to be targeted. Hence, the commercial interests of several domestic firms (some of which were already testing the French and British systems), in combination with the changing regulatory situation, may have been what aroused AT&T's interest in becoming actively involved.[21] What is more, AT&T would have hardly felt very comfortable seeing a foreign system invade its home market.

AT&T's PLP development was presented alongside the first announcement of the CEPT videotex standard in 1981. With it, AT&T adopted a position that was a hybrid of the German "arbitration" position and the competing endeavors to internationalize domestic technological developments. AT&T promoted PLP as an attempt to marry the existing systems proposals to a unified approach, which was meant to lead international standardization out of the very unsatisfactory state it was in at the end of the first study period in 1980 (Wetherington 1983). This "neutral" position was also institutionally induced: having participated in videotex standardization from the beginning, the United States was expected to base any national development on the agreements achieved via that process. Although the proposal was similar to that for Telidon, the existing leeway had been used so that the differences between PLP and Telidon were sufficient to keep Canadian companies from gaining any advantage. And the adoption of elements of the alphamosaic systems allowed AT&T to tout PLP as a significant improvement.

AT&T's involvement in videotex significantly affected all the participants in the standardization game. Videotex activities in the United States increased, thereby jeopardizing foreign endeavors to dominate the US market. The interests of US firms did not last, however. Different use patterns, the early diffusion of personal computing, and the complicated and changing regulatory situation made videotex an even less profitable and less promising undertaking in the United States than it was in the other countries.

In Japan, a close national coalition of the major players in the telecommunications industry developed a highly sophisticated videotex system that utilized photographical coding. The CAPTAIN system was a joint project of the Ministry of Posts and Telecommunications (MPT) and Nippon Telephone and Telegraph (NTT), involving from the beginning such important manufacturers as NEC, Hitachi, Matsushita, Fujitsu, Oki, Sony, and Toshiba (Kashiwagi and Taniike 1981, p. 98). Japan's videotex was closer to the technical environment of telefax, using a very similar coding method (Run Length Limited Code). Exclusively technical arguments were used to justify this separate route to videotex development. Owing to the peculiarities of the Japanese alphabet, the coding method was deemed essential. Hence, it received recognition as an option by the CCITT without obstacles, and in return for this recognition Japan never attempted to compete head-on with the other approaches. Like most of the other national systems, the Japanese system aimed at compatibility with teletext. For geometric coding, adopted in 1983, an approach similar to NAPLPS was chosen (Shibui 1984). By adopting this approach of simple coexistence, Japan remained on the fringe of the standardization controversies.

Minor and Major Conflicts

In many respects, the whole process of the standardization of videotex was a process of conflict—at least until the mid 1980s, when the failure of the service became more acute and interest died down. Rather than analyze the dynamics and the outcomes of specific incidents, therefore, we will examine the entire standardization process in the light of the progress of the conflict, the composition of the contending parties, and the building of coalitions.

At the outset of standardization, the UK and France were the two strongest contenders. The initiation of videotex standardization at the international level was even triggered by the conflict when France started work on teletext in the CCIR that went against the UK's objectives. The CCIR then drew the CCITT's attention to the topic. The UK lost its domestic flexibility in international negotiations quite early and could henceforth only defend its system. Committee standardization became for the UK more like an arena, similar to the market, where systems competition was being pursued. But the general uncertainty regarding videotex development and use brought with it the possibility of benefiting from cooperation. France, having much less videotex experience and not yet bound by imminent implementation decisions, was more cooperative. The early presence at the talks of Germany, which lacked a national development of its own, also reinforced the opportunities for technical coordination on the basis of consensus.

At this stage, most of the issues concerned serial versus parallel coding and other display aspects (number of rows, etc.). The direction of technical embeddedness was also contentious. The UK had decided on compatibility between teletext and videotex, and France was also inclined in this direction. Germany, however, pushed strongly for compatibility between teletex and videotex. But similarities of the national positions in Europe, most of all the common choice of alphamosaic coding, actually outweighed these issues. In view of this commonality, there was the initial belief that an agreement just might be possible.

The introduction of Telidon made standardization much more difficult. Canada's choice of alphageometrical coding implied greater technical divergence. Moreover, Telidon adhered to another service concept. A sophisticated graphics capability was deemed necessary because advertising—a use not emphasized at all in Western Europe—was believed to be a major application. The display-independent coding necessitated a very expensive decoder. Telidon was based on an altogether different assessment of the pace of technical development because it was more oriented toward computing.

Despite these differences, France sought closer cooperation with Canada. The needs of the French-speaking Canadians meant that Canada's system had to provide for the display of English and French, and this represented a possible solution to the problems between France and the United Kingdom. Cooperation between France and Canada was also facilitated by Canada's relatively unconfrontational stance: Canada merely wanted its system to be treated on a par with other countries' systems; it did not plan to overtake them. Furthermore, the other half of Canada's strategy—to try to set the North American de facto standard—was not common knowledge, so Canada's proximity to the United States made it an important partner for France (Savage 1989, pp. 202–203).

By this stage, the UK lacked the flexibility for cooperative moves. The appearance of the very different Canadian approach had demonstrated that achieving any cooperative gains without individual costs—the pure-coordination game—was impossible. And the UK's domestic commitment was simply too great to permit bearing any individual costs. Seen from this perspective, videotex standardization appeared to be a zero-sum game lacking any incentive for a compromise solution. Consequently, at the last meeting of SG VIII in the study period until 1980, the UK suggested deleting Telidon as an option in order to arrive at a CCITT recommendation (S.100) that had as few options as possible (Ford 1979, p. 5.3).

Such confrontational behavior is exceptional in the CCITT, and it contradicts the dominant notion of a common search for the best technical solution to coordination problems. The other participants did not acquiesce to the United Kingdom's hostility: the United States and Japan decided to support Canada, and France, seeing itself in direct competition with the UK, immediately backed Canada too. To broaden the coalition and further isolate the UK, France lobbied Germany and Sweden to give their support, threatening otherwise to veto recommendations for teletex, which these two countries urgently wanted to realize. In consequence, the UK's position was considerably weakened by the broad support for Telidon (Savage 1989, p. 204), and Telidon was recommended alongside Prestel, Antiope, and CAPTAIN.

Although the conflict was now manifested in S.100, with no system able to triumph over the others, many participants regarded this recommendation as a new impetus to continue standardization of videotex. At least, S.100 obliterated the most confrontational "my system only" strategy. The recommendation could serve as a basis both for advancing the competing economic interests through systems approved by the CCITT and for continuing negotiations in order to strengthen the individual position vis-à-vis the necessary future modifications and refinements of the standard.

Concomitantly, the European countries switched their attention to the CEPT, where work was accelerated. The need for a shared technical specification, no single position having won the support of the committee, had shifted the emphasis to the German strategy. Further pressure was exerted by the ongoing development of AT&T's systems and by signs of closer collaboration among the United States, Canada, and Japan. In this climate, it became easier for Germany to mediate the British and French positions, and to find acceptance for its compromise position, under the threat of otherwise advancing a national German standard. Despite significant difficulties and tensions, which led some external observers to liken the negotiations to a war, a CEPT agreement was signed in May 1981.

Much like S.100, the CEPT standard brought together Prestel, Antiope, and the Swedish version of Prestel as Profiles 2, 3, and 4. Moreover, these different systems were given a common "umbrella" with Profile 1, the technical specification used for the German Bildschirmtext (Gabel et al. 1984). For Profile 1 it had been decided, after long technical discussions, to use the composition method for diacritical marks. Originally developed for teletex, this method allowed videotex to satisfy the demands of all European languages. Profile 1 combined serial and parallel coding and was said to be downwardly compatible with Prestel and Antiope. Nevertheless, this did not mean that compatibility for European videotex was henceforth established. On the one hand, the standardization dealt only with a small part of the overall system (neglecting, for example, network interconnection); on the other hand, the existing Prestel and Antiope systems needed to implement certain additions if they were to be compatible with Profile 1. This approach toward agreeing on a common standard was called the "onion strategy": a common core was surrounded by different options, whose implementation was voluntary (Morganti 1982).

In 1981, when the CEPT standard was presented in conjunction with the first demonstration of AT&T's PLP, a new phase in the standardization of videotex began (Heys 1981). Each of the second-generation specifications was advocated as an eventual compromise between the disparate videotex developments. Thus, at the announcement of PLP it was claimed that compatibility among Telidon, Antiope, and PLP had been achieved. But by virtue of their coexistence the PLP and CEPT standards demonstrated that the standardization conflict had merely been transferred to another level. The balance of the different national coalitions, however, was significantly changed by this event. Shortly after the parallel announcement, the CEPT attempted to initiate bilateral negotiations with the United States. In November, the US Department of State, the official delegate to the CCITT, was contacted with the proposal that if PLP were to integrate the CEPT standard as a subset the CEPT would in turn adopt the geometric coding of PLP. However, the United States did not respond to this attempt to coalesce these developments until October

1982, when the United States officially submitted PLP to the CCITT. This was almost immediately followed by another cooperative venture launched by the CEPT: the WWUVS initiative.

In the meantime, the various countries sought to strengthen their positions by individual action. For Canada, the British inroad had demonstrated that international standardization required multinational support. Simultaneously, AT&T was venturing into the other part of the Canadian strategy, the domination of the North American market, with PLP. In the light of this situation, Canada made strong overtures toward the United States on behalf of a joint US-Canada development. The outcome of this cooperation was NAPLPS, which was completed in parallel with the WWUVS endeavor. In June of 1982 ANSI and the CSA finished their work on this specification, which was publicly backed by 80 North American companies (including IBM, Control Data, Hewlett-Packard, DEC, Texas Instruments, AT&T, and Bell Canada). Furthermore, Japan adopted the NAPLPS geographical coding.

Europe, where videotex had originated, now seemed even more isolated, but here too the positions of various countries changed in the period 1981–1984. The UK's involvement in international standardization declined. The adoption of Prestel had been mediocre, partly because of too expensive terminals.[22] The privatized British Telecom now targeted the business market but withdrew its support from Prestel. With market liberalization, competing services were introduced, and Prestel soon lost its distinction and importance (Schneider et al. 1991, pp. 190–191). Another problem with Prestel's diffusion was more closely associated with standardization: the common decoder with teletext implied much commonality between the services. This put Prestel in the difficult role of a private good competing with a public one (Larratt 1981, p. 25).

Germany, in contrast, was still at the initial stage of videotex development, with strong political support. That country continued to play an active role in the CEPT, strongly pushing the revised videotex standard that was eventually agreed to in 1983. This standard included such advanced features as photographical representation and geometric coding. After the failed negotiations between the United States and the CEPT, the graphic kernel system (based on an ISO standard rooted in German developments) was used for geometric coding. Germany still hoped that the various existing videotex systems would finally migrate to the CEPT standard, at least within Europe. In that case, Germany, having implemented CEPT Profile 1 early on, would be a late winner in the standards competition.

France continued to push its position on standardization after the 1981 decision to implement Antiope on a large scale with the backing of heavy government subsidies. Much as the UK had done several years earlier, France had now lost all flexibility as to compromise. However, to advance its chances of exporting Antiope, France needed to make sure that Antiope remained a part of the internationally recommended specifications. That is why, at the revision of the CEPT recommendation, France pushed hard for an addendum that exempted it from implementing Profile 1.

At the 1984 Plenary Assembly of SG VIII, held to consider the Red Book recommendations, France's position led to a conflict that demonstrated the political relevance of videotex. After the failure of the WWUVS initiative, the 1984 T.101 recommendation included only CEPT, NAPLPS, and CAPTAIN as options, and it restricted the validity of the old S.100 (which contained Antiope explicitly) to one more study period. In a highly unusual step, France vetoed this recommendation at the Plenary Assembly, although in earlier meetings of the Working Party and on other occasions the French delegates had raised no objections. Only two other countries, Belgium and Cameroon, voiced opposition; thus, T.101 was "recommended" rather than "recommended unanimously." France's hope in this instance was that other countries, which at an earlier point might have consented to a difficult compromise, would now join in the dissent, opening up the debate for another study period.

What kind of general pattern can be discerned in this conflict? All countries were encouraged by the extreme expectations raised by the prognoses of videotex diffusion and by the endeavors to realize return on investments. International service considerations were secondary or even less important. For a short period, owing to the significant uncertainty in the early stages of systems development (relating, for example, to future use patterns), it looked as if the different participants were quite willing to cooperate, as a common technical solution still seemed possible. But this view soon came to appear unrealistic. A joint technical specification was not really sought. Countries wanted to keep their domestic systems and achieve de facto domination of exports. Committee standardization complemented market strategies, and the debate in the CCITT was unusually full of political overtones.

As France's opposition at the Plenary Assembly indicated, this sometimes led to a reversal of the division of labor between technical and political actors. Normally the work of the technical study groups is only rubber-stamped at the Plenary Assembly, to the extent that the study groups formulate the questions which the PA will later allocate to them (Irmer 1990). This time, the political aspects of the CCITT also played a dominant role outside the Plenary Assembly. Members of technical study groups were constrained in their work on possible compromise solutions by the rather firm national positions.[23] In the case of recommendation T.101, where rudiments of a joint technical position were nevertheless achieved at the technical level, the result was weakened by the political participants at the Plenary Assembly.

Once the cooperative spirit had vanished at the technical level, tactics determined the positions of the various countries to a high degree. Although the primary goal was to win the zero-sum game, the secondary goal was to keep another country from succeeding. The UK began in this vein by openly trying to eliminate Canada's Telidon. AT&T's PLP was based on Telidon, but not to such a degree that it would have given Canada the competitive edge. The maneuvers of France, Germany, and the UK were subtler. Because Germany pursued success by acting as a mediator, tactics and changing coalitions prevailed over head-on collisions. Whenever the UK or France seemed to dominate the conflict, Germany tried to counteract this. For instance, after the French decided to support large-scale implementation of Antiope, the Germans defended the position of the information providers who had incorporated in Prestel about a thousand pages from the German pilot system.

Finally, videotex provides many examples of how controversies can be open or hidden to differing degrees. France's opposition to videotex at the Plenary Assembly was the extreme form of evident political disagreement with a technical compromise that had been worked out (in collaboration with French delegates) at the technical level. But such blatant means were used only as a last resort. Two of the more frequently used means were bilateral intergovernmental lobbying outside the official meetings and a politically determined national mandate restricting the scope of negotiation.[24] Thus, political and economic considerations were crucial, though less obvious, determinants for the preference of specific technical solutions. Yet this does not mean that technical reasoning played no role. Several physical constraints existed, such as the number of signs that can be displayed using a seven-bit code or the possible size of a page on a TV terminal. Technical predilections—rooted, for example, in different engineering traditions and system concepts—also played a part. However, it is sometimes difficult for observers, and even for the participants in standardization, to judge whether nontechnical arguments are being disguised in a discussion on (say) direct versus composition coding, or whether the technical differences are really the crux of the issue.

Conflict Resolution (Termination)

How was the standardization conflict resolved? The CCITT could not formally decide not to pursue standardization, nor did it have the institutionalized means to openly discuss different political and economic interests and to solve them via package deals, side payments, or turn-taking. Distributive issues of a battle-of-the-sexes situation can only be discussed informally in the CCITT. Moreover, in many instances the videotex standardization process had the structure of a zero-sum game. Standardization thus had to continue suffering harsh conflicts. A multifaceted solution, which included the various approaches as options, was the only way to achieve any result from standardization. To stabilize a stalemate in such a way is a common solution in committees once compromise is impossible. Either it provides the opportunity for further progress in the next study period (because the different contenders at least acknowledge their mutual existence), or the committee may be supplemented by market forces (since the implementation of the system may lead to the dominance of one option). From an institutional perspective, standards with incompatible options embody a failure in disguise, because standards officially aim at preventing or overcoming incompatibility.

With videotex, competing economic interests typically contravened the willingness of different participants to compromise. The comparatively late start of standardization provided a valid excuse, since sunk costs had already been incurred. Moreover, after the start of standardization, additional systems were proposed which were only partly legitimized (as in the Japanese case) through specific national conditions. That competition does not necessarily lead to the emergence of increasingly contradictory standard proposals is suggested by the Canadian effort to merge Telidon with AT&T's PLP into the NAPLPS standard. Here, the need for a large market with economies of scale encouraged compromise over conflict. This was aided by the very similar applications environments and the tradition of shared telecommunications standards in Canada and the United States. Furthermore, Germany stressed the need for compromise for the sake of a large market but, at the same time, preferred a "new" standard instead of backing one of the existing proposals.

In this constellation, a common technical position could only be established as part of the future course of technical migration. Technical enhancements of the different systems were to take the same approach. The consequence was a peculiar mixture of ex post and ex ante standardization. A range of possible future extensions to videotex were standardized and were integrated as options. The CEPT standard resulted from this strategy, but the problems of this approach became manifest in the German implementation: the common developments were too sophisticated and forward-looking to overcome existing differences. While harmonization could thus be approached, problems still arose in implementation. An immediate implementation was required to capture a market segment and to increase the opportunities for wider diffusion, but the technology necessary for the sophisticated extensions could not yet be built. German videotex diffusion suffered considerably from this contradiction (Schneider 1989, pp. 140–150).

Multiple constraints had been imposed on the new videotex service by the commercially motivated decision to start this innovation, mainly through combining existing widespread technologies embedded in heterogeneous technical environments (TV, computing, telecommunications). Additional complications arose from the desire to make videotex compatible with teletex and/or teletext. The standardization of the future extensions was also affected by the heterogeneity of the three technical domains. This time, however, not the constraints but rather the positive technical capabilities of the constituent technologies guided the actors in their efforts. A common belief in the future feasibility of many features made it easier to compromise. The features of the TV were most dominant here. With the standardization of 4096 colors, of blinking, and of sound, the attempt was made to profit from the capabilities of the medium to the utmost extent. The different application environments of videotex and TV, although already visible, were largely ignored.[25] A comprehensive service concept was missing.

The way videotex standardization was caught between competition and coordination requirements did not leave much scope for a successful conclusion. One possibility could have been the implementation of a multi-option standard to be activated selectively, according to the national context. This option—which can be compared to selling all electrical equipment with an all-purpose adapter for all different national networks but restricting normal use to a single network context—would have provided a basis for economies of scale in terminal manufacturing while allowing for nationally specific applications contexts. Yet here too videotex suffered from a lack of simultaneity in different developments: by the time technology was advanced enough for such a concept (e.g. multi-standard decoders), videotex was already outdated.[26] Indeed, part of its failure had been caused initially by the absence of stable international standards and by the incoherence of applications concepts.

8

Facsimile (Telefax)

Some technologies affect our lives significantly within a relatively short period of time. Suddenly, it is strange to think that something we rely on quite naturally was not around just a few years ago. Even more amazingly, this change need not be the result of a recent major scientific breakthrough that has provided a technical solution to a long-felt need. It may be a familiar technology that has suddenly become widely diffused, highlighting the whole array of social preconditions needed for the successful diffusion of a technology.

Facsimile—a technology that has undergone a remarkable diffusion since the early 1980s—has its roots in the nineteenth century. Unlike videotex, facsimile left all prognoses as to its diffusion far behind in its actual development, giving another illustration of the difficulties of forecasts. Facsimile's success was dependent on many preconditions, among which standardization was of considerable significance. The agreements of the 1960s and the 1970s on international facsimile standards created a basis for compatible international service. Before, interoperation had been restricted to machines produced by the same manufacturer. Equally relevant was the lifting of regulatory constraints. Without permission to use the public telephone network for this service, facsimile's diffusion could not have been as widespread.

Against this background, facsimile is an impressive example of the limits of market-based standardization, the significant potential of committees, and the benefits of technical coordination.

Overview of the History of Facsimile Standardization

Although it was only after 1980 that facsimile developed into the most widely used service for text telecommunications, the technology is based on one of the earliest principles of electrical communications over distances. It was initially patented in 1843, but it existed very much in the shadow of Morse telegraphy.[1] However, news picture transmission developed into a relevant application at the beginning of the twentieth century. Initially relying on private-line networks, the service became more relevant when it was discovered that the general telephone network could support transmission over long distances too.

The desirability of standardization was felt relatively early. In 1929 the German Post Office proposed at the Second Plenary Assembly of the CCIT (the precursor to the CCITT) that those parameters of phototelegraphy necessary for an international operation should be standardized. First specifications were agreed upon in 1929, with some refinements following in 1936. But thereafter no more changes could be achieved for the next 20 years or so (Bohm et al. 1979, pp. 176–177). Applications during this time extended to include the military. After World War II the transmission of weather maps became important and office facsimile gradually developed.

In many countries, office facsimile was restricted to use with connections that were privately leased or owned and could not be transmitted on the public switched telephone network (PSTN).[2] These common regulatory restrictions may be regarded as important feedback on the incompatibilities that existed between the proprietary facsimile machines of different manufacturers until well into the 1970s.[3]

As the technology improved and as the differentiation of uses increased, standardization in the CCITT picked up again. In 1956 the United Kingdom and the United States submitted contributions for the standardization of "document facsimile telegraphy." However, no real progress was made on the specifications in 1956, in 1961, or in 1964. The first steps toward standardization of office facsimile were taken in the United States, in the newly established technical committee (TR-29) of the Electronics Industry Association (EIA). A first standard, published in 1966, included many imprecisions due to problems of cooperation, which continued into the following implementations: "Each manufacturer stayed within the wide tolerances required to get the 'standard,' but didn't convert to the 'recommended for eventual standard' items." (McConnell et al. 1992, p. 299) Nevertheless, interoperation between facsimile machines of different brands became possible for the first time, although it was fraught with errors.

Regulatory changes in the United States were behind these more determined steps toward standardization. From 1963 on, greater and greater use of the PSTN for office facsimile was permitted. Direct connection was approved in 1978 (McConnell et al. 1992, pp. 300–301). When it became possible to link machines outside private networks, the lack of interoperability started to become a constraint.

At the CCITT, standardization of office facsimile was soon taken up by Study Group XIV, and in 1968 the Group 1 recommendations for the first generation of standardized facsimile equipment were approved.[4] The recommended specifications differed fundamentally from the American TR-29 standard in the frequencies used for black-and-white (2100 Hz rather than 2400 Hz, and 1300 Hz rather than 1500 Hz), making interworking almost impossible. Within Europe the new Group 1 facsimile standard led to only partial compatibility between different machines, with the result that the standard was no substitute for the previous brand-specific compatibility within private networks.

The low effectiveness of the Group 1 standard was due to a variety of reasons. Although use of the PSTN became progressively more possible, most facsimile traffic was still conducted on private networks, and in that context proprietary standards were sufficient. A public facsimile service for the PSTN, which would have made the lack of appropriate standards less problematic, was still missing. Moreover, direct long-distance dialing—an important precondition for an attractive service—was only slowly being implemented on the PSTN. With these comparatively low benefits of compatibility, standardization for office facsimile was slow to progress in the CCITT. Although manufacturers were interested in working on standardization, their low commitment made agreement on a common specification difficult. As a result, the recommendations in 1968 included some ambiguities, lacked certain parameters altogether, and in other respects were much too complicated to warrant implementation.

In sum, the Group 1 standard was not very successful in achieving more than very basic interoperability. Group 1 consisted of the two recommendations, T.2 and T.4. Recommendation T.2 specified the modulation of information. As with the connection of computers to the PSTN, the facsimile information generated by the terminal had to be converted (modulated) when it was handed over by the sending terminal for transmission in the analog network. In the receiving terminal the information had to be reconverted (demodulated) in order to be reproduced in its original form. Recommendation T.4, which spelled out the procedure for signaling between terminals,[5] was never

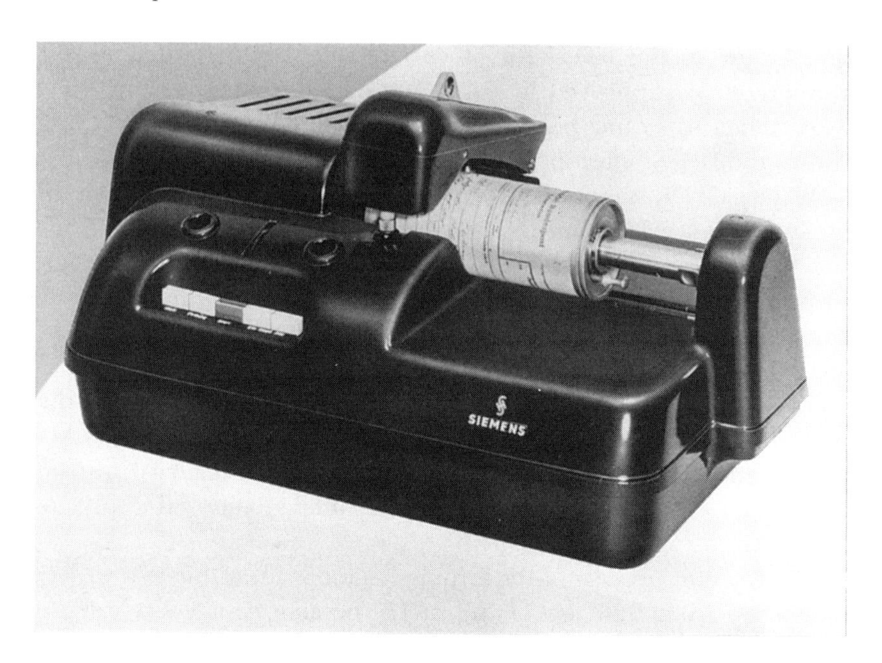

Figure 8.1
A Siemens-Hell machine of the 1950s. (Courtesy of Siemens Forum, München.)

implemented by manufacturers; it was deemed too complicated. Since a Group 1 terminal was operated entirely by hand, it was possible to neglect specifications for signaling, although the latter would have facilitated communications. Facsimile thus required a prior telephone call to indicate that a transmission was being planned. The scanning and receiving mechanism worked with a revolving drum, on which the paper had to be placed by hand. After manual synchronization, transmission ensued. The end of the transmission was signaled via a telephone call (figures 8.1, 8.2).

Recommendation T.2 included the main specifications of Group 1: an analog machine with a transmission time of 6 minutes per page. However, it did not specify whether amplitude modulation (AM) or frequency modulation (FM) was to be used; thus, equipment using both methods existed side by side, all partially following the Group 1 recommendation. Only after some time did FM turn into a de facto standard, subsequently implemented in many facsimile machines (such as the one shown in figure 8.3).

Figure 8.2
Parallel telephone control of the machine shown in figure 8.1. (Courtesy of Siemens Forum, München.)

The study period (1969–1972) that followed the adoption of the Group 1 specifications was characterized by a significant lack of contributions, despite the ongoing progress in the development of faster facsimile machines and despite the introduction of the first digital facsimile machine (by Dacom, in 1969). When the entire Study Group XIV met for the first time, shortly before the Plenary Assembly, many questions went unaddressed; work on other questions was hampered by the need to collaborate with other study groups. This low interest coincided with the retirement of the study group's chairman toward the end of the period. As a result, the dissolution of SG XIV was

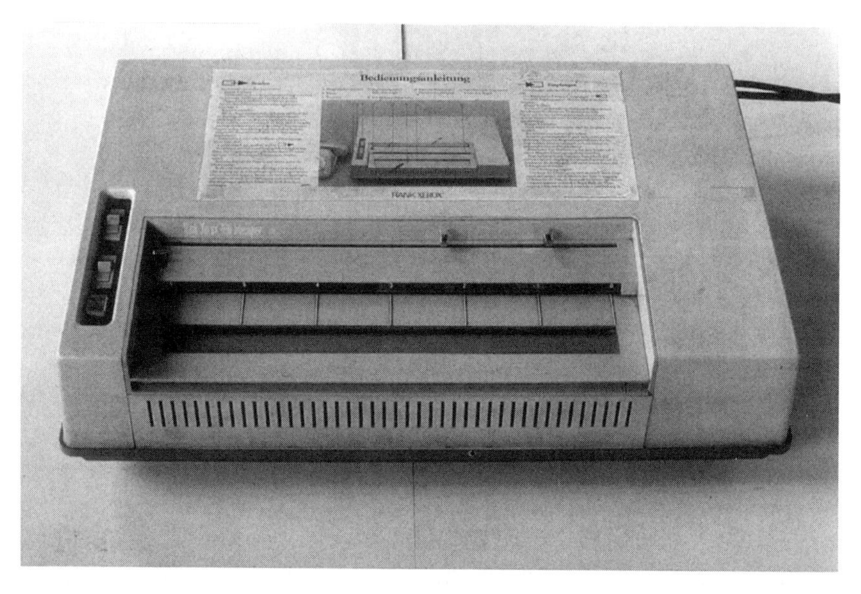

Figure 8.3
The Rank Xerox Telecopier 400, one of the most widely used quasi-Group 1 facsimile terminals of the early 1970s. (Courtesy of Museum für Fotokopie, Mühlheim/Ruhr.)

seriously considered. The CCITT is always trying to keep the number of study groups down, and in this case the evidence of low interest combined with the impending retirement of the chairman provided a perfect opportunity for the dissolution of SG XI. Facsimile standardization narrowly avoided this fate.

The low level of attention was a transitory phenomenon for SG XIV. What is interesting about this study period is that many matters that would become pertinent much later (including color facsimile, digital facsimile, and a standardized facsimile service) were already on the agenda. The issue of a standardized facsimile service was related to the low degree of compatibility achieved so far. As specifications intended to ease interworking were sought, it was in the context of study question 7 that most of the later work for the Group 2 and Group 3 specifications was done.

In the next period (1973–1976) the work of SG XIV intensified considerably. It seems that the first TR-29 and Group 1 standards had demonstrated the potential benefits of compatibility, which were increasing with regulatory changes that opened the PSTN for facsimile. At the same time, the need for cooperation had been revealed, and

	transmission time/page			modulation/ redundancy reduction		major recommendations	year
Group 1	6 min	analog networks	analog terminal	AM/FM	manual operation	T.2 (modulation) T.4 (signalling)	1968
Group 2	3 min			AM-PM VSB		T.3 (modulation)	1976
Group 3	1 min		digital terminal		automatic operation	T.30 (signalling; revised T.4 for Group 1-3)	1980
				Modified Huffman Code (Modified Read Code)		T.4	1980
Group 4: class 1 (basic) class 2 (teletex receive-only) class 3 (two-way teletex inter-working)	seconds	digital networks		Modified Modified Read Code		T.5, T.6	1984

Figure 8.4
Main features of various facsimile groups.

the manufacturers were willing to take an active role in standardization despite the fact that standardization often lagged behind the development of machines. Work on the Group 2 standard, and later also on the Group 3 standard, dominated this period. Moreover, the division of standards into Groups 1, 2, and 3, as laid down in recommendation T.0, was itself an achievement of this period. It is important to note that the different groups of facsimile machines by themselves are incompatible and cannot interoperate; they use different methods for coding, processing, and transmitting information. Communication between machines in different groups is therefore, by definition, impossible. Communication can be realized only if machines embody the standards of different groups in parallel. Figure 8.4 summarizes the main features of the various groups.

The membership of SG XIV was dominated to an unusual degree by manufacturers, mirroring the range of existing machines and the focus of standardization on terminal aspects. At the meeting of the whole study group in 1974, for example, 26 of the participants were national delegates from postal, telephone, and telegraph administrations (PTTs) or from recognized private operating agencies (RPOAs); another 28 were from scientific or industrial organizations (SIOs), almost all of them affiliated with major facsimile manufacturers. In many cases, manufacturers submitted contributions describing aspects of existing products, so that agreement very much depended on a high commitment to standardization.

The Group 2 standard employed a more efficient approach to modulation—one that came close to the theoretical optimum and allowed for twice the transmission speed. Although several alternative methods were proposed, the main confrontation took place between Graphic Sciences and Xerox, each of which already had an installed base of machines. Other parameters for Group 2 machines could simply be copied from the existing T.2 recommendation for Group 1 equipment (COM XIV (35E: 14), 1973/76). Recommendation T.3 for Group 2 equipment was issued in 1976.

Work on Group 3 facsimile also began in the period 1973–1976. In contrast to Group 1 and Group 2, which use analog techniques, Group 3 works digitally. Greater speed (1 minute per A4 page) is achieved because digital encoding allows redundancy to be reduced. Hence, the search for the right type of redundancy-reduction technique was one of the main problems studied in this period, albeit without result. Other parameters of Group 3, including type of modem to be used, degree of resolution, and number of picture elements, had already been agreed upon.

Other important work of the period 1973–1976 concerned revising the T.4 recommendation for signaling, which had been too complicated to implement in the context of the Group 1 standard. A complete reexamination of T.4 resulted in a new recommendation, T.30, for signaling of Group 2 and Group 3 machines, work on which continued into the next study period. T.30 could also be used to achieve a late compatibility for Group 1 machines. It had also been decided to adopt for Group 1 a proposal by the British company Muirhead and the British Post Office inverting the frequencies for black-and-white in the FM system, as recommended in T.2, starting at the beginning of 1975.

Study Group I, responsible for service aspects of facsimile, was not very active in this period; its work was restricted to definitional issues. This changed in the next period (1977–1980), when many administrations were considering the introduction of facsimile service. The foundations had been laid by the successful recommendation and implementation of the Group 2 standard, which had provided worldwide facsimile compatibility for the first time. The greater possibility of facsimile communication increased the demand for the equipment. In 1975 the number of terminals in the United States exceeded 100,000. In SG I, the question of facsimile services was now regarded as urgent, and among the issues discussed were a centralized directory for facsimile (to be held in Geneva) and the opportunity for "bureaufax" service between post offices. Other problems of lesser importance, including the obligatory identification of subscribers, also needed to be solved. Interestingly, the service recommendations that were agreed upon for facsimile largely reproduced recommendations for telex service.

Later in the 1977–1980 study period, once the question had arisen in SG XIV, SG I turned its attention to the standardization of Group 4 facsimile. Here the main issues were whether only the public data networks (PDNs) or both the PDNs and the PSTN would be used, whether Group 4 should be designated separately as "datafax," and to what extent compatibility between different groups of facsimile equipment was to be obligatory. Because compatibility could be achieved in terminals or in networks, it was decided to leave the final decisions to the network operators. Some farther-reaching issues were equally relevant—for instance, the question of coherence among the new generation of non-speech services and the desired degree of interworking between them. Study Group I supported the attempt to define common protocols for teletex (the upgraded telex for electronic word processors), and Group 4, involving also the possible "mixed-mode" operation between the services, in the context of the ongoing work on layered architectures, which was later to become the open systems interconnection (OSI) reference framework.

For SG XIV, the completion of the Group 3 specifications (T.4) was the most urgent matter in the 1977–1980 study period. Manufacturers in particular were very keen on this. The choice of a redundancy-reduction mechanism was the central problem in standardization, but the completion of T.30 for signaling in Group 3 equipment was also on the agenda of this study period. Those who were planning immediate introduction of national telefax service hoped to build on Group 3 equipment, and this created a strong incentive for manufacturers to have standardized equipment ready.

Throughout this study period, the dominance of manufacturers in SG XIV continued. In the middle of the period, at a meeting in Geneva, the ratio of delegates of organizations categorized as SIOs to delegates from PTTs and RPOAs was 52:38. It was not until the period's last meeting, held in Kyoto, that the ratio was reversed to 41:63. At that meeting the Japanese had a significant share of the delegates: 8 representatives of Japanese SIOs participated, and 23 of the delegates were from Nippon Telephone and Telegraph and

Kokusai Denshin Denwa. This indicated the gradual rise of the Japanese, although input from North American and European sources remained dominant. In terms of the market, according to a monograph on facsimile written by Americans active in standardization (McConnell et al. 1992, p. 304), manufacturers in Europe and North America had already begun to lose their position with the Group 2 standard: "CCITT standards opened up worldwide markets for the Japanese equipment."

Japanese delegates were also heavily involved in the standardization of Group 4 machines, a process that had already been initiated in the first study group meeting of the period. This new question appears to have been triggered, in part, by the incipient work of SG VII, the study group responsible for data networks. Guided by study question 35 of SG VII, this group focused on facsimile in public data networks, an undertaking that threatened to violate the established domain of SG XIV.

In the early stages, substantial work on Group 4 came from the United States. Recommendations T.5 and T.6 were, however, not finalized until 1984. A substantial part of the work dealt with the question of common protocols with teletex (COM I (104E: 33), 1977/80). Teletex was expected to succeed and replace telex in the era of electronic word processing. Hence, initially it was IBM Europe more than the PTTs that demanded these standards.

In this context, far-reaching changes were initiated. The work on common protocols was to occur jointly with the work of SG VIII, which was responsible for teletex. Links were also set up with SG VII (mainly because of its work on OSI, with which the work on common protocols was affiliated). Two important consequences arose from these interlinkages: the Group 4 standard became interrelated with teletex, and the collaboration with SG VIII became the basis on which the 1980 Plenary Assembly decided to dissolve SG XIV (again coincidentally with the retirement of a chairman). Consequently, SG VIII was now solely responsible for work on standards for teletex, videotex, and telefax. The intention was that it would increase the efficiency of standardizing the new telematic services.[6]

In the 1981–1984 study period, work on facsimile was concerned with matters arising from the implementation of the Group 1, 2, and 3 equipment and facsimile service. The main activity, however, was the completion of the Group 4 recommendations. The work continued to follow teletex closely. Three classes of Group 4 equipment ultimately resulted, specifying a stand-alone facsimile machine and machines with one-way or two-way interoperation with teletex. The background to this work was the overall belief in the growing importance of integration and interoperation in telecommunications.

With the Group 4 specifications, facsimile standardization had changed greatly. It now had a clear ex ante character, whereas before manufacturers had often brought in existing technical solutions to determine, via an extended series of tests, which one was to be recommended by the CCITT ex post as standard. However, incremental improvements to existing implementation-based recommendations continued, so the more obvious change for the moment was the very different composition of SG VIII relative to SG XIV. Now PTTs and RPOAs were dominant—for instance, at a Geneva meeting in 1981 there were 117 delegates representing them and only 66 representing SIOs. In addition, a lot of time had to be spent on issues not relevant (or only slightly relevant) to facsimile.

In this period, SG I worked on improvements to the existing facsimile service recommendations. With regard to Group 4, possible complementary services[7] and the question of interworking between facsimile and other services were addressed. Moreover, the bureaufax service (which followed the idea of telegram service) was now an important focus of work conducted in collaboration with the United Postal Union (UPU).

For both SG I and SG VIII, the 1980s were an era of incremental and stable work on facsimile, now that the greater decisions had been taken. For SG VIII, there was only one slightly contentious decision that had to be made, and that was in 1986 on the question of an error-correction mode (ECM) for Group 3. In addition, the incremental work profited greatly from a specific characteristic of the Group 3 standard: the protocols employed in "handshaking." Handshaking is the procedure used to build up a connection between two terminals, including the alignment of relevant features by way of mutual agreement of the two terminals that is specified in recommendation T.30. Handshaking allowed manufacturers to enhance equipment with proprietary features, since information about specific capabilities was exchanged between terminals before transmission (McConnell et al. 1989, pp. 25, 77). One new standard was note-fax for private households in A5 or A6 paper sizes, which could also be fitted conveniently into cars (in the radio slot, for instance).

While standardization progressed smoothly, the 1980s saw an unprecedented and very much unexpected growth in the number of facsimile machines. In 1988, a year with very high growth rates, Group

3 machines increased by 75 percent worldwide, by 81 percent in North America, and by 156 percent in Western Europe. This growth was accompanied by significant price reductions: "The lowest priced Group 3 units in 1988 cost about 5 percent of those introduced at the end of 1980!" (McConnell et al. 1989, p. 181) During these years, Japanese manufacturers established their dominance in facsimile production; by 1988 they supplied about 90 percent of total production.[8] Their close national coordination helped Japanese manufacturers to homogenize many standards that included multiple options for implementation. The manufacturers settled some disagreements (concerning, among other things, specific "handshaking" conventions) internally. As a result, facsimile messages were exchanged more efficiently between Japanese machines of different brands than between non-Japanese terminals.

Considerable market success was almost completely confined to Group 3 machines. Group 4 facsimile did not take off at all. Large Group 3 production facilities existed, with very fast equipment life cycles that exploited add-on features of this standard's equipment. The different careers of Group 3 and Group 4 terminals, as well as the actors' diverging strategic visions of the telecommunications technology for the future information society, induced an unprecedented upheaval in the smooth standardization process during the 1989–1992 study period. A small group of actors submitted a proposal to modify the Group 3 standard for use with digital networks as a cheaper and more pragmatic alternative to Group 4 terminals, which were dedicated for use in highly integrated ISDN telecommunications systems. This move precipitated a prolonged and serious controversy. A persistent vagueness of actors' positions characterized this conflict; together with the highly relevant technical and economic factors, this made resolution in the CCITT by the usual means difficult.

Central Technical and Economic Problems

Facsimile is designed much less as a systemic technology than videotex. It is virtually a terminal technology, and the early decision to specify it for use in existing networks supported this component character. For transmission, routing, and addressing, facsimile does indeed rely on existing networks—typically the PSTN. It has to meet the requirements of these networks, which sometimes is regarded as a constraint but which most significantly makes the necessity of providing recom-

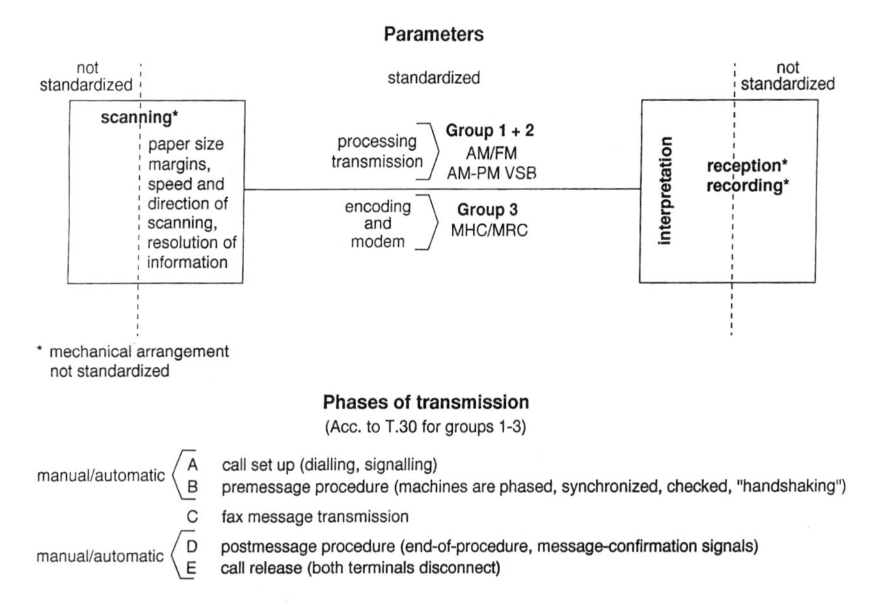

Figure 8.5
Technical aspects of the facsimile procedure.

mendations for these functions redundant. Thus, from the start of facsimile standardization, many of the difficulties faced by other standardization projects could be avoided—in particular, all decisions as to where to realize which functionality. Facsimile is, therefore, a clearly structured and comparatively easy technology for which only a limited set of functions need be realized in the sending terminal and the receiving terminal—provided, of course, that the network can accommodate the service in the first place. Among the necessary functions are scanning, processing, and transmission of information, on the one side, and the reception, interpretation, and recording of information, on the other (Perabo 1981, 1982). The parameters that have to be standardized can be deduced from these functions (Grallert and Hammer 1979; figure 8.5 here).

To explain the different steps of a facsimile transmission, it may be useful to start with distinctions drawn and standardized in T.30 concerning the five different phases of a transmission. In phase A the call is set up by dialing the number and receiving the signal that the other machine is not busy. In phase B, the pre-message procedure is activated. Both machines have to be phased, or synchronized. Each machine signals its capabilities—for instance the speed of the modem involved or the availability of an error-correction mode. This pre-message "handshaking" phase is the provision that creates the great flexibility of the Group 3 standard: it is possible to implement optional or proprietary features in machines as the precise capabilities of communicating terminals are exchanged before transmission. Phase A and phase B are the two phases which in the mainly manual operation of Group 1 and Group 2 terminals were usually harmonized by way of a telephone call. Next is phase C, the transmission of the message. Phase D, the post-message procedure, includes the exchange of end-of-procedure and message confirmation signals. The final phase, E, represents the release of the call, when both terminals disconnect. With older terminals, again, the latter two phases were coordinated in manual operation through the aid of an extra telephone call.

The technical functions of facsimile can be presented against the background of the above-mentioned steps in the call procedure: Facsimile is a *raster scanned graphics* system. An image is decomposed into a set of raster lines. On digital systems, the image is further decomposed into a series of evenly spaced samples along each raster scan line. These samples are called picture elements or "pels" (Schmidt 1983, p. 1147). The scanning of the message starts in the sending terminal. The precise mechanical arrangement required for the scanning or for the reception and recording of information is not part of the standard.[9] For Group 3 and Group 4, a flat-bed mechanism is used. A drum scanning system had been common in Group 1 and Group 2, where the sending paper and the receiving paper had to be wrapped around a drum. However, it is necessary that scanning be standardized for the size of the paper, including the minimum margins, the speed, the direction, and the scanning density or resolution (i.e., the number of lines and the number of rows).[10]

In the processing and transmission of signals, then, the important differences are between the analog Groups 1 and 2 and the digital Groups 3 and 4. Table 8.1 summarizes the technical parameters for all groups. In Group 1, amplitude modulation and later, increasingly, frequency modulation were used for both processing and transmission on the analog telephone network. Group 2 terminals managed to double the speed and thus reduce the transmission time from 6 to 3 minutes by using band-compression technology. The modulation procedure, called amplitude modulation–phase modulation vestigial sideband (AM-PM VSB) technology, a method close to the theoretical optimum for this type of facsimile, was standardized for Group 2.

Group 3 machines work digitally, achieving their one-minute-per-page transmission time through the encoding of information in binary signals, which reduces redundancy. Modems are used to convert the digital signals for transmission on the analog telephone network. More precise, two different modems are used: a 300-bits-per-second modem for the pre-message procedure (Phase B) and a separate high-speed modem, with speeds of either 9.6 and 7.2 kilobits per second or 4.8 and 2.4 kilobits per second, for the sending of image data (the choice of speed being part of the pre-message procedure). SG XVII standardized the modems for data transmission over telephone networks separately, specifying V.27 or V.29 modems. Group 3 facsimile, realized by means of faster modems, was originally standardized for 4.8 kilobits per second.

The core of the Group 3 standard is the redundancy-reduction code. A fundamental distinction exists between one-dimensional and two-dimensional codes. A one-dimensional code reduces redundancy (duplication of information) by exploiting the strong correlation between adjacent pels in a horizontal direction: a code is designated for a number of adjacent pels sharing the same information instead of transmitting the same information several times. A two-dimensional code employs the same principle, but along vertical and horizontal lines, thereby reducing redundancy much more at the cost of becoming more susceptible to errors (Rallapalli 1981). This danger is particularly relevant to analog networks, so two-dimensional coding is only optional in the Group 3 standard. The modified READ (relative element address designate) code was chosen, and the modified Huffman code (MHC) was recommended as the one-dimensional, regular code for Group 3 in recommendation T.4 (Hunter and Robinson 1980).

Group 4 facsimile is very different from the other three groups. It is standardized for public data networks. In contrast to the idiosyncratic protocols and codes of the other groups (which are dedicated exclusively to the exchange of facsimile messages), Group 4 is seen as part of the comprehensive OSI world, and it is designed to be integrated into a comprehensive network (such as the Integrated Services Digital Network) that allows for a great variety of applications.[11] Three classes of Group 4 are defined. Class 3 offers two-way mixed-mode interoperation with teletex, Class 2 offers a one-way (receive-only) mixed mode, and Class 1 is a pure facsimile mode. Group 4 uses a further modification of the modified READ code for a two-dimensional

Table 8.1
Outline of standardized parameters for each facsimile group.

	Apparatus					
				Group 4***		
Parameter	Group 1	Group 2	Group 3	Class 1	Class 2	Class 3
Apparatus recommendation	T.2	T.3	T.4	T.5	T.5	T.5
Network	PSTN	PSTN	PSTN	PDN (PSTN, ISDN)****	PDN (PSTN, ISDN)****	PDN (PSTN, ISDN)****
Transmission time/A4 (min.)	6	3	Approx.1	-	-	-
Number of pels per scan line	-	-	1728	1728, 2074* 2592*, 3456*	1728, 2074** 2592, 3456*	1728, 2074** 2592, 3456*
Scanning density	3.85 (1 p/mm)	3.85 (1 p/mm)	3.85, 7.7*, (1 p/mm)	200, 240** 300*, 400* (1 p/i)	200, 240** 300, 400* (1 p/i)	200, 240** 300, 400* (1 p/i)
Modem	FM (1700 ± 400Hz)	AM-PM VSB (fc: 2100 Hz)	PM (V.27ter), AM-PM (V.29)*	-	-	-
Data rate (kb/s)	-	-	2.4, 4.8, 7.2*, 9.6*	2.4, 4.8, 9.6, 48	2.4, 4.8, 9.6, 48	2.4, 4.8, 9.6, 48
Coding scheme	-	-	MH MR*	Modified MR (T.6)	Modified MR (T.6)	Modified MR (T.6)

Control procedure, protocol, recommendation	T.30	T.30	T.30	T.62, T.70, T.71, T.73	T.62, T.70, T.71, T.72, T.73	T.62, T.70, T.71, T.72, T.73
Remarks					Reception only for teletex and mixed mode	Transmission and reception for teletex and mixed mode

*Option; **required for teletex and mixed-mode reception; ***first recommended end of 1984; ****further study. Source: Kobayashi 1985, p. 33.

redundancy reduction. This modification of the Group 3 two-dimensional code is the only commonality with Group 3 machines. Otherwise, the teletex protocols for handshaking procedures and signaling take the place of the former T.30 protocols, which could be used for Groups 1, 2, and 3 (figure 8.5). With the transition of Group 4 to the integrated telematics world, however, facsimile standardization has become much more complicated. Decisions for Group 4 are now interrelated with specifications for other applications, just as other standards may have repercussions on Group 4. From this point of view, the isolated realizations of the facsimile Groups 1, 2, and 3 have been less burdened with compatibility requirements than Group 4.

For the sake of compatibility, all parameters that bear on the signals being transmitted have to be standardized. This comprises resolution, method of modulation (Groups 1 and 2), bit rate for transmission (Group 3), encoding (Group 3), pre-message harmonization, interface with transmission channel, geometry of scanning and recording, and so on (BMFT 1977, pp. 155–156). An important characteristic of facsimile standardization "before Group 4" is that, although the parameters are not unrelated, they are not tightly coupled; that is, no hierarchy of parameters exists where the choice of one central variable predetermines the shape of the others. Rather, two-dimensional interrelations or simple tradeoffs are typical, and many parameters have inherent limitations. An example of the former is the interrelationship, given a certain transmission capacity, between resolution and transmission time. The possibility that the speed of transmission is constrained by the speed of scanning might also be mentioned. Moreover, the naturally given limitations of parameters are quite numerous. There are biological limits to the human eye, which it is not desirable to exceed technically. More than 256 shades of grey cannot be perceived, and a resolution of more than 8 rows or 8 pels per millimeter is more than the eye can accommodate (BMFT 1977, p. 13). Physically, there are clear theoretical limits to the conventional methods of modulation and encoding. Both the modulation approach to Group 2 and the one- and two-dimensional encoding of Group 3 are close to the bounds of theoretical possibility, so further improvements in this trajectory could hardly be expected.

A consequence of the fact that the different parameters are definite and loosely coupled is that standardization can proceed incrementally. Moreover, it is relatively easy to determine a set of criteria with which to evaluate existing options (e.g., quality and time of transmission).

Thus, many decisions could be taken using quasi-scientific methods-e.g., on the basis of test rows of different options.

Important in this respect was the decision to standardize in such a way that facsimile could build on existing networks. Consequently, the relatively common question of whether to realize functionality in terminals or in networks came up only when compatibility between different groups that used different networks (predominantly Groups 3 and 4) was being discussed in SG I. Because a gateway between networks is necessary anyway, conversion for compatibility may be realized either centrally in the gateway or in each terminal. Apart from this, the choice of the PSTN meant that facsimile had to adapt to a preexisting structure. On the one hand, this implied constraints because the telephone network is adapted to and optimized for voice communications, as with the relatively long setup of dialed connections and the sometimes high rate of minor transmission errors which can be tolerated (only) with a telephone service. Moreover, at the international level, these and other difficulties are exacerbated because network parameters differ. On the other hand, the telephone network offers tremendous advantages as the most ubiquitous network there is, providing worldwide basic compatibility. On this basis, the use of facsimile was dependent only on the availability and diffusion of facsimile terminals, and not on further additional infrastructural provisions and investments. This benefit included the fact that many problems had already been solved for telephony, and these solutions were then regarded as directly valid for facsimile too. Examples are tariffs and international accounting procedures, as well as the whole international numbering plan. In addition, more detailed service specifications were adapted from the existing telex recommendations for the subscriber service and from the telegram recommendations for the bureaufax service, minimizing again the need for new coordination.

Group 4, in contrast, was not just a loosely coupled "appendix" to the telephone service; because of its orientation toward OSI, it was embedded in a more principled system of protocols, whose requirements had to be taken into account. At the same time, another hitherto-important prerequisite for the relatively smooth and incremental standardization process, the lack of overtly significant economic stakes, had been experiencing gradual change. In contrast to the expectations of videotex, the expectations of facsimile diffusion had always been relatively slight. It was, after all, an old, well-known technology, not part of a newly emerging vision of the information society. Economic stakes developed parallel to the incremental standardization, and for a long time they were sufficient to let standardization proceed but not so high as to be a basis for conflict. It was mainly with the surprisingly fast diffusion of facsimile in the mid 1980s, following a somewhat disappointing record on the part of European and North American manufacturers, that economic stakes and market competition started to bear on facsimile standardization.

Actors and Interests

Throughout the long-term and incremental standardization of facsimile, the constellations of actors and interests contrasted with those found in the videotex case. A row of relatively independent, low-profile decisions characterized facsimile; accordingly, the actors adopted various positions of favored options or showed lack of interest in each particular case. Because the technical problems were loosely coupled, few major technical decisions ("technological guideposts" in Sahal's (1981) sense) had to be taken; thus, larger controversies, concerned with limitation of future choices, could be avoided. In view of the fact that interest constellations were issue-specific and did not display generalized features, it is not possible to offer a comprehensive overview, as has been tried in the videotex case. Only certain major characteristics need be highlighted.

From the start, facsimile standardization was dominated to an unusual degree by manufacturers. Network operators participated only because they expected that new terminals (and services) would bring additional traffic into their networks once they were opened to facsimile applications. Additionally, standardized equipment had the side effect that operators could forgo the installation of conversion equipment for incompatible terminals in the network (Mao and Hummel 1981, p. 748; Sung 1983, p. 132). Beyond this broad economic interest, operators were not called upon as political agents to defend more comprehensive national interests. As a consequence, facsimile standardization was rarely characterized by pronounced national positions. One notable exception was Japan. Interestingly, Japan's national facsimile coordination was modeled on the British Facsimile Industry Consultative Committee (BFICC). Although facsimile was often addressed in national preparatory meetings, the problems of aligning a fixed national position with a flexible strategy at the international level of the CCITT, and other two-level game complications of the sort that came up in the case of videotex, were not really significant in the case of facsimile. Rather, individual contributions were often made to the CCITT meetings directly, without having been coordinated in advance at the national level.

Because facsimile had existed for so long, sufficient experience had been gained with respect to the limits of proprietary standards. Moreover, once committee standardization had started, the limited compatibility that had been achieved with Group 1 machines and with the American TR-29 standard drove home the benefits of compatibility and the absolute necessity of cooperating across company boundaries to achieve this. As a result, firms were very cooperative and generally committed to standardization, despite the parallel competition they pursued on the market. Because actors had learned to perceive facsimile standardization as a positive-sum game in which everyone could benefit, work on Group 2 and Group 3 was largely characterized by a cooperative interaction orientation. A strong motive for this orientation was the fact that facsimile was not particularly regarded as a growth industry.

A central factor in this positive constellation was that the technical problems to be resolved were only loosely coupled. This allowed for a highly incremental standardization process involving many small decisions; at the same time, it implied that no single decision could prejudice the future course of the work. Moreover, the positions on any one issue could not reveal the overall corporate strategy. Far-reaching interests were difficult to deduce from minor issues, and the success of a single decision did not undermine the chances of others in the next round of decision making. This interpretation may at least explain why, in decisions on Group 2 and Group 3—for example, in the important questions of modulation and coding—it was no secret which manufacturers supported which solutions. The loose coupling of technical problems also made it relatively easy to establish commonly accepted general criteria by which to judge various company-backed options.

Typically, decisions were closely related to applications already existing or being developed. Thus, facsimile was standardized ex post or in parallel to marketing plans. Moreover, the issues were not far-reaching; they concerned only rather isolated components, not the architectural level. This changed with the Group 4 standardization, which was soon associated with ISDN (the long-term standardization project for the next generation of integrated public telecommunications networks). With the Group 4 standardization, facsimile took on a long-term character and, at the same time, was transformed into an issue closely connected to more basic architectural considerations. The aim of integration finds one expression in the idea of mixed-mode applications, which allow users to communicate between different kinds of applications (i.e., to receive a facsimile message with a teletex terminal). This implies the need to structure different applications along the lines of common principles. In this respect, the ISDN reveals similarities with the OSI model, under whose umbrella ISDN standardization was soon placed.

The work on Group 4 facsimile coincided with a change in the constellation of actors. Work on protocols began in the 1977–1980 study period, first in cooperation with SG VIII and later under its auspices, following the decision to develop joint protocols with teletex and thereby replace the highly idiosyncratic technical base of facsimile. This technical integration into the telematics world was an important factor in the absorption of SG XIV in SG VIII (Staudinger 1985). Here, facsimile manufacturers had lost their dominance. Manufacturers with interests in teletex or videotex were present in equal number; most important, however, network operators were the dominant group in SG VIII. Because network operators—owing to their much larger sunk investments—plan over much longer periods than terminal manufacturers, the new proclivity for ex ante standardization became entrenched in the social dimension. Another side effect of the merger was that, although work on telematics as a whole was seen to become more efficient, the single participant interested in facsimile could only progress more slowly than before: discussions were no longer focused exclusively on facsimile; indeed, decisions on facsimile had to be assessed in terms of their implications for applications other than facsimile.

As a result, Group 4 facsimile standardization proceeded in a much more "theoretical" way than previous standardizations, not being directly connected to ongoing development work and marketing plans. Actors' interests in Group 4 consequently were much more principled, and, instead of specific technological solutions, the general technical approach was at stake. The integration of facsimile into the general trajectories of telecommunications and computing technologies also implied that facsimile technology might become feasible for firms that had not been active in the field. For example, very early in the 1977–1980 period, IBM Europe began to promote the idea of common protocols for teletex and facsimile. However, parallel to the standardization of Group 4, facsimile production had moved almost exclusively to Japan, from whence very few technical features were promoted vigorously. Japanese actors tended to emphasize the testing and refinement of new proposals put forward by others, for instance as to their interoperability with the installed base.

But why were European and North American actors still active in standardization after Japanese manufacturers had gained dominance with Group 3 equipment? Their contributions were predominantly motivated by two categories of interest. First, Western manufacturers were still active in the marketing of so-called OEM facsimile machines, the original equipment manufactured in Japan (McConnell et al. 1989, p. 180). As distributors of this equipment, they had to adapt the machines to their national technical network environment; moreover, they had a direct interest in offering their customers units with new standardized features which responded to the perceived needs of communications. Both elements could elicit contributions to the standardization process. Second, the whole aspect of facsimile-based services and overall technical development clearly increased in relevance after facsimile's unexpected success in the 1980s. Furthermore, there were still hopes that Europe and the United States might recover lost ground in manufacturing, although it is difficult to trace this desire to specific incidents. Indications here, however, are gleaned from interviews with network operators who thought, for example, that German manufacturers had a fair chance of entering the market with Group 4 equipment after the Group 3 development had been misjudged entirely. It was implied here that operators had put some effort into the attempt to make the standards suitable for domestic manufacturers, but that they had grasped this opportunity incompletely.

Owing to its ex ante nature, Group 4 standardization does not really offer examples of decisions guided by pronounced commercial interests. Moreover, because of its marginal market relevance (attributable to the very slow introduction of the ISDN, on which Group 4 primarily relied), Group 4 did not attract much attention once the basic specifications had been recommended. Therefore, only Group 3 offers instances in the 1980s of discernible market or service interests' lying behind certain decisions. But again, these interests were always very issue-specific. In the middle of the 1985–1988 study period, for example, Britain proposed standardizing an error-correction mode, motivated by OEM suppliers who saw a need for this feature among their customers. Japan, however, was against it, because manufacturers already had proprietary methods for ECM and were not interested in a standard. Another example of a Group 3 option concerns technical specifications for small A5 or A6 machines. These were strongly promoted by Japanese manufacturers hoping to capture the consumer market. Most other delegations did not share their economic expectations. In both cases, standards were eventually approved because they were not seen to seriously threaten the course of facsimile development.

In the last study period (1989–1992), a protracted and heated debate characterized much of facsimile standardization. The issue at stake was whether or not to standardize a digital Group 3 terminal as a cheap alternative to Group 4. The difficulty experienced in bringing this question to a firm conclusion—analyzed in greater detail below—indicates that participants were far from indifferent to the point. Their precise business interests, however, were difficult to detect in the course of technical argumentation. Hence, in contrast to previous disagreements, it was not possible to draw up a list of different positions. Thus, the features of this situation were the reverse of the typical features of conflictive situations in the 1970s. Two reasons for this were certainly the different composition and the sheer number of participants in SG VIII, which made it much more difficult to differentiate positions. Contributions from countries instead of single manufacturers were more frequent than in earlier conflicts, and so the problem of two-level coordination of negotiation processes arose. It was sometimes very difficult to detect precisely who advanced ideas in the national delegation. The high degree of speculation surrounding this debate resulted from the unusually broad scope of the issue: with a digital Group 3 standard, the still-dominant idiosyncratic approach to facsimile was again put forth as a viable future alternative to a Group 4 facsimile standard that was meant to be a crucial element in a comprehensive telematics architecture. Hence, the debate was also about architectural and strategic choice regarding telematics. Did the new structured approach of OSI and the ISDN offer the greatest technical and economic advantages, or was it a mistake, to be rectified sooner rather than later? Comprehensive technical interests regarding the future architecture of the telecommunications system were affected, as were direct economic interests in the market for facsimile machines and services.

The configuration of the contesting groups indicates that the conflict did not split manufacturers from network operators or service providers. Not just the whole Japanese delegation, but also France and Germany were against the standard—Japan apparently because of its commitment to the development and production of Group 4 equipment, the other two countries because of the operators' concern to avoid variety in standards and to safeguard a compatible telefax service. The promoters of the Group 3 digital option were two British subsidiaries of Japanese firms (Matsushita and NEC), which had secured the backing of the whole British delegation and particularly of AT&T in the US delegation. One reason quoted for the support for Group 3 digital was the interest of the United States in breaking Japan's dominance in facsimile equipment, because a digital Group 3 terminal could be manufactured more easily by more diverse suppliers. Because the precise interests were difficult to attribute in this case, personal motives were often cited. Were the two main proponents—one from the United Kingdom and one from the United States—only aiming to increase their standing by demonstrating that they still had ideas? These questions, which imposed difficulties on all the participants, complicate our analysis of the conflict's process and its termination.

Minor and Major Conflicts

The debate over Group 3 digital standards was the only major conflict facsimile standardization ever had to face. Before we turn to it, let us examine some of the minor conflicts that arose in regard to the standardization of Group 2 and Group 3 equipment.

The standardization of Group 2 equipment began at a time when such equipment already existed in the form of proprietary corporate implementations. From an analytical point of view, late standardization is prone to conflict. Past investments may have to be sacrificed and internal developments abandoned if a joint solution is to be adopted. At issue in this particular case were the modulation and demodulation methods and the scanning line frequency. Other elements of Group 2, such as scanning track and density, could be taken over from Group 1.

Different procedures exist for the conventional modulation procedures from a theoretical point of view, all coming close to the theoretical optimum and thus making further enhancements impossible. However, not all methods work equally well on all the different networks that exist. Thus, testing the different methods, with the help of specific "test charts," became important in decision making (table 8.2).

Table 8.2
Test results and preferences concerning facsimile modulation.

Rank Xerox	Tests on two FM systems using alternate encoding and one AM VSB system; no decision; further testing required.
Hell GmbH	Tests on FM with alternate encoding (Hell-Muirhead design), AM-PM VSB system of Graphic Sciences, FM system with alternate encoding (Hell, Xerox); AM-PM VSB was superior.
SECR	AM VSB proposal; tests on AM-PM VSB; further testing required.
MGVS (Fax Communications) Co Ltd.	Tests on AM-PM VSB; further testing required.
Xerox Corporation	Tests of the FM and AM-PM VSB system; no clear preference; further testing required on real lines.
NTT	NTT's AM-PM system; an advantage of the system was easy compatibility with the AM Group 1 machines used in Japan.
Cable & Wireless Ltd.	Were disappointed that so many delegates were advocating further testing since they doubted that it would produce more conclusive results; Graphic Sciences system should be adopted.
IPTC	Failing a clear technical superiority, the decision should be taken on other grounds.
Swedish Administration	Preference for AM on grounds of smaller bandwidth requirements and greater immunity to impulsive noise.
French Administration	Further time to assess tests on its own network.
Graphic Sciences	Promoted AM VSB modulation; machines working in 40 countries including international connections; further testing under control of independent body.

Source: CCITT, study period 1973–1976, COM XIV (30-E: 4–6).

Such tests had been used previously. Different available methods were backed by different firms. As we only want to analyze the decision-making procedure between different alternatives, we need not consider in detail all the differences between the modulation techniques. In the modulation controversy we can concentrate—without explicating their technical content—on the positions of the two dominant antagonists. The Xerox Corporation and Rank Xerox (its British section) stood in one camp, proposing an FM approach; on the other side, Graphic Sciences pleaded for an AM-based vestigial sideband system.[12] Xerox was forced onto the defensive quite early. Although tests were already being carried out jointly by Germany and the United Kingdom, Xerox proposed at a study-group meeting in 1974 that a Special Rapporteur Group (SpR) be commissioned "in order to study evaluation criteria and test conditions which will lead to the selection of an acceptable Group 2 modulation system" (COM XIV (18E: 10), 1973/76).

The new group consisted of more than thirty participants. Although the testing of modulation methods on national and international circuits was intensified, conclusive results were not reached. It was found that the Xerox "duobinary FM with center frequency 'black' shows distinct advantages provided the level of listener echo is low [though] it would not be acceptable with the echo levels currently encountered in the UK" (ibid., 30E: 2). At the next meeting, Rank Xerox countered this finding (obtained by Hell and Muirhead), claiming that there was "no conclusive evidence to decide either on FM or AM and that further testing, particularly on international lines, was required" (ibid., 30E: 4). The Xerox Corporation reported similar results for AM and FM systems. Most of the actors (with the exception of Cable & Wireless, which was in favor of the Graphic Sciences approach, already being used in 40 countries) wanted further testing; nevertheless, Xerox suddenly faced difficulty in keeping its FM approach among those being tested. A show of hands had resulted in "a majority in favor of AM but a substantial number in favor of further testing. There was no support for FM. In view of this it was agreed to start drafting a recommendation based on AMVSB with alternative parameters with the intention of finalizing it at the next meeting." (ibid., 30E: 5) However: "On the second day of the meeting the wisdom of the previous day's decision to concentrate on drafting a recommendation for an AM system to the exclusion of FM was questioned. Xerox pointed out that there were 4 AM systems still under consideration so why exclude 2 more systems just because they were FM." (ibid., 30E: 6) This received some support in another show of hands. Though the PTTs and the RPOAs wanted to adhere to the earlier decision, it was decided "after much heated discussion" that FM should be subjected to further testing. However, Graphic Sciences urged that this time testing should lead to a definite decision, for which some guidelines were issued. At the next meeting, after further testing, "agreement was rapidly reached that a vestigial sideband (VSB) amplitude modulation—phase modulation (AM-PM) modulation system should be adopted" (ibid., 32E: 1).

The question of carrier frequency posed the next problem. In this case, Rank Xerox was supported by other manufacturers, and Graphic Sciences had a large installed base. "The choice was rapidly reduced to two frequencies: 2048 and 2100 Hz. There was prolonged discussion of the merits of the two frequencies and a number of technical issues were raised. Eventually a position of stalemate was reached with Graphic Sciences supporting 2048 Hz because of its use in their existing machines and with Rank Xerox as the chief supporter of 2100 Hz." (ibid., 32E: 2) A suggested compromise was deemed technically unfeasible, and the views of the PTTs and the RPOAs produced no decision either. The matter therefore had to be referred to the SG meeting. Here, views were again sought among PTTs and RPOAs only. Although many expressed no opinion either way, 2048 Hz received no support at all, so 2100 Hz was finally agreed upon. As a result of the two decisions, the installed base of proprietary (Graphic Sciences) Group 2 machines was incompatible with the new standard (McConnell et al. 1992, p. 303).

The standardization of Group 3 equipment gained momentum parallel to the work on Group 2 in the period 1973–1976. One important decision centered on the choice of a digital redundancy-reduction method for one- and two-dimensional coding; this led to similar problems as the decision on the modulation technique for Group 2. Again, various redundancy-reduction techniques were proposed and studied, evaluated, compared, and further assessed as to their vulnerability to transmission error (COM XIV (18E: 22), 1973/76). In contrast to the Group 2 decisions, there was no conflict between two main opponents, although it was again possible to clearly denote the different positions of the actors (table 8.3). At the final meeting the choice between codes could not be made: " . . . the efficiency of all the one-dimensional codes proposed was approximately the same; the two-dimensional codes . . .

are more efficient than certain one-dimensional codes; while it was not yet possible to take a final decision between one-dimensional codes and codes extending over several lines, because the criteria of choice (efficiency, simplicity of design, vulnerability to errors, transmission time, patent rights, etc.) have not been sufficiently studied, there was, nevertheless, a large majority of views in favor of simple one-dimensional codes (modified Huffman type)" (ibid., 34E: 5).

Another open question was the choice of resolution (table 8.3). The question of resolution is linked to that of transmission speed. A resolution of 1728 elements per line necessitates the use of 4800-bits-per-second modems for the target transmission speed of 1 minute per page. However, "a number of countries [could not] yet contemplate a service at that speed on switched telephone networks" (AP VI (35E: 8)), and in any case the modems were not yet standardized. It became possible to reach a partial decision on both issues when Japan, a strong supporter of two-dimensional coding, proposed a compromise, with their favored higher horizontal resolution of 1728 elements per line, but with the one-dimensional run-length coding as the standard and the two-dimensional coding only as an option. Thus a preliminary agreement for the modified Huffman code (MHC) as one-dimensional coding could be drawn up (ibid., 35E: 9).

In the next study period, the MHC was further specified in a joint contribution by several firms.[13] Although the choice remained contentious for a while (figure 8.6), this version was later accepted with slight modifications. But the question of two-dimensional coding remained problematic, although it was not entirely divorced from one-dimensional coding. Whereas for Group 3 two-dimensional coding was only an option, for Group 4 it was considered to be the standard mode. An important aspect in the choice of the coding system was the assurance that "all corresponding patent rights were to be available and this free of charge" (COM XIV (7E: 3), 1977/80). Japan had initially reserved these rights in its proposal of the READ (relative element address designate) two-dimensional coding scheme. IBM submitted an alternative proposal. The urgency of the matter was emphasized at a meeting in December 1978, but agreement on the method of selection of a coding scheme was already proving very difficult. AT&T and the US delegation had proposed a list of criteria comprising compression factor, error sensitivity factor, cost of implementation, and compatibility with other facsimile codes (ibid. 66E; 70E: 20–22). In this situation, the "CCITT Secretariat stated that in the case where the proposed

Table 8.3
Facsimile resolutions and systems proposed by manufacturers and PTTs. Adapted from CCITT, study period 1973–1976, COM XIV (30-E, annexes 1 and 2).

Organization or company	Horizontal resolution		Vertical resolution (line/mm)	System proposed
	elements/line	points/mm		
BNR	1536		3.85 and 7.70	One-dimensional coding (modified Huffmann, elimination of branch lines)
Cable and Wireless	1536 or 1728[a]		3.85 and 7.70	Simple code, more complicated options according to manufacturers' needs
CIAJ (Japan)	1536/1728		3.85 and 7.70	
CIT Alcatel	1536		3.85 and 7.70	
Graphic Sciences	1536		3.80 and 7.20	One-dimensional coding (similar to Huet 9 run lengths and determined by algorithm)
Hasler AG	1728		3.85 and 7.70	One-dimensional coding (B1 if Xerox assigns its patent gratis)
Hell		8	3.85 and 7.70	One-dimensional coding (modified Huffmann)
KDD high definition	1728			RAC coding
Kalle Infotec	1728		3.85 and 7.70	Coding on 2 lines
Muirhead				One-dimensional coding (modified Huffmann for black-and-white runs)

NTT				
normal	1728:2		3.85 and	Further studies must be made
high definition	1728		7.70 (square resolution only)	
Philips	1536		3.86 and 7.71	
Plessey	1536	7.14	3.85 and 7.70	One-dimensional coding (modified Huffmann)
PTT France	1536 or 1728[a]		3.85 and 7.70	One-dimensional coding (possibility of optional extension to a compatible two-dimensional code)
PTT Germany	1728	8	3.85 and 7.70	One-dimensional coding (modified Huffmann with end-of-line code)
Rapifax	1728		2.7 4 and 8	Coding on 2 lines
Swedish PTT	1728 points		3.85 and 7.70	No preference at the moment
General Post Office	1536 or 1728[a]		3.85 normal 7.70	It is essential to reach an agreement at the meeting in January 1976. We support the French position.
Xerox standard	1024 1536		3.85 or 5.13 3.85	One-dimensional coding on white runs only (modified Huffmann)
high quality	1536 or 2048		7.70	
3M	1024 option 1536		3.86 and 5.13	

a. Only one value should be retained.

Figure 8.6
Proposed facsimile coding schemes. Source: CCITT, study period 1977–1980, COM XIV (7-E: 11–12).

PTT Germany	No final decision yet. Further studies on reconstruction possibilities.
Hell/Siemens	Open. Coding scheme which allows reconstruction not yet investigated sufficiently.
Kalle Infotek	AD Code of document no. 15* or for further study.
Philips	ITC document no. 4.*
Rank Xerox	Modified Huffman (EIA-BFICC) now.
Xerox Corp.	Modified Huffman (EIA-BFICC) now.
AEG-Telefunken	Code EIA-BFICC.
PTT France	EIA-BFICC provisional with modified EOL in document no. 13* and further studies.
CIT-Alcatel	Provisional modified Huffman.
PTT Finland	Choice not yet possible; one year before approval.
NTT	Simple and systematic one-dimensional coding scheme easily extensible to two-dimensional code, for example WYLE's code.
KDD	Simple one-dimensional code easily extensible to two-dimensional system, for example WYLE's code.
CIAJ	Simple and systematic one-dimensional code extensible to two-dimensional coding, for example WYLE's code.
Rapidfax	Document no. 7.*
Dacom Inc.	No sufficient information. Modified Huffman code after further study.
Graphic Sciences	EIA-BFICC modified Huffman.
MGVS	EIA-BFICC modified Huffman.
3M	Modified Huffman run-length encoding.
General Post Office	No strong preference. Fully defined, adequately tested and freely available code.
Cable & Wireless	Document no. 7.*
IPTC	One-dimensional modified Huffman run-length with an option for bi-dimensional coding.
ITT-Creed	Use the available time until November 1977 for further study.
Muirhead	Document no. 7*—now.
Plessey	Document no. 7*—now.

Figure 8.6 (continued)

PTT Sweden	Too early for decision. Prefer a coding scheme which enables reconstructing from the right-hand side.
Hasler AG	Modified B.1 for example—possibility of error-correcting facilities—efficiency same range as modified Huffman—and no patent.

*—Document no. 4 proposed by Philips: a redundancy reducing run-length code for digital-facsimile signals with line-correcting capabilities.
—Document no. 7 (joint document).
—Document no. 13 proposed by French Administration: method for evaluating the error sensibility of coding systems for digital-facsimile machines.
—Document no. 15 proposed by Rapifax: comparison of coding schemes by computer simulation.

codes are approximately equal with respect to all evaluation criteria, then a code should be adopted as the standard which is by that time the most used in facsimile machines, such procedure having been applied in the CCITT work during the past for many kinds of techniques to be standardized" (ibid., 70E: 22). Furthermore: "Both IBM and Japan agreed, that it would be very difficult to produce and agree upon a compromise code. They both agreed that the codes were very similar." (ibid., 70E: 27) But the criteria of diffusion put Japan in a much stronger position: the READ code was already a compromise among several manufacturers, RPOAs, and PTTs, and because Group 3 machines were being manufactured with this code the existing base of 1500 machines was increasing at a rate of 6000 machines per year.

Other actors entered the conflict. In March 1979, Germany submitted an alternative two-dimensional code, originally proposed by Hell, with a duty-free license assured. Xerox, 3M, the GPO, and AT&T followed with their own proposals, the one from 3M being explicitly promoted as a compromise position. Extensive testing of the various codes followed in a combined effort involving France, Germany, the National Communications System (a US government agency), IBM, and the CCITT secretariat. At the next meeting of the study group, held at the end of 1979 in Japan, a decision had to be made to avoid a proliferation of different techniques. The United States, unable to arrive at a national position, only put forward a set of principles for the group to consider in making its choice. France opposed the whole idea of a two-dimensional coding standard on the ground that such a standard could not guarantee the necessary transmission quality on the French telephone network.

A decision was reached in the consecutive meetings of an ad hoc group formed at the final meeting of SG XIV. The study group proceeded by gathering all points on which agreement already existed, next to those still in contention. At the second meeting of the group, Japan agreed to a modified version of its READ code proposal in the interest of compromise. Several options for modified READ codes were discussed, a major problem being that 3M, Xerox, AT&T, and KDD preferred a vertical reference code extending to 3 pels while IBM proposed 2 pels simply because "IBM stated that they preferred the IBM code" (ibid., 129E: 62). The next meeting started with an assurance by the Japanese delegation as to the availability of free licenses should the Japanese proposal be unanimously selected as the recommended option. This was received favorably, but IBM continued to support its own proposal. On the next day, however, an agreement was achieved on a modified READ code, under the compromise that the uncompressed mode wanted by IBM and others in the US delegation should be included as a recognized optional extension. Accompanying this was an addendum, insisted on by Germany and France, reserving the right of PTTs to require terminals to make it apparent if two-dimensional coding had been used in a transmission. "A spirit of compromise and a desire to reach a conclusion had enabled a solution to be found." (ibid., 129E: 55) Thus Japan finally succeeded in having its proposal agreed to in a modified version. For the Group 4 coding scheme, yet another modified version (the modified, modified READ code) was later accepted.

In the mid 1980s, a conflict arose with regard to the error-correction mode.[14] The ECM was originally promoted by two British subsidiaries of Japanese firms (Matsushita and NEC) in the hope of selling more reliable equipment more easily. Error correction by means of a specified mechanism verifies incoming information and initiates retransmission in the event of error. This slows transmission by 10 percent. Obviously, such a mechanism is attractive when errors are likely to occur because of poor network performance, or when a certain quality of service is to be guaranteed by the operator. Germany was quite a strong supporter of the option, the Bundespost being one of the few national administrations offering a "telefax" service. The US delegation was split on this issue because of its background of supposedly more reliable networks.

The Japanese delegation, wanting to rely on existing proprietary ECMs, resisted such a standardized mode fiercely. In contrast to other enhancements of Group 3, where proprietary options gave way to standardized ones, standardization was rejected in this case. After some time, Japan's unwillingness to cooperate was weakened through some political lobbying by British government officials in Japan, and an ECM standard was drafted that—it was commonly perceived—Japanese manufacturers supplying the relevant markets could not help but implement. Often the assumption is made in standardization committees that if Japanese firms do not actively try to influence many decisions, it is because they can produce efficiently whatever the standards and the markets request. However, this time the Japanese opposition was also interpreted as a sign that Japan already regarded the facsimile market as its own market and did not want others to define the shape. In addition, Japan, like the United States, feared a loss of efficiency through longer transmission times.

Another problem was that an alternative technical solution for handling errors had arisen. The Union of Soviet Socialist Republics had proposed an error-limiting mode (ELM) aimed at limiting an error's impact on the rest of a document instead of initiating the retransmission of damaged lines. Despite some initial opposition, both the earlier ECM proposal and the USSR's alternative were quickly given unanimous approval, and in 1986 (in the middle of the study period) they were added as options to the Group 3 standard. As things transpired, the ELM was never implemented in machines for the relevant markets, nor had its implementation ever been expected. Thus, error correction provides an example of the acceptance of an option that obviously had no significant support, as well as showing how overt political lobbying may at times determine a standardization process. It appeared to be easier to pass an irrelevant option than to shoulder the responsibility for drawing up a compromise. The ELM posed no threat to a coordinated development of facsimile.

The recent conflict over Group 3 digital is generally deemed to have been the worst confrontation in the history of facsimile standardization. The conflict started at the first meeting of SG VIII in the last study period in 1988 with the British proposal—clearly attributable to a delegate working for Matsushita in the UK—to standardize a Group 3 option for 64-kilobits-per-second networks.[15] A parallel idea for such an option had come up in the United States, and here too one individual—a consultant working for AT&T—was closely connected. Besides the British and American delegations, the delegation of the USSR backed the proposal, albeit much less forcefully. France and Germany immediately rejected it. Japan was to become another strong opponent, but only after the first few meetings. Other countries were mainly indifferent, either not getting involved in the conflict or marginally supporting one of the two sides. Austria and Denmark were seen to support France, Germany, and Japan out of political allegiance; the Netherlands once argued in favor of the proposal, which was seen as resulting from operator interests in possible conversion services between different facsimile groups.

The idea behind the Group 3 digital proposal was very simple: since Group 3 terminals work digitally, it would be very easy to virtually take the modem out of the terminal, allowing it to run on digital networks and the ISDN. Only very minor amendments to existing protocols were thought necessary by the proponents. However, since with the Group 4 terminal a facsimile standard for digital networks had been in existence for some time, other countries did not readily regard the proposal as only one of the many standardized options to the basic Group 3 standard. Rather, Group 3 digital was seen as an attempt to replace the existing Group 4 standard.

The defenders of Group 4 saw no reason for such an unusual step. In their argument they not only invoked the institutional rules of the CCITT; they also referred to cost considerations. Standards recommended by the CCITT cannot easily be abandoned. This informal rule protects investment and increases the certainty of action based on these recommendations. Only if recommendations are quite old, and clearly not in use any more, are they sometimes deleted. This was obviously not the case with Group 4, which was seen by its defenders as an optimal solution of future coordination problems. From their point of view, the slow diffusion of Group 4 until that time had been entirely due to the unexpectedly slow diffusion of the ISDN. Accordingly, the high costs of Group 4 terminals were interpreted as a transitory phenomenon, caused by the initially low production runs and not by the overly complicated protocols (whose costs would also decrease once the software was developed and put on a chip). Since the standard differed from Group 3, the defenders of Group 4 argued that a Group 3 digital terminal would have to contain the complete Group 3 technology to be compatible with the installed base, thus causing much higher production costs than was being claimed by its proponents. To be truly compatible with all existing machines, Group 3 digital would also need to integrate the Group 4 technology, thus making it even more expensive.

Supporters of Group 3 digital saw things from a very different angle. For them the rivalry between two standards was not problematic, because it offered a choice for the user. Group 4 had not sold well, and in view of its complicated protocols and expense this was not unexpected. With the speed and reliability of existing Group 3 machines, users would not see the point of the transition to a much more expensive generation of new equipment. If new terminals were to be sold, it was because they offered new capabilities incrementally, at little cost, such as a Group 3 terminal's ability to run on digital networks. From the start, Group 4 had been connected with the teletex development and with demands for interoperation with this service, which accounted for its technological complexity. Teletex was regarded as dead, which made it preferable for standards to specify an alternative path for the technical development of facsimile, divorced from the failures of the past and from the risk of having no viable future standard at all.

The argument about teletex was not accepted by the opposition. Group 4 included three different classes, and the discussion so far had focused on class 1, which could not interoperate with teletex anyway. Another major argument supporting Group 3 digital—the longer transmission time needed for the handshaking procedure with Group 4 compared to Group 3 digital—was believed to be irrelevant. Because Group 4 was adapted to the OSI environment and the ISDN, more terminal capabilities had to be checked. This handshaking procedure took 6.35 seconds, compared to only 1.55 seconds for Group 3 digital. Given that on a 64-kilobits-per-second channel the transmission of four pages only takes about twelve seconds without the handshaking time, this was not insignificant. The handshaking overhead even multiplies in networks that include several satellite links (presumably a transitory problem in view of the increasing importance of international fiber networks). On the assumption of sufficiently long time intervals in the tariffs of service providers, however, the longer handshaking procedure would not generate additional costs—so the proponents of Group 4 argued.

The debate on handshaking, in particular, shows that the conflict was in good part an architectural one. Group 4 represented the long-term plan of the ISDN and OSI, the principled technical approach that allowed for future network and service transitions, and the rather unrestricted integration of new service features and new technical components. Group 3 digital embodied the old pragmatic and rather unidirectional facsimile philosophy and its past success. However, the strategic question of architectural choice was not debated explicitly. The promoters of the Group 3 digital standard did not openly oppose OSI; rather, they argued for an incremental transition to digital networks. Moreover, the old dispute over the closure of SG XIV and the related conflict between manufacturers and operators were partly implicit in the conflict between architectural technical interests. Group 3 digital appeared to be the solution that SG XIV would have chosen in the absence of dominance by SG VIII, by teletex, and by operators. Informally it was argued that a principled layered architecture might be advantageous for long-term network investment because it allowed for modular upgrading. Such an architecture was considered misplaced for terminals, however, because terminal manufacturers wanted to sell and replace whole terminals rather than only to add modules.

Such were the main actors and arguments involved in the conflict. The aspects that made it so memorable were the extreme difficulty experienced in trying to solve it within one study period and the lack of any transparent reason why this was so. One main problem was the great insecurity of the interests behind the Group 3 digital proposal. For the promoters of Group 4, there were so many open questions behind the arguments for Group 3 digital that it seemed to them very much a matter of idiosyncratic interests on the part of two prominent individuals, who had managed to get the support of their respective delegation (the British and the American). But it was not clear whether this was merely a case of individual predilection and disposition or whether there were corporate interests behind it. Though it seemed understandable that a consultant to AT&T might push a topic connected with a contract, there was some perplexity felt as to why Matsushita UK should promote an issue that Matsushita Japan opposed. Important in this context was the perceived lack of detailed technical work backing the "digital" proposal. While the promoters of Group 3 digital said that production could start just as soon as there was a standard, because the development work could be paid for out of petty cash, the opponents lamented that there were fundamental problems with the protocols and that essential questions (concerning, e.g., the necessary interworking between digital and analog networks) had not even been addressed.

This uncertainty provided scope for speculation that the US Department of State had paid the AT&T consultant to put Japanese manufacturers into a state of uncertainty; that Matsushita Japan had to take the route it did via its British delegate because it could not openly confront the Group 4 standard that Matsushita had once promoted, in particular since the Japanese vice-chairman of SG VIII was the "father" of the Group 4 standard; that the US Department of Defense strongly supported the standard and wanted to place a large order; that the United States was interpreting Japan's resistance as an attempt to maintain a virtual monopoly in the facsimile market; and that this was an entirely meaningless conflict, kept going only by two individuals. Yet it was always clear that these were speculations. Though they were used to account for the conflict, they were never cited as a basis for action. Rather, each of the concerned parties continued to try to resolve the conflict in their own favor: the proponents of Group 3 digital pushed for that standard; the proponents of the Group 4 standard opposed Group 3 digital in the hope of gaining time.

In contrast to other conflicts, there was no attempt made in this case to find a more acceptable middle ground by splitting the problem into more manageable parts and/or establishing criteria for choice. Instead, proposals were constantly rewritten, arguments were continually restated, and the conflict tended to spill over to questions not directly related to the basic issue. As no progress could be achieved in substance, the existing procedural means were turned to. A Special Rapporteur from "neutral" Italy was commissioned to gather the various arguments and to point out possible ways to find a solution. But this measure failed, and only a list of the various incommensurable arguments emerged. Furthermore, SG I, the group responsible for services, was asked to give direction to SG VIII and to point out if a Group 3 digital standard was desirable. Yet this procedural safeguard for maintaining the functionality of study groups was handicapped by the presence of the two promoters of Group 3 digital from SG VIII among the ranks of SG I. After a while, the defenders of Group 4 noted that the same two individuals were asking SG I questions from SG VIII, and that in the next meeting they were supplying to SG VIII the answers from SG I.

Conflict Resolution (Termination)

Within the formal provisions guiding the CCITT's work, explicit mechanisms for overcoming conflict are scarce, although a strong institutional expectation exists that consensus should and can be reached. There are some procedural provisions to enhance decision-making capacity: for instance, that SG I be asked to give guidance, or that a dedicated rapporteur group be set up to study the problem. Otherwise, there is only the formal decision rule requiring a simple majority among national delegations, which is precisely the non-desired means for ending disagreement.

The early conflicts in the choice of modulation technique for Group 2, and of one- and two-dimensional redundancy-reduction technique for Group 3, show similarities. In each case, different technical options were promoted by identifiable firms and agreement on an option was achieved in a similar way. Each time, once various solutions to the technical question were proposed, an attempt was made to draw up a common set of criteria for the evaluation of the options. The prevalent criteria were efficiency, simplicity of design, vulnerability to errors, transmission time, and patent rights. In the choice of a Group 2 modulation technique, a special rapporteur's group was established "in order to study evaluation criteria and test conditions which will lead to the selection of an acceptable group 2 modulation system" (COM XIV (18E: 10), 1973/76). A series of tests were conducted in different parts of the international network, using the same test chart. Because of a lack of time, the results underwent a "simple visual comparison" (ibid., 30E: 3). Xerox, with the inferior method, did not readily concede defeat—a common reaction, and similar in IBM's case of the choice of a two-dimensional coding technique. Nevertheless, having agreed on a "scientific" decision-making procedure comprising the establishment of criteria, testing, and assessment, the supporters of the inferior method could only extend the testing until they had to give in.

Likewise, the choice of a Group 3 redundancy-reduction mechanism was based on an agreed-upon procedure for decision making, including study of various redundancy-reduction techniques, evaluation of performance, comparison of quality, and assessment of vulnerability to transmission errors. The difficulties experienced in agreeing on vertical and horizontal resolution for Group 3 equipment provide a clear example of how decision-making capacity can be improved through a subdivision of the problem into smaller "chunks" to be managed individually. Thus, it was agreed to separate the choice of horizontal and vertical resolution from the choice of the redundancy-reduction technique, and, in addition, to have the choice of the vertical resolution not "strictly determined" by the choice of horizontal resolution. Furthermore, a set of criteria (legibility of characters, quality of reproduction, simplicity of equipment design, transmission speed, compatibility with Group 2, and effects of transmission errors) was approved for the decision taken on resolution (ibid., 34E: 3f).

An important background to this common sequential decision-making process is the loose coupling of technical problems, which permitted the unbundling of decisions on relatively independent issues in line with the concept of decomposability of systems in Simon's (1962) "architecture of complexity." While being affected by given physical opportunities and constraints, the case of facsimile demonstrates how the structure of a problem may be due both to design and to interacting social and technical factors. This is exemplified most clearly by a central feature of the Group 3 standard: the provision that proprietary or standardized options may be implemented beyond the basic compatibility. This possibility is based on the handshaking procedure, in which the specific capabilities of the interacting terminals are checked. In this way it has been possible to keep the Group 3 standard very flexible, introducing new features gradually and keeping the basic standard as the lowest common denominator. This has not only been a basis for the incremental development of more advanced yet compatible facsimile Group 3 machines; it has also aided standardization, in that the specification of options has become a feasible possibility.

Options can be seen as enhancing the decision-making capacity. Proposals which are strongly promoted may be recommended as an option, providing for a standardized development in a technical area not deemed relevant for ubiquitous use but deemed relevant for a more specialized field. Options thereby allow one to respond to more specialized, designated communication demands within overall compatibility. Because it is not necessary for a proposal to become relevant for the whole installed base, it is possible to evade an otherwise-likely conflict. This possibility of delimiting the influence of options may explain why, in the case of the ECM conflict, the ELM option was readily agreed upon alongside the ECM option, instead of going through a period of conflict until unanimous support for one option could be achieved. Still, there are cases where some actors strongly desired a feature not deemed relevant by others (for example, the case of the note-fax) and cases where agreement on an option did justice to both sides while managing to evade conflict.

Not all contentious issues in the standardization of facsimile could be solved cooperatively in a sequential decision-making process. The

ECM again is an example where political lobbying on the part of the British government in Japan was seen to have an influence. Within the context of the institutionalized provisions of the CCITT, the formal majority rule is the principle employed as a last resort, for cases where major actors are uncompromising, cases where different positions cannot be combined, or cases where there have been difficulties in subdividing a problem into manageable parts. The choice of frequency for Group 2 was such a case: Xerox, having just lost the conflict on modulation, was not prepared to make concessions. Although most PTTs and RPOAs expressed no preference either way on this matter, a majority decision taken at the last meeting of the period led to the choice of 2100 Hz, the frequency promoted by Xerox, despite the fact that there was an installed base of Graphic Sciences machines using 2048 Hz, which now would no longer be fully compatible although their modulation method had been chosen.

Majority decisions are usually regarded as problematic. They document the failure of consensus and thus imply that a recommendation does not have the broad backing desired for voluntary implementation. Clearly, technical arguments have not sufficed to convince others of the need to adopt a common position. Most of all, majority decisions are slightly illegitimate, because the official position of the national delegations is predominantly influenced by the national PTTs and RPOAs. Other members of the study groups are formally excluded even if they have committed resources to the technical work and have rather high stakes in the eventual decision. Also, less binding majority decisions in the form of indicative voting, in which all members can participate with a single vote, are generally regarded as inappropriate. The choice of resolution for Group 2 demonstrates how majority decisions may turn against significant invested stakes, although the informal principle exists to decide in favor of the installed base whenever other criteria are absent. The resulting strong reluctance to end a conflict with a majority vote exerts pressure to compromise on a common position, so that the threat of the "shadow of the majority decision" can be seen to facilitate agreement.

The Group 3 digital conflict differed from earlier disagreements in two respects: the constellation of opponents and promoters was not precisely clear, and there were no attempts made at problem-solving (either via an agreed procedure of establishing criteria and testing or through subdivision into smaller components). Endeavors to bring about a solution through procedural means failed as neither SG I nor the SpR could bring about a clear compromise. In SG I there were some initiatives—for instance, the one advising SG VIII to simplify the existing Group 4 standard. SG I also considered deleting the Group 4 class 2 and class 3 specifications, since mixed-mode operation with teletex had become irrelevant. But this study group was hampered by the inclusion in its number of some of the key participants in the SG VIII conflict. Only the proponents of Group 4 made a compromise proposal for a Group 3 standard using Group 4 protocols (called UDI, for "unrestricted digital information"), which would be less harmful to the Group 4 developments than the Group 3 digital. Yet this attempt to oust the Group 3 digital proposal was not, and obviously could not be, accepted as a compromise by the other side. Indeed, the two camps adhered to genuinely incompatible architectural models of the telecommunications network. The principled OSI/ISDN architecture, with its analytical distinction between different communication layers, must be seen in sharp contrast to the traditionally built-up architecture of the telephone network, which mixes control information about signaling, routing, etc. with the content of a telephone call or a facsimile message in a rather idiosyncratic way. Thus, it turned out that—in contrast to past experience—there was no basis for agreement on common criteria for assessing the technical performance and efficiency of the competing proposals, a prerequisite for which would have been a consensus on the overall architecture.

The conflict over Group 3 digital claimed much of the attention of the 1989–1992 study period, at the expense of other important work. As a result of the constant regurgitation of the same vacuous technical arguments, it often proved impossible to break the conflict down into manageable parts. In view of the deadlock, the chairpersons of study groups I and VIII felt in the end that they had to act. Initially in SG I, then later in SG VIII, the same procedure took place: the chairperson gathered the heads of delegations to discuss the issue and reach a conclusion on how to proceed. Both times, the result was in favor of adding digital options to the established Group 3 standard. The compromise resulted in the inclusion in Group 3 of both the Group 3 digital and the UDI counterproposal, which had emerged during the conflict within the committee. It was not even necessary to take a vote, since a consensus for this solution could be obtained among the delegation leaders. Thus the UDI proposal—which had been introduced by the opponents of Group 3 digital with the aim of limiting the variety of standards by replacing the Group 3 digital with an alternative more in line with Group 4—was now accepted as an additional incompatible option. However, this was part of a favorable compromise: each party was granted its proposed option. Although some delegates felt that the delegation heads had given in too readily (despite all the valid arguments against Group 3 digital, so that two individuals had managed to sway the whole study group through their national delegations), there was no doubt that this result had to be respected. In the end, it was a relief that the conflict was over.

9

Message Handling (X.400)

The convergence of computing and telecommunications has unleashed a tremendous technological potential, on the basis of which the traditional telephone service has been complemented by an array of new telecommunications services. The new technological capabilities are accompanied by significant expansion of the range of actors relevant to telecommunications and by liberalization of the former monopolies.

The increasing diversity of actors and technological options causes significant problems for standardization. Indeed, the conflict over the Group 3 digital standard for facsimile (discussed in chapter 8) has already demonstrated the new diversity implicit in the possibility of embedding technical options in the new, principled OSI architecture or of following a more incrementalist approach.

Similar problems and difficulties with the emerging structure of the market in telecommunications accompanied the work on X.400, a standard for electronic mail. The greater the number of actors and the greater the number of technical options for standardization, the higher is the risk that distributional issues will obstruct the building of a consensus in institutionalized standardization. The Comité Consultatif International Télégraphique et Téléphonique is maladapted to dealing with problems of this type; however, such problems can be avoided if standardization starts early. The standardization of X.400 is an impressive example of an attempt at early, comprehensive standardization. As we will see, a head start allowed X.400 standardization to progress smoothly, without major internal difficulties or conflicts. However, X.400 has not managed to take on the Internet, and its diffusion has been rather disappointing. Thus, X.400 demonstrates the pitfalls of ex ante standardization: if the work of the standardization committee is protected from the influence of markets, and if the marketing of products is years away, a comprehensive system of standards may remain largely theoretical.

Overview of the History of X.400 Standardization

Of our three case studies of standardization, X.400 is the most recent, the most ambitious, and the most comprehensive. Initial work toward this standard started in 1978 within the International Federation for Information Processing (IFIP), an organization involved in prestandardization work (see chapter 3 above). The IFIP set up a new working group, WG 6.5, for "international computer-based messaging." Within WG 6.5 three subgroups were relevant to work on message-handling systems (i.e., electronic mail systems): the subgroups focusing on user environment, systems environment, and policy environment. Each of these subgroups had North American and European co-chairpersons and held separate meetings on both sides of the Atlantic. The North American Systems Environment Subgroup was particularly relevant. The initiative behind this new project can be traced back to several individuals working at Canadian Bell Northern Research[1] who believed in the need for an international standard for electronic mail in the context of research on computer networks and the "office of the future" (Uhlig et al. 1979).

Several proprietary computer mail systems were in existence at the time. However, they were restricted to a host computer shared between the interacting partners, so that the underlying technological principle was very similar to distributed computer processing. As regards a long-term plan for the office of the future, therefore, these systems were technically too constrained. Later the X.400 series of CCITT recommendations resulted in an attempt to overcome these limits by specifying a generic, nonproprietary, and international system to which all existing proprietary electronic mail systems could be connected for use in intersystems communications (Cunningham et al. 1982). From its inception, this project was based on the newly defined layered OSI standards architecture, for which it represented the first realization of a standard on the top application layer (Maciejewski 1989).

A network already existed: the ARPANET, created by the US Defense Advanced Research Project Agency (DARPA) connected heterogeneous computer systems at particular universities and research organizations by means of nonproprietary and generally unclassified protocols. Mainly designed for remote log-in and file transfer, the ARPANET was also used for message handling by computer engineers and the research community. The basic tools required were already in place in 1975 (Denning 1990). Despite its military background, the ARPANET was not restricted to military traffic. Civilian use predominated, and in the early 1980s the military segment of the ARPANET was separated organizationally and transferred into the MILNET. However, the ARPANET was not designed for public use. It was, in fact, not even accessible to most members of the academic community in the United States. Thus, several initiatives were set up to provide research networks for those not connected to it (Rogers 1996; Quarterman 1990; Abbate 1994b). At first the ARPANET's protocols, though quasi-open (unclassified), were perceived as US Department of Defense standards, not as international standards. The network also lacked many of the features of traditional telecommunications services (Hart et al. 1992). Thus, the ARPANET was not regarded as a prototype for a network of e-mail services for the general public (although Telenet, an early offshoot, offered such service in a few US cities). Therefore, when several persons working at Canadian Bell Northern Research (BNR) initiated the standardization of message-handling systems, it was a new and ambitious undertaking from the viewpoint of those associated with the established scene of international standardization. Because of its precedents (proprietary systems and the ARPANET), the project met with a lot of interest among academics and people in computing firms in North America.

Because the idea for the standardization project had arisen in the middle of a CCITT study period, work on it started in the IFIP as a search for an interim solution. The IFIP was particularly well suited for initial work on a new project. It was less exclusive in its boundary rules than the CCITT, and work could progress much more easily because it was defined as "pre-standardization" and because it was not expected to produce official documents. Several meetings of the North American subgroups of WG 6.5 were held, with many people from computing firms and universities in attendance.

In the IFIP, a conceptual model based on a layered model of computer communication and following the basic features of the postal mail system (figure 9.1) was developed that already included most of the basic characteristics of X.400 (Cunningham 1983, p. 1426). Parallel to this conceptual work, standards for message-handling systems (MHS) had to be prepared. In principle, the CCITT and the

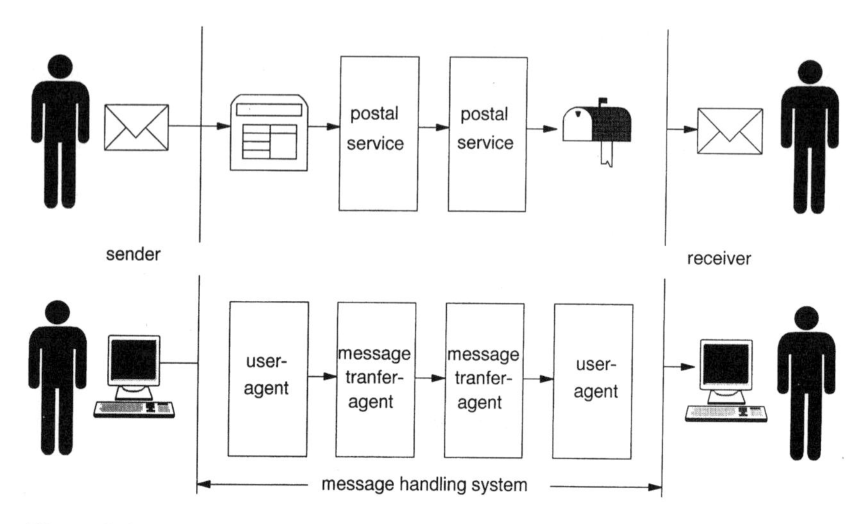

Figure 9.1
Conventional mail and electronic mail.

International Organization for Standardization were appropriate organizations, the former for the telecommunications and network aspects and the latter for the computer and terminal aspects. From the beginning the CCITT was preferred, both in light of BNR's background in telecommunications and as a result of a strategic decision taken with a view to improving the future chances of the new technology. As the committee of the national network operators, the CCITT was the organization best suited to securing the commitment for implementation from the relevant actors. It was to the CCITT, therefore, that the initiators of the project at BNR turned for standardization.

Before the end of the CCITT's 1977–1980 study period, it was necessary to choose a study group and to seek the acceptance of a study question. (Traditionally, the study groups formulate the questions to be studied; these formulations are then subject to acceptance by the Plenary Assembly.) Each of these needs required some strategic action. Either SG VIII or SG VII could have been chosen, depending on whether MHS was regarded more as a medium for text communication (similar to teletex and facsimile) with an emphasis on terminal aspects (SG VIII) or whether data communications and the systems rather than terminal aspects were emphasized (SG VII).

In the case of MHS, SG VII offered clear advantages. More computing firms were present in that group than in SG VIII. Also, SG VII had done the work on packet switching in data networks and was involved in open systems interconnection (OSI) on the part of the CCITT. Moreover, this study group had a Canadian chairman, which made it easier for a Canadian (Ian Cunningham) to initiate work on a new topic. Furthermore, because the main opposition to MHS came from the supporters of teletex, which as a point-to-point text-communications service was perceived as a direct competitor to a planned store-and-forward MHS, it was deemed better not to choose SG VIII.

Nevertheless, the formulation of a study question for MHS faced opposition from teletex supporters in SG VII too. This problem was managed strategically. Cunningham, the initiator from BNR, began by informally seeking the support of several countries. Having gotten the support of the United States, Japan, France, the United Kingdom, the Netherlands, and some other countries, he formally proposed the new question. This immediately drew on the support already obtained; it also took the potential opponents (Sweden and Germany, both of which were highly involved in developing teletex[2]) by surprise, finding them without prepared counterarguments in the face of such broad approval.

Having pushed through the question (figure 9.2), Cunningham attempted to institute a special rapporteur group (SpR) for the interim, to start work before the beginning of the next study period; his argument was that MHS was a particularly complex and pressing issue. But Germany and Sweden mounted opposition, thereby causing the proposal to fail, although five such interim SpRs were instituted for other urgent points of study, mainly in the context of OSI (COM VII (R1-E :4), 1981/84). Cunningham then pursued an alternative plan and distributed documents on the work done so far in the IFIP, along with a proposal for holding a series of informal meetings in the interim period to discuss the issues. Many actors were interested, and three "interregnum meetings" were convened in 1980 and 1981 so that much of the functional model and other work done by the IFIP could be discussed and confirmed in a larger and more heterogeneous circle of actors—including the firm of Bolt, Beranek and Newman (which was highly involved in the ARPANET) and the firm General Telephone and Electronics (owner of the ARPANET's commercial offspring Telenet, whose network had been developed by BBN).[3]

The support of the Canadian chairman of SG VII was quite important for these meetings. In the face of criticism, he emphasized their

Figure 9.2
Initial study question for message handling. Source: CCITT, Study Period 1981–1984, Com VII (1-E: 19–20).

Question 5/VII—message handling facilities

considering that

a) A number of administrations are planning to provide message services
   - i) which will provide message preparation facilities (e.g. editing capabilities),
   - ii) which will provide sending, transporting and receiving facilities,
   - iii) which will provide message storage facilities,
   - iv) which will permit the transfer of messages of different types having a large variety of formats,
   - v) which will provide message to message type and format conversion facilities,
   - vi) which will be accessible to users via a variety of terminal equipment,
   - vii) which will be accessible by a variety of other services;

b) i) High level protocols are being defined by Study Group VIII for Teletex, facsimile (e.g. Group 4 apparatus) and Videotex and it is expected that these services will be provided on public data networks by several administrations,
   - ii) Teletex, facsimile and Videotex terminals may require message handling facilities for enhanced communication requirements;

c) Manufacturers require support in the form of standards for the manufacturing of equipment;

1. What new message handling services and facilities are technically suitable for provision in public data networks?
2. What technical conditions (e.g. protocols) should be recommended concerning such services and facilities?
3. What are the effects on high level protocols developed by Study Group VIII?

Note 1: The purpose of this question is to provide one or more recommendations that complement or modify existing recommendations and which give direction to future CCITT studies.

Note 2: This question should be studied in close collaboration with Study Group I for the operational aspects and Study Group VIII for the terminal aspects.

unofficial status. He permitted the distribution of IFIP documents. He also made it possible for the documents of the interim group to be distributed when official CCITT work started. It was then decided, with the support of Germany, to use these documents as a basis for the subsequent work. Consequently, a rather small and (for the CCITT) unrepresentative group at the IFIP managed to influence much of the later CCITT standard by feeding its results gradually into this forum.

In the 1981–1984 study period, SG VII worked on the technical details and protocol specifications of message handling in the framework of the IFIP's functional model. About thirty participants were present at the meetings, and there were enough active members to have a first version of X.400, consisting of eight different specifications defined in parallel, finished by the end of the study period. In this very productive period, the group did not have to cope with any major conflicts or problematic decisions, and the very competent SpR group chairman—the initiator from BNR—remained very influential in the achievements of the group, as he managed to keep a balance between allowing enough time for difficult decisions and keeping to a tight schedule. SG I, responsible for services, did not work on MHS standardization in this period. This was another strategic element of the standardization initiation: to try to keep SG I uninvolved, in view of the significance of European operators in this study group. The Europeans had no experience yet with electronic mail communications and were more interested in the alternative teletex service.

Problems for standardization were less attributable to work in the CCITT than to insufficiently separated domains of organizational competence.[4] Parallel to the CCITT's work in SG VII, the International Organization for Standardization (ISO) started standardization of MHS in the early 1980s, its work based similarly on the IFIP's.[5] This duplication of workload was caused in part by the different membership rules of the IFIP, the CCITT, and the ISO, so that IFIP members not represented in the CCITT carried the work over to the ISO; on the other hand, MHS standardization as such fell between the areas of competence of the CCITT and the ISO. While the CCITT was concerned because of the international communications aspect of MHS, the ISO was involved because of its activities in the standardization of OSI and in the area of computing and private networks. Until 1984, however, the ISO did not manage to make much progress. Besides the ISO, the European Computer Manufacturers Association had also started work on MHS. From the CCITT's point of view this was less problematic than the parallel work of the ISO, because the ECMA, which generally feeds its standardization work into the ISO, is not an official standardization organization. Fortunately from the viewpoint of the CCITT, overlapping membership with the ECMA made it possible to influence this organization and align its MIDA (Message Interchange Document Architecture) specification with X.400. This was done by Siemens, which sent a single rather active delegate to the ECMA, the ISO, and the CCITT, partnered with a specialist from Xerox who was "on loan" to Siemens for this task. Because a unified international standard for MHS was in the interest of both firms, they carried out much of the adaptation of MIDA to X.400. This standard was then submitted in 1984 to the ISO, where it became a relevant part of the slowly developing work on MHS. The parallel involvement of the ISO and the CCITT was a long-term problem, demanding interorganizational coordination to avoid the specification of two different standards focused loosely on public versus private networks and also to avoid direct duplication of the work.

Because the ISO had progressed much more slowly in the beginning, the first draft standards of MOTIS (Message-Oriented Text Interchange Systems) did not appear until 1986, by which time it was obvious that incompatibilities with X.400 existed. The ensuing efforts to align the work of the two organizations saw the CCITT in the advantageous position of being further ahead and having four or five times as many participants as the ISO. An attempt to hold joint meetings, with the chair alternating, failed after two meetings, and the groups then tried to align their work via liaison persons.[6] The CCITT, having already recommended a first set of specifications and being the organization with the greater investment, was not willing to respond to the ISO's requests for modifications. Once the CCITT has issued recommendations, it is, as an institutional rule, bound to use them as a basis for future work. The ISO, on the other hand, was not content with parts of the 1984 version of X.400. It was said that the ISO would never have recommended such a preliminary set of standards—which shows that the rather long CCITT study periods, in fact, result in quite early recommendations, for fear of otherwise being delayed for another study period of four years. Moreover, the ISO was, of course, very reluctant to be treated as a junior partner. Not having recommended its own standard, it did not want to compromise on technical perfection. In particular, the 1984 X.400 did not follow the OSI framework strictly, because some work on message handling and OSI had been done in parallel. For instance, X.400 (84) did not clearly separate layers 6 and 7 of the OSI frame of reference. Moreover, enhancements to OSI had been necessary: from the original peer-to-peer protocols to distributed systems relevant for MHS (Craigie 1988–89). The ISO, however, felt obliged to promote a strict OSI solution for MHS.

In 1980, while the CCITT and the other well-established standardization organizations were still at an early stage of their work on MHS and at the same time about to get involved in jurisdictional issues, the Department of Defense declared the protocol suite TCP/IP its official network standard. Subsequently TCP/IP became the single family of protocols employed in the ARPANET. Like OSI, TCP/IP is based on the principle of layering, although the architectural structures are different. Furthermore, TCP/IP did not arise from the work of official standards committees; it arose mainly from research commissioned or funded by DARPA (Comer 1991, pp. 139–158; Edwards 1996, pp. 259–271). At the time, TCP/IP was a series of successfully tested protocols to be implemented in a network, whereas OSI was an abstract (though ambitious) architectural system. Even more significant in the specific context of the work on message handling systems was the publication of specifications for the exchange of mail between computers in the ARPANET/Internet Requests for Comments (RFC) series in 1982. The Simple Mail Transfer Protocol (SMTP) is described in RFC 821, and the exact format of mail messages is specified in RFC 822. From the X.400 perspective this simple protocol is incomplete, because it deals with how a system passes messages across a link from one machine to another but leaves the user interface and other functions unspecified (Comer 1991, p. 399). On the other hand, SMTP indicates that work outside the international standardization organizations, on a central component of MHS in heterogeneous technical environments, had already reached a degree of maturity that was missing from X.400 at the time.

Technically, some outstanding extensions to X.400 were agreed upon in the 1985–1988 study period of the CCITT: new security features, interworking of MHS with other telematics services, the physical delivery of messages, and so on. The "message store" was another addition made to allow for the storing of messages which the terminal was not ready to receive. Another topic was the "implementer's guide," whereby experience made in the early implementations was fed back into the standardization work with a view to correcting mistakes or facilitating improved solutions. All in all, there was significant interest in the work, and the membership of the SpR group increased to approximately seventy. Besides SG VII, SG I (responsible as it was for services) started to be involved in MHS in this period. However, SG I never assumed a prominent role in the course of standardization; rather, it adapted the 1984 version of X.400 to a service recommendation (F.400).

Before the CCITT recommended its revised specifications in 1988, there had been several steps toward a joint approach with the ISO, none of which had been a complete success. In particular, differences still remained on the question of compatibility with the 1984 version. While the CCITT aimed at maintaining backward compatibility with the 1984 version, the ISO in its 10021 draft standard series opted for greater divergence, in favor of stricter compliance with the OSI model. In the ballot procedure for the standard, however, several national bodies—among them Australia, Canada, France, Germany, Sweden, and the United States—pressed for compatibility with the 1984 X.400, bringing the ISO closer into line with the CCITT.[7] However, some differences persisted in the version adopted in October 1988 (figure 9.3).[8]

Between 1985 and 1988, in the United States, the National Science Foundation (NSF) became involved in building up a national education and research network. The year 1986 saw the creation of the NSFNET, a high-speed network connecting a multitude of regional and university networks. DARPA had encouraged many of the providers of these networks to use TCP/IP (Norberg and O'Neill 1996). With this protocol suite also implemented in the NSFNET, the foundation was laid for a large network of networks to evolve in the United States. By the time the 1988 version of X.400 was finalized, the network of networks (today called the Internet) already comprised several hundred networks. However, the CCITT and (to a lesser extent) the ISO either did not take notice of this development or tended to underestimate it as confined to military or other use outside the domain of commercially utilized public networks and as technically inferior to OSI. This position was indirectly reinforced by the US government, which declared that public procurement of network technology should follow rules laid down in specifications called Government OSI Profiles (GOSIP). This, of course, required OSI-conformant products.

In the TCP/IP community, there was a move afoot to get the ISO to consider approving selected TCP/IP specifications as OSI standards. Assisted by the National Bureau of Standards (NBS), these moves partly succeeded when the ISO started standardizing protocols for local area networks (Abbate 1994b, pp. 139–143; chapter 4 above). But neither the CCITT nor the ISO reacted officially to the growing popularity of e-mail communication in the Internet based on SMTP. In view of the potential competitor, however, both organizations were concerned to streamline X.400 and to add some functionality. Only one individual actor, who mostly observed standardization and only marginally participated in e-mail-related activities in the ISO and the CCITT, tried to bridge the gap between SMTP and X.400. He did not do this in one of the established standardization organizations; rather, he carried his idea into the Internet community. There he proposed a "mapping" procedure between X.400 and RFC 822, and a first version of this proposal was approved for publication by the Internet committees as RFC 987 in 1986. In addition, a broader move to establish interoperability between SMTP and X.400 through a gateway originated in the Internet context. Under contract to the US Defense Communication Agency, the NBS developed this gateway, which was declared from both sides to be a step toward migrating from Department of Defense standards to OSI protocols within Department of Defense networks (Tang et al. 1988).

In the 1989–1992 study period, SG VII's SpR group for X.400 and the ISO's SC 18 WG 4 held "co-located" meetings for their work on MHS, sharing a venue for ease of coordination and also in view of overlapping membership. In this period, refining and correcting mistakes in the existing recommendations were important issues. In particular, the 1988 recommendations contained many orthographic errors. In a complex standardization project such as X.400, feedback of experience from implementation is important, since, in contrast to the IETF procedure in the Internet (see chapter 3 above), it is not explicitly required that protocols written for the standard be tested in advance. Moreover, there were moves (particularly in the ISO) for additional specifications in the context of the message store, to allow incoming messages to be filed in different folders. Such a new addition, however, was strongly resisted, and its recommendation was therefore postponed. Other new technical work to be completed concerned provisions to use X.400 as a basis for electronic data interchange (EDI) and voice messaging.

All in all, the standardization of X.400 proceeded relatively smoothly, despite the significant complexity of the issue. It started in a very exclusive circle in the IFIP, which succeeded in defining a conceptual framework that was later to be adopted as the basis of

Figure 9.3
The 1988 ISO and CCITT standards for message-handling systems and message-handling functions. Source: Tietz 1988, p. 513.

Recommendation or standard	Joint MHS		Joint support		CCITT only	
	CCITT	ISO	CCITT	ISO	System	Service
MHS: System and service overview	X.400	10021-1				F.400
MHS: Overall architecture	X.402	10021-2				
MHS: Conformance testing					X.403	
MHS: Abstract service definition conventions	X.407	10021-3				
MHS: Encoded information type conversion rules					X.408	
MHS, MTS: Abstract service definition and procedures	X.411	10021-4				
MHS, MS: Abstract service definition	X.413	10021-5				
MHS: Protocol specifications	X.419	10021-6				
MHS: Interpersonal messaging system (IPMS)	X.420	10021-7				
Telematic access to IPMS					T.330	
MHS: Naming and addressing for public MH services						F.401
MHS: Public message transfer service						F.410
MHS: Intercommunication with public physical delivery services						F.415
MHS: Public IPM service						F.420

Recommendation or standard	Joint MHS		Joint support		CCITT only	
	CCITT	ISO	CCITT	ISO	System	Service
MHS: Intercommunication between IPM service and Telex						F.421
MHS: Intercommunication between IPM service and Teletex						F.422
OSI: Basic reference model			X.200	7498		
OSI: Specification of abstract syntax notation one (ASN.1)			X.208	8824		
OSI: Specification of basic encoding rules for abstract syntax notation one (ASN.1)			X.209	8825		
OSI: Association control: service definition			X.217	8649		
OSI: Reliable transfer: model and service definition			X.218	9066-1		
OSI: Remote operations: model, notation, and service definition			X.219	9072-1		
OSI: Association control: protocol specification			X.227	8650		
OSI: Remote transfer: protocol specification			X.228	9066-2		
OSI: Remote operations: protocol specification			X.229	9072-2		

standardization in the CCITT, the ISO, and the ECMA. The greatest difficulties resulted from the overlapping organizational competence. Then, in the last study period, the X.400 recommendations slowly stabilized. The implementation of X.400, however, continuously lagged behind expectations. For a long time the significant time lag between the recommendation and the implementation was explained by the issue's significant technical complexity. The base standards recommended by the CCITT and the ISO have themselves never been a sufficient basis for implementation, but needed to be further specified in functional standards and profiles. The need for functional standardization has to do with the very large amount of options specified for X.400. From the user's perspective the elements can be grouped into "basic," "essential optional," and "additional optional" features. In practice the first two categories are considered to be mandatory and the last to be optional for implementation (Manros 1989, p. 38). Early implementations of the 1984 standard thus only appeared around 1988.[9]

The actual implementation and diffusion of X.400 has been rather disappointing for those who invested time, mindshare, and money into this innovative and technically ambitious project. One reason is that for a long time X.400 remained in a transitional state. The 1984 version of X.400, which did not offer much more functionality than existing, less complicated alternatives like SMTP or proprietary e-mail products, was technically outdated four years after its approval. The 1988 version, on the other hand, could not be implemented immediately after its adoption because functional specifications were still missing. Even more critical has been the fact that significant initial investment was necessary if use was to be made of X.400. As this standard was the first realization of OSI functionality on the application layer, it was to bear the burden of implementing all lower layer OSI protocols in order to make use of e-mail. OSI in general has high initial costs (Law 1989), certainly too high for one single application. Thus, X.400 diffusion was contingent on the availability of other applications in the OSI world. These appeared rather late on the market, if at all. Moreover, for those potential e-mail users who were predominantly interested in a medium facilitating fast transmission of messages in public networks, facsimile (which had taken off by the end of the 1980s) offered a simple-to-use alternative (Borenstein 1991).

Because firms are reluctant to undertake immediate implementation of more advanced versions of X.400, and because the existing implementations have not been sophisticated enough to offer incentives for widespread transition to X.400, the future prospects of this standard have remained ambivalent. The extension of the X.400 series in the last study period, which made it applicable to electronic data interchange and voice mail, held out new hopes for the future relevance of X.400 (Scheuerer 1990, p. 31; Saunders and Heywood 1992). The recent development of the market for computer network technology, however, does not support these aspirations for OSI products in general or for X.400 in particular (although X.400 no longer must be run in an OSI environment; it can now be implemented in local area networks as well as in TCP/IP-based networks). One consultancy firm that has observed the US messaging market for many years and has tended to sympathize with X.400 has now come to the following conclusion: "Internet messaging products now dominate the residential/consumer messaging market and are working their way up tier by tier towards the large enterprise level. Established vendors in all segments are adopting the Internet open systems technology as rapidly as possible in order to blunt its impact on their market share." (*Rapport Messaging Review* 3 (1996), no. 5, p. 13)

Because many basic Internet protocols are public-domain software, they are increasingly regarded as open system standards. In the area of electronic messaging, these open standards include SMTP, its Multipurpose Internet Message Extensions (MIME), and the so-called Post Office Protocol (POP). In combination with proprietary software, these standards have enhanced the functionality of electronic mail to a degree that makes it difficult for any other e-mail system to keep pace. Moreover, the public commitment of governments, large vendors, and many standardization committees to open systems standards no longer necessarily implies a confession to the OSI frame of reference, but is in all probability targeted at Internet standards. If there has ever been a "religious war" (Drake 1993; Malamud 1993) between TCP/IP and OSI, the latter seems to have surrendered, and X.400 may end up as the most prominent victim of this war.[10] The war, as will presently become clear, was certainly not waged within the confines of the CCITT. Institutional boundary rules prevented the enemies and their ideas from invading the arena of X.400 standardization.

Central Technical and Economic Problems

The X.400 complex is a highly complicated series of standards defining a complete global message-handling system based on the OSI framework. Its degree of abstraction from existing technical systems

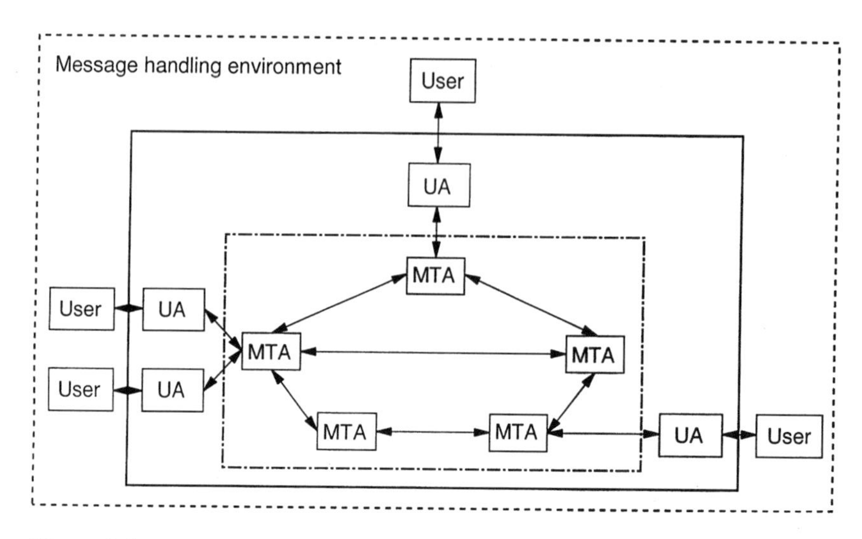

Figure 9.4
Functional view of the X.400 model. MTA: message transfer agent. UA: user agent.

and e-mail implementations is, at the same time, based on a relatively simple general model of electronic messaging.

From the start, X.400 was conceptualized as a store-and-forward service, in contrast to point-to-point connections, where direct real-time interaction occurs, as is the case with teletex, facsimile, and telephony. The work of the IFIP had already resulted in another major characteristic of X.400: the conceptual separation between the user agent (UA) and the message transfer agent (MTA) (figure 9.4). The UA is the part of the system closest to the user, normally a personal computer or a workstation. However, X.400 does not prescribe the shape of the UA. The UA feature provides the functionality to receive and send messages; however, it does not specify the precise user-machine interface, which can be specified separately.

The UA is connected to one MTA, which together with several other MTAs constitutes the message-transfer system (MTS). Messages thus originate in a UA and are routed through the MTS until arriving at the MTA connected to the receiving UA. This simple model was conceptualized at the IFIP, the important new step being the separation between UA and MTA, which had not been part of existing proprietary mail systems. This separation, the basis for the high degree of abstraction of X.400 from different existing technical implementations, is fundamental for the standardization project. This conceptual innovation, which allows looser coupling of system components and thus more flexibility, finds a clear parallel in the ordinary postal system (figure 9.1). The MTS is the backbone of international interconnection, based on standards that were intended to be kept stable after the 1984 version. The MTS also provides for content conversion between different capabilities of UAs, including other services (Cunningham et al. 1982, p. 158). The MTA-UA separation is also the basis for the possible integration of proprietary e-mail systems into X.400. Only the specifications for an MTA have to be implemented as an interface to an X.400 system; internally, private protocols for MTA-MTA interaction may be used, ignoring the whole level of the UA (Kille 1983).

The interaction between different UAs is organized in the interpersonal messaging system (IPMS). The relation of the MTS to the IPMS can be again illustrated by analogy to the mail service. Whereas the MTS transfers the envelopes without regard to their content, the IPMS refers to the messages, which are split into a header and a body. The header may then contain specifications as to the type of content: normal ASCII, EDI, voice-messaging, and so on. In contrast to the stable MTS, the IPMS is more flexible for future additions and different uses. The IPMS includes facilities for access: the access unit (AU) to other telematic services, such as telex or facsimile, and to the physical delivery of messages.

The whole message-handling system builds on a variety of standardized protocols comprising the main part of the standard. These protocols manage interaction and message transmission between UAs and MTAs (figure 9.5). Protocol P1 connects the different MTAs and constitutes the MTS. P2 contains the necessary information exchanged between the UAs in the IPMS; it differentiates the envelope of a message, including the address, from the message content, and it also includes information on the type of message, such as the transfer of EDI and voice-messaging, as extensions of X.400.[11] P3, the final protocol standardized in 1984, specifies the UA-MTA interaction. Of the 1984 specifications, the P3 was one of the weakest items, being quite complex while providing only little functionality. For example, under P3 the MTA determines the transfer of messages to the receiving UA, which was problematic because the UA (say, a personal computer) in a store-and-forward system is not always ready to receive incoming messages.[12] As a consequence, P3 was rarely implemented, and in 1988 an alternative specification, P7, was standardized for the message store.

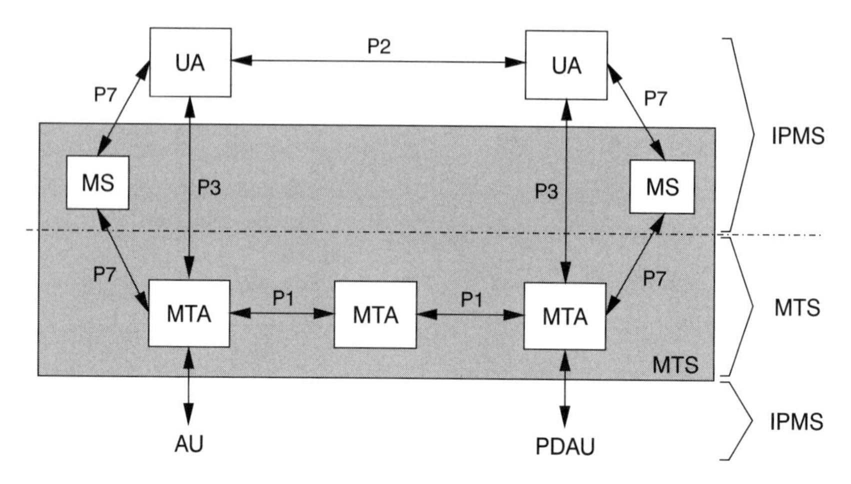

Figure 9.5
Protocols controlling interaction and transmission in X.400. AU: access unit to other telematics services (e.g., teletex). IPMS: interpersonal messaging system. PDAU: physical delivery access unit (postal service). P: protocol. MS: message store. MTS: message-transfer system. MTA: message-transfer agent. UA: user agent.

The MS is located in between the MTA and the UA. From here, messages may be retrieved by the UA. Work toward the MS had originated in the ECMA, and it was taken from here to the ISO and the CCITT.

The routing of messages in X.400 has not been standardized separately; it is part of the address. The initial idea for addressing was that it should be relatively self-explanatory, using only personal attributes (surname, organizational affiliation, etc.) that could be easily taken from a business card (Cole et al. 1985). However, for a number of reasons—not least the various existing proprietary mail systems employing different addressing conventions—this plan could not be realized (Kerr 1981). As a result, directory standardization was started in 1985 (the X.500 series) to make routing and addressing information internationally available by electronic means. Meanwhile, it was agreed to organize MHS into different "management domains," each consisting of at least one MTA responsible for the correct routing of messages within its realm (Cunningham 1983, p. 1429). With regard to these domains, a most important distinction between the Administration Management Domain (ADMD) and the Private Management Domain (PRMD) was introduced in the 1984 recommendation.[13] PRMD designates a private corporate network and ADMD the public network of a postal, telephone, and telegraph administration (generally only one per country) (figure 9.6). Each address has to indicate the ADMD to which it belongs. Here the necessary information is held as to the appropriate MTA and UA the messages have to be routed to. Although the preferential treatment of ADMDs was problematic from the start, it did have a technical rationale: compatibility was likely to be much better between ADMDs, since public networks had less technical variety than private ones.

Being a standard series for the two upper OSI layers, X.400 may be based on any technical implementation of the lower layers, despite the fact that in general X.400 is built on X.25 data networks—as recommended by SG I—because of the ubiquity of these networks and the interests of the PTTs in generating additional traffic in these networks.

The X.400 series contains many more features, but these do not concern us here. What should be noted is that a lot of work done in the context of MHS standardization was later transferred to other applications. Such was the case with the Abstract Syntax Notation 1 (ASN.1), originally developed for X.400 and recommended as X.409 in 1984. ASN.1 was also applicable to the reliable transfer and the remote operations, which were then, consequently, recommended in the X.200 series dedicated to OSI (X.208, X.209, X.218, X.219, X.228, X.229). Being the first standardization project for the application layer, the scheme still required some work in the X.400 context to supplement the newly established OSI framework. This was primarily in the area of enhancements based on the fact that OSI had been specified for peer-to-peer communications and therefore did not fit the requirements of distributed systems such as MHS, where communications between dissimilar systems such as MTA-UA had to take place (Craigie 1988–89).

Since work on OSI had continued into the 1981–1984 study period, some of the 1984 X.400 recommendations fitted the model imperfectly. For instance, the 1984 version could not take account of OSI's presentation layer because this work was being concluded in parallel, and P2 turned out not to fit the OSI protocol definition. The subsequent adaptation to OSI was problematic, as it required divergence either from the 1984 recommendation or from the OSI model. Further technical problems were a consequence of the very ex ante nature of standardization. Only the ensuing implementations could reveal mistakes and technical problems in the standard, leading to revisions

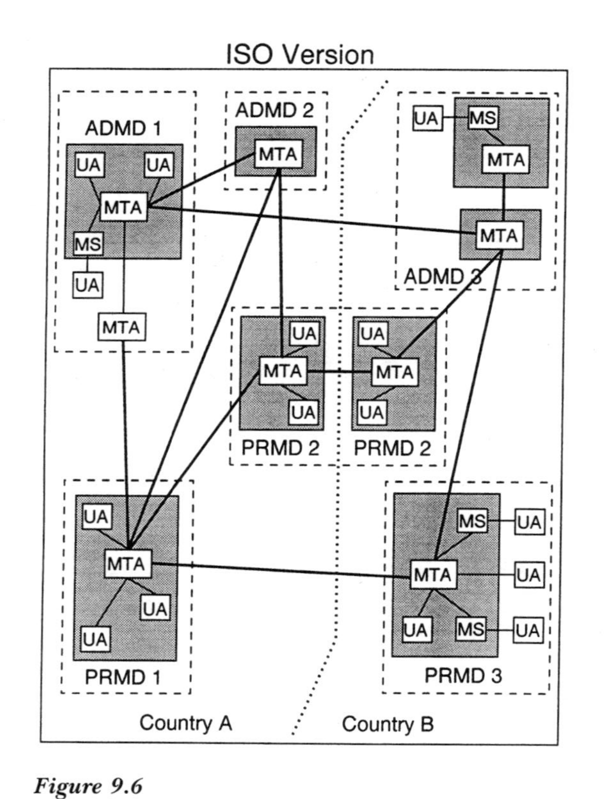

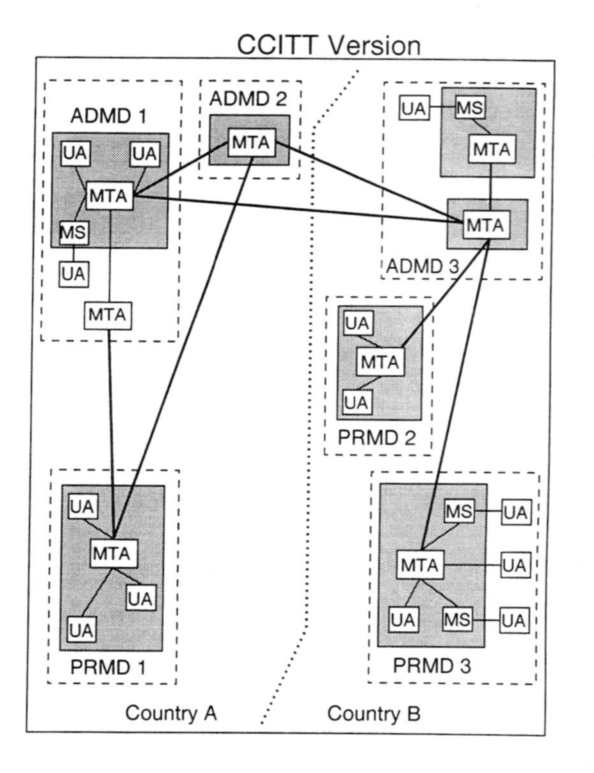

Figure 9.6
X.400 links between management domains. UA: user agent. MTA: message-transfer agent. MS: message store. ADMD: administration management domain. PRMD: private management domain.

in the implementer's guide. Such a late exposure of technical problems also occurred in the 1988–1984 interworking, where implementations revealed at a later stage that compromises made for the sake of compatibility in the 1988 version had failed to provide complete compatibility with the 1984 version.

The loose coupling achieved through the separation of MTA and UA and the very ex ante nature of the standardization project were instrumental in circumventing many potential technical and economic problems. Only the first implementations introduced economic constraints, which were fed back into ongoing standardization. Because MHS was the first implementation of the OSI application layer, and because of the very abstract and complex nature of OSI in general, MHS required significant initial investment. This made the stability and the backward compatibility of MHS standards even more important.

Owing mainly to the far-reaching conceptual work already completed in the IFIP, the standardization of MHS progressed surprisingly smoothly in view of the technical complexity of the subject. "The work on Message Handling has progressed far faster than I, as Rapporteur, could have possibly hoped for," wrote Ian Cunningham (1983, p. 1430). Nevertheless, several fundamental problems remained part of MHS standardization, giving rise to the few conflicts encountered in the process. Not only was X.400 the first standard for the upper layer of OSI; it was also a project involving the converging technologies of telecommunications and computing. Moreover, it was probably the first standardization process equally concerned with public and private networks and with the resulting need for the CCITT and the ISO to manage a joint standardization process. And this process of organizational adaptation took place in an unstable environment where increasing liberalization altered the demands imposed on the shape of the technical system.

Actors and Interests

The background to MHS standardization was provided by the development of plans for office automation and for the office of the future, which merged with the interests in nonproprietary, global electronic mail systems. With these underlying ideas, X.400 paralleled videotex standardization somewhat; however, in contrast to the very short-term expectations held of an internationally extended use of existing systems in the case of videotex, X.400 was conceptualized as a long-term project. From the beginning, it was planned as an implementation of the newly developed OSI model, and there was a consensus among participants on the need to specify a totally new system (since all existing proprietary systems were too limited). In a foreword to a book on X.400 published by one of the standardization activists (Manros 1989), James White, another activist, distinguishes a tactical and a strategic aspect of X.400 within this long-term perspective. The tactical aspect refers to the potential to interconnect proprietary systems via the MTS (message transfer system); the strategic considerations aim at the complete implementation of the protocols of the X.400 series, including the IPMS (interpersonal messaging system).

Owing to the very long-term character of MHS standardization, only indirect commercial interests were present initially. Standardization was regarded as an aspect of research and development, and consequently most of the people who were sent to assist its development were from research departments. At the more open IFIP, and later at the ISO, many of the participants were academics who had gained their initial experience with electronic mail in the context of ARPANET and who believed strongly that only a nonproprietary electronic mail system could achieve the necessary critical mass to make the service a success. Long-term commercial interests were coupled with significant elements of scientific and professional interests in working on the project.

The beginning of MHS standardization was thus characterized by a very favorable constellation of actors and interests: a long-term, ex ante standardization based on the joint insight of the insufficiency of existing systems, together with the need for a clearly new system, based on OSI, which no company or country would be able to realize by itself without prior cooperation in standardization. There was enough long-term commercial interest to delegate experts to standardization; however, until the first implementations of the 1984 X.400 version appeared, toward the end of the next study period, delegates were given little precise direction. Standardization was only loosely connected with firms' strategies, being legitimized predominantly by long-term interests, alongside the knowledge that other important firms were similarly involved. Delegates had plenty of scope to realize individual and professional interests in standardization within the broad confines of the delegating firms. Experience gathered from existing proprietary systems was used as a basis for contributions, but this was restricted to overall architectural knowledge and never went as far as to push for specific protocols.

Besides the shared desire to cooperate in a long-term project, few of the firms adopted identifiable positions. The firms favoring proprietary systems, for the most part, did not involve themselves at all in standardization, as it was obvious that the choice would not be one of a proprietary system competing with X.400, but rather whether a limited proprietary "niche" strategy or the global OSI strategy would be more profitable. IBM was an exception here in that it did participate passively, interested in observing the course of the work, but not interested in contributing actively toward furthering the obsolescence of its proprietary system, as one delegate put it. This underlines that IBM saw X.400 not as a complementary, but as a substitutive system to its own. More noticeable than in the technical work, IBM became active at a higher level, when it tried at a plenary meeting of SG VII, toward the end of the 1985–1988 period, to push through the agreement to transfer MHS standardization from the CCITT to the ISO, where IBM's own influence was much stronger. However, in view of the fact that standardization had progressed this far, IBM's late activity was in vain.

Similar opposition to IBM's had occurred earlier in the CCITT from the side of the teletex supporters. Initially, Germany and Sweden were hostile to MHS standardization. But Germany soon switched to a policy of cooperation, and the Deutsche Bundespost delegate was said to have helped the Canadian SpR chairman with his diplomatic skills and to have shared some of the knowledge he had gained through his longer experience with the CCITT. Sweden, however, articulated renewed opposition toward the end of the study period, when the eight recommendations for MHS were endorsed. Here Sweden reserved its position on MHS: "The Swedish Administration does not see Message Handling Services as completely separate new services, but as complementary services to other CCITT services such as telex, telephone or telematic service. It is therefore our position that the Message Handling Services should be defined by the CCITT Study Groups responsible for the basic services. . . . We also think that further work in this area should be done in one Study Group, preferably Study Group VIII, within the present structure of CCITT." (COM VII-R 22(A)-E: 13, 1981/84)

Besides these examples of attempts at hindering MHS standardization, there were very few instances where specific actors had certain positive stakes to realize in the standardization process. France Telecom, one of the few remaining carriers organizationally merged with a postal service, was highly influential in the specification of the physical delivery option for X.400 in the 1985–1988 period, owing to its interests in the automation of postal services. Other participants being indifferent to the subject, France Telecom was successful with the proposal. A more recent example concerns AT&T, which in the study period 1989–1992 was instrumental in the standardization of a voice-messaging content-type addition. Again, other participants were disinterested. This demonstrates, of course, a general feature of standardization: it is necessary to either receive support or be met with ignorance for successful standardization to occur.

The last examples illustrate exceptions to the rule that joint problem solving was the dominant characteristic trait of the standardization process, where all those active developed a common world view on global MHS. As the first SpR noted, "At one point the group almost became stronger than any outside influence." Maybe this strong group cohesion had also restricted the attention for developments of message handling in the ARPANET.

Among the actors there were significant differences as to the degree to which they influenced standardization. The most important actor up to 1988 was Bell Northern Research, which had already managed to optimize its influence through the decision to start work within the IFIP (a small circle in which ideas could be developed and propagated). BNR initially combined Northern Telecom's interest in office automation and the newly converging technologies with Bell Canada's interest in message handling. A proprietary system, ENVOY, had already been implemented but was regarded as insufficient in the long run. Soon, Northern Telecom withdrew its plans to become active in office automation and concentrated instead on its core business, telecommunications. From then on, Bell Canada was the driving force behind BNR's activities in MHS standardization. This interest declined once Bell Canada had implementations operative, in addition to the general slight disappointment with X.400 diffusion in the late 1980s.

Another central actor was a delegate from Xerox who became a special rapporteur for MHS in the 1985–1988 study period and who was the originator of the ASN.1 notation syntax and some other important contributions, for instance on the principles of remote operation. Xerox at the time had a proprietary message system, and one goal of standardization work was to see that the new MHS did not develop in such a way that it would be difficult to combine with Xerox's XNA architecture. However, the fact that the delegate in question changed jobs from Xerox to 3COM while continuing his work on X.400 points to the involvement of more indirect strategic corporate stakes, and to his significant personal commitments. The definite instrumentality of Xerox for some provisions of X.400, moreover, never led to it adopt a significant position in existing implementations, a point equally true for Bell Canada and Northern Telecom with regard to BNR's involvement.

A further example of the impact of specific actors is Siemens, which engaged in the definition of the message store (MS), initially in the ECMA and then (in the 1985–1988 study period) in the CCITT. Again, there was certainly commercial interest in having the MS addition to X.400, but for Siemens it might have sufficed for any other participant to do the work on the standard. France Telecom, in this case, was a vociferous opponent of the work, on the basis of a very early implementation of the 1984 version.

One notable feature of X.400 was an unusual degree of heterogeneity among participating actors. Telecommunications and computer manufacturers represented divergent stocks of technical knowledge from the different technological fields. Moreover, private networks were just as relevant to standardization as the public networks of participating PTTs and RPOAs. This unusual constellation rarely resulted in explicit confrontation between the respective views on MHS standardization. One significant reason for this success was the early achievement of a comprehensive systems concept, elaborated in a homogeneous group at the IFIP, and its fortunate transfer to the CCITT. In the ensuing work, an orientation toward problem solving and cooperation generally overcame the potential for conflict, and different groups of actors seemed, to some extent, reluctant to push their particular interests. Thus, administrations were interested in the participation of computing firms as a way of attaining a common standard for public and private networks. Telecommunications manufacturers such as Siemens, combining interests in both fields, were hesitant to become too prominent because of their peculiar relationship to the German network operator, which had primary importance in company policy. Moreover, there was some understanding among the groups as to their different spheres of competence: network operators were more influential in designating the functions to be provided, manufacturers were more prominent in doing the necessary technical work.

Within the heterogeneous actor constellation, the main basis for antagonistic interests was the difference between the academics predominantly working in the ISO and the participants with commercial backgrounds. In cases of disagreement, the latter group disputed the idea that academics could have legitimate interests to advance, in view of the fact that the academics' interests were professional or scientific more than economic. Yet these differences only appeared at a time when initial implementations represented a basis for more specific commercial interests, which were difficult to combine with the technical perfectionism of the academic side.

With the wisdom of hindsight, many of the participants in the X.400 effort doubt that such a large project could be repeated in the present climate of ongoing liberalization and internal organizational restructuring into profit centers, where investments in standardization have to be directly accountable. Standardization of X.400, as a facsimile manufacturer who was also involved in message handling put it in one of our interviews, was regarded at the time as something to be undertaken "as a good citizen of the international community," akin to what was noted in the context of the facsimile case, namely that "to have an SpR is just like having a race-car in the automobile sector."

Minor and Major Conflicts

In view of the size and the novelty of the MHS project, there was a noticeable lack of conflict during standardization, commensurate with the scarcity of specific strategic interests, the dominant cooperative orientation, and the ex ante nature of the work. The few conflicts that did occur were relatively minor and failed to reflect the intensity of the videotex and the Group 3 conflicts. All but one of these conflicts were, to some degree, connected with the organizational co-involvement of the CCITT and the ISO.

The first conflict occurred during the 1981–1984 study period and evolved out of an internal US context. The National Bureau of Standards, developing standards in the area of information processing, had started work on MHS standardization (Wilson 1984, pp. 40–45). The NBS (today known as the National Institute of Standards and Technology) is an agency of the US Department of Commerce. The NBS issued the Federal Information Processing Standards (FIPS), which were to be used by other federal agencies in order to achieve systems compatibility (des Jardins 1983). FIPS corresponded to national or international standards, and they often specified features that were left open in those standards. Preparation of FIPS in message handling in the early 1980s was aimed at feeding them into the CCITT process. The conflict arose from work done for the NBS by Bolt, Beranek and Newman—a company that, as we have already mentioned, had been much involved in the ARPANET work, also on the basis of government contracts. A standard related to message formats and encoding, already approved by the NBS as FIPS 98, ran counter to the work done on ASN.1 by Xerox in the CCITT context. The difference between the two proposals lay chiefly in the degree of abstraction. The work commissioned by the NBS was already related to specific applications, whereas the Xerox contribution was, roughly speaking, concerned with overall architectural principles. Within the US delegation, these different proposals led to a memorable confrontation during a CCITT meeting in Virginia hosted by Telenet in the fall of 1982. The BBN project leader who represented the NBS in that meeting found that the Xerox approach, which was very much in line with earlier work done in the IFIP and the CCITT, received much stronger backing from the other members of the national delegation. Political rearguard action later in Washington did not alter the situation. Rather, it mobilized the newly created Electronic Mail Association, in which Telenet and some other commercial carriers officially supported the majority position in the CCITT. Eventually the NBS withdrew its proposal, and in consequence the agency preferred to resign completely from standardization in the CCITT rather than to change its position on this specific subject.

The difference between the ISO and the CCITT manifested itself in all the other conflicts. The parallel standardization of message handling by these two organizations created problems of coordination. Coordination was essential in avoiding the promulgation of two incompatible international standards for MHS, which could affect the future diffusion of the system beyond the mere duplication of effort. In practice, the coordination of a joint strategy leading toward standardization was not easy to achieve. There were organizational differences in working procedure, and particularly in formal decision making, because the CCITT recommends a standard after 4 years whereas the ISO holds a ballot whenever the standard is ready (see chapter 3 above). Moreover, the two organizations approached MHS standardization from very different perspectives, and were supported in this by their different memberships. The CCITT standardized with a view to telecommunications, public networks, and international services; the ISO came from the side of computing and private networks and tended to focus on terminals and products instead of communications. This implied differences in regard to the need for standardization. For the CCITT, standards were needed where international telecommunications depend on compatibility. "The work of CCITT on Message Handling Facilities concerns only the transfer of messages across Management Domain boundaries." (Cunningham et al. 1982, p. 157) For the ISO, the rationale of standardization was not as focused; this was exemplified most clearly in regard to the question of a standardized extension to the message store, which we will discuss below.

Manifest difficulties started soon after each of the organizations took active notice of the other's work (around 1986, when the ISO took the first ballot on certain draft standards of its Message-Oriented Text Interchange Systems).[14] These difficulties grew out of the finding that some protocols resulting from the ISO's work were incompatible with the 1984 version of X.400. All conflicts arising from the parallel work reflected different aspects of the problems connected with the standardization project.

The first conflict concerned backward compatibility with the 1984 version of X.400. Having recommended a standard, the CCITT was bound to maintain compatibility with it in the versions that followed. The ISO had no such existing commitments; what is more, it voiced its concern about shortcomings of 1984 X.400, such as the lack of proper consideration for the OSI presentation layer (to which the ISO was committed) and the problems resulting from the barely adequate attention paid to future extensions (notably P1 in the 1984 version). Each of these shortcomings gave way to a compromise on technical perfection in the 1988 version of X.400, which the ISO was not prepared to consent to (Craigie 1988–89).

Consequently, there was the danger that two sets of incompatible international standards might exist should no compromise be reached.[15] Hence, the ensuing discussions between the ISO and the CCITT were difficult. From the CCITT's point of view, a "latecomer" was suddenly putting several years of effort into jeopardy, while the ISO group was not content to be treated like a junior partner. In view of existing implementations, it was inconceivable for CCITT participants that representatives of commercial firms at the ISO could question the need for backward compatibility, since the avoidance of uncertainty is a major factor in the successful start of large technical developments. Early revisions and incompatibilities, it was believed, would kill off both the expectations of and the existing commitment to the system. The CCITT's resistance to change was interpreted by the other side as typical PTT arrogance. The controversy did not only center on the question of existing commitment versus technical perfection; it was closely linked to the different memberships of the two groups.[16] The university people in the ISO were not seen to represent legitimate commercial interests, and the ISO participants representing firms were believed not to act in the best interest of their delegating organization.

Often the conflict transcended national boundaries. In the United Kingdom, many academics participated in the ISO, preferring to change the standard immediately rather than live with technical mistakes forever, whereas British Telecom was strictly against deviating from the 1984 recommendations. As a result of the difficult internal British discussions, the UK coordination groups for the CCITT and the ISO were merged in order to avoid any future controversy. Moreover, conflict characterized the balloting on the MOTIS standard that took place in 1988. Several parts of ISO standard 10021 did not receive immediate approval, and in the majority of cases the demand for compatibility with the CCITT recommendations was reflected in the comments received by the national members. In the end, there was no solution to the conflict. Instead a compromise was reached: the CCITT recorded the 1984 recommendation as mandatory and the 1988 version as optional, and the ISO wrote down its 1988 MOTIS version as mandatory and the 1984 X.400 as optional (Willmott 1989, p. 6).

Another conflict took place in the 1989–1992 study period, in the now "co-located" meetings. This concerned proposals for additions to the new message-store standard. It was mainly participants from the ISO side who worked on the respective contributions to extend the message store (MS) by providing it with the capability to file incoming messages in different folders and thereby allow the receiver to selectively retrieve messages according to subject instead of on a "first come, first read" basis. Academic participants from the United Kingdom contributed greatly to the development of this proposal, which also received the backing of delegates from Norway, the Netherlands, Australia, and Israel. The chief opponent of the proposal was the delegate from Siemens, who had originated the work on the simple MS for the ECMA and the 1988 X.400 version. As a special rapporteur, he was in a good position to block the work. Some IBM delegates backed him, as did the German and US delegations; France, Japan, and SG VII in general were rather in favor of the new development. The supporters of the MS extension identified IBM and Siemens as the main opposition and attributed this to vested interests of both firms in an existing standard for document filing and retrieval (DFR) that was not part of X.400 and could be used as an alternative. However, the CCITT participants tended to see the conflict as a further example of different approaches to defining the desired scope of standardization.

The CCITT, which was fundamentally interested in telecommunications services, did not deem the MS extension for DFR relevant to an international standard. The interest in this issue was considered to reflect the ISO's focus on terminals and private networks rather than on communications. However, it did not matter for international network compatibility whether these problems were solved in a standardized or a proprietary way, either on the network side or the terminal side. A proprietary solution was preferable in the interest of the utmost stability of the standard. Continued enhancements to the recommendations put existing implementations in jeopardy, since users were likely to require implementations to follow the most recent set of recommendations, and these commercial interests were only reluctantly renounced in favor of additions that were unnecessary from the point of view of international service. Therefore, for the CCITT, this conflict again involved aspects of misjudgment by the ISO group as to the commercial requirements of standardization.

The last conflict we wish to discuss focused on the distinction between the Administration Management Domain (ADMD) and the Private Management Domain (PRMD). The designation of the ADMD as part of the address necessary for correct international routing imposes considerable disadvantages on international private networks. It implies that all transborder communications should be routed via the respective national ADMD rather than handled directly by the transnational private network, or between two PRMDs, thereby incurring the cost of unnecessary use of the public network (figure 9.6).

This simple approach to the routing problem reflected the view of the prevailing public telecommunications monopolies during the first study period (1981–1984) of work on X.400 in SG VII. At the same time, there was a broad consensus in SG VII that the group's work should not predispose regulatory positions. The members of this group knew that the standard was meant for both public and private networks, under very differing regulatory conditions, and that otherwise implementation would be hampered. In view of pending work on directories (X.500) and other constraints, the specification of an ADMD indication was thus chosen as a relatively simple and temporary solution (Cunningham 1983, p. 1429). Simplicity obviously does not signify that nobody was to gain from it. For the PTTs this was certainly a beneficial solution. But there was also the technical reason that internationally compatible public networks (ADMDs) were likely to be more easily interconnected than different PRMDs.

With ongoing liberalization, growth of private networks, and the ISO's involvement, such an approach could not go unchallenged. In its work on MOTIS, the ISO objected to having to distinguish ADMDs from PRMDs; it preferred to speak only of MDs (management domains). Thus, the ISO did not include any structural differences in the terminology of ADMDs and PRMDs, but allowed for all interconnections. Such a liberal approach had problematic consequences for the CCITT, since the ADMD specification did have a function. As a result, SG VII vehemently opposed the ISO's design in the 1985–1988 study period. For the ISO, as for many non-PTT members of the CCITT, this opposition was a matter of incumbents' resisting the loss of privileges, since that the ADMD provision guaranteed a lot of traffic. With increasing national liberalization and with changes in national coordination, opposition in the CCITT to direct transnational interconnection could not be maintained by the PTTs.

Although opposition declined, the problem was not directly tackled by the CCITT participants. Most contributions in this regard originated from the ISO side of the co-located meetings. Proposals to standardize routing in MHS, in order to release addressing from the routing function, implied too fundamental a change of existing concepts and implementations and was thus unacceptable. Another proposal forwarded by the ISO group, and by the UK university delegates in particular, was the "single space solution," which provided for leaving the space in the address for the ADMD blank. Messages would then be routed over different MDs, without favoring the incumbent.

This proposal was implemented in the UK (apparently after large petrochemical users had lobbied to break British Telecom's opposition), but it did not solve all the technical problems. The information normally given by the ADMD indication would now have to be provided in another way, requiring all PRMDs to have a list of all other PRMDs in the country—and, at worst, where no PRMD was indicated, possibly even to have a list of all registered users in other PRMDs—in order to be able to route the messages to the right addressee.

In the CCITT, where the Deutsche Bundespost had earlier threatened that it would not transfer any messages originating or terminating in the Federal Republic of Germany without specification of the ADMD, opposition to the ISO proposals declined. In this context, it should be understood that for the big national operators the privilege of having all the ADMD traffic was not the only stake: there were the steadily rising interests of transnational carriage of MHS, for which the ADMD provision was a disadvantage. However, no attempt was made to solve the problem; rather, it was stated that these were regulatory matters to be solved outside the CCITT in the national context. In view of the differing regulatory positions of member states, it was impossible to find a consensus in the CCITT. Therefore, the CCITT preferred to claim no responsibility for the direct interconnection of private networks and PRMDs, since it only standardizes public networks.

Yet the conflict did not relate only to different areas of competence between private and public networks. It was again interpreted as an issue where the ISO did not correctly assess the requirements the CCITT had to meet, since tariffs were not considered at all—and here the question of whether to route among different PRMDs or ADMDs was indeed relevant. This type of problem was ignored by most of the ISO delegates, whose experience stemmed either from academic networks or from local-area networks in which tariffing (and, often, complex routing) was no issue at all. The whole debate about ADMD and PRMD, in fact, nicely exemplifies the fact that the incumbents in telecommunications often pursue exclusive solutions to coordination problems—problems that, with liberalization, necessarily become subject to more comprehensive standardization.

Conflict Resolution (Termination)

There were very few conflicts in the case of X.400 standardization, partly because of favorable circumstances: whereas videotex was embodied in different national systems, MHS was seen international from the start. If it was to realize global communication and interoperability, it could not be designed by a single company or a few companies. The standardization process was also smoothed by conscious efforts to evade opposition: the way the IFIP conceptual model was introduced in the CCITT in the face of initial resistance was strategically planned by the initiator from BNR. Consequently, it was possible to start work in the CCITT on the basis of an existing consensus on fundamental principles of X.400, which included the common perception of the work as highly abstract and ex ante. The internal US conflict with the NBS illustrated the dominance of this approach.

There are also more specific examples, in the case of X.400, of potential opponents' being approached strategically. From the beginning, teletex-X.400 interworking was placed on the agenda as a way to counter the disapproval of the teletex supporters. Interworking facilities could be interpreted as an indication that MHS was meant less as a substitute for teletex than as a complement to it.[17] The apparently conscious efforts not to involve SG I had the similar aim of evading potential difficulties with the European PTTs, which had a monopolistic perspective on service and were rather unfamiliar with computer communication. Likewise, the CCITT group was not unprepared for the incipient co-involvement of the ISO. Fear of a competing committee standard, rather than of a proprietary firm standard, was one reason behind the attempt to have a first set of recommendations ready as early as 1984. Existing standards were a legitimate basis from which to be relatively uncompromising vis-à-vis the ISO's suggestions, and they were a structural advantage for the CCITT in all discussions. For the successful management of the standardization process, the choice of the special rapporteurs was important. Although rapporteurs changed every study period, continuity in their work was preserved as each SpR managed to be succeeded by someone who had a proven interest and capability in MHS standardization and who shared a similar outlook.

Another important concurrence in the productive standardization process was the maintenance of a small and homogeneous group of active participants who were highly committed to the progress of the work. Steadily broader support could be gained for their developments, since they would not be impeded by the heterogeneity of perspectives and interests the standardization process touched upon. Even in cases where the differences between public and private networks or telecommunications and computer manufacturers became relevant, the factions put forth predominantly moderate positions.

Of the few conflicts that did occur, the one over backward compatibility with the 1984 recommendation of X.400 was the most serious. Identical texts of the theoretically quite compatible CCITT 1988 recommendation and the respective ISO standard could not be achieved, but significant incompatibilities were prevented because the CCITT played out its dominant position regarding competence and legitimacy in this technical field. Although unfamiliar interorganizational negotiations were required, the way the conflict over the MS extension was dealt with was a typical CCITT approach. The actors who opposed the extension tried to gain time by simply deferring a decision. Although they could not smother the initiative for additional specifications, they succeeded in delaying the work into the next, post-1992 study period, hoping that supporters would lose interest and discontinue the work.

Interpretation and Generalization

10

Standardization as Coordination

After introducing our theoretical approach of actor-centered institutionalism, we began our analysis of the processes of international standardization in telecommunications with a discussion of the links between standards and technical development. We then examined the institutional, organizational, and procedural aspects of standardization and developed a general model of the standardization process based on our approach. We continued by taking a closer look at the CCITT and its relation to regional and national standardization bodies, and then turned to our case studies of the standardization of videotex, facsimile, and message handling. In this chapter we shall try to selectively conflate the different elements of our work into a more comprehensive picture of technical coordination through institutionalized international standardization.

Telecommunications technology can be regarded as a prototype of a network technology whose components can be assembled to form a worldwide system. No central authority exists to design, construct, and operate such a system in a coherent way. The components must, however, be able to interoperate if communication between two terminals at any place in the world is to be possible. Consequently, a multitude of actors need orientation in shaping their activities in order to comply with the system's requirements. This does not mean that the requirements must be taken as given and unchanging. Neither does it necessitate a central plan or some kind of common spirit among the actors interested in building up a universal telecommunications system. Rather, self-interested actors—as designers or producers of technical components and subsystems, as contractors constructing parts of the network, or as operators, carriers, service providers, regulators, or users—are capable of contributing to the development of a large technical system. Compatibility—a collective good—plays a key role in getting the technical system to work without friction. Achieving compatibility requires coordinated action.

Technical standards provide clues to the design of the relational properties of technical components that are relevant to compatibility. Thus, standards help coordinate the development and operation of technical systems. Standards have gained significance with the growing demand for telecommunications services (especially at the international level), with the growing number of actors involved, and with the new technical opportunities arising from the basic innovations in computer technology and optoelectronics. As long as telecommunications was nationally contained and monofunctionally confined to telephony and telex (basically electromechanical technologies), it was hierarchically organized and coordinated. Standards were of little importance.

The costs of coordination are transaction costs—that is, costs which are not directly related to the production of goods and services but which involve the costs of reaching and safeguarding agreements on contracts along with more general costs of information (Williamson 1985, pp. 15–42; Alchian 1984). Producers, vendors, and users of general-purpose technology face lower transaction costs than those committed to special-purpose technology. Many components of telecommunications systems are essentially special-purpose devices that can be used only in combination with complementary components. Thus, knowledge is needed about the features of the components that are relevant to their interoperation. Compatibility standards will provide and generalize this knowledge and will, as a means of coordination, reduce transaction costs.

Parts of these costs, however, are only transferred from the stage of production, operation, or use of technical components to the process of standard setting, thereby affecting perhaps a different circle of actors, who for their part enjoy the privilege of directly influencing the shape of a standard. Standards organizations with routinized working procedures and stable rules of membership and decision making provide an environment that reduces the transaction costs of standardization because the basic conditions of each standard's production do not need to be renegotiated. Standardization organizations thus coordinate the process of standard setting. However, as with the standards themselves, the opportunities and constraints established by the institutionalized rules of these organizations do not affect all actors in the same way.

Standardization as a Special Case of International Coordination

Our case studies focus on international standardization. Traditionally, in many countries, telecommunications was operated by public authorities and enjoyed a monopoly position. However, the old order has been eroded, and national governments are beginning to open telecommunications markets to competition, confining themselves to a regulatory function. Still, the political significance of telecommunications in the areas of infrastructure, technology, and industrial policy has obviously remained high. The International Telecommunication Union's Standardization Sector (ITU-T), formerly the Comité Consultatif International Télégraphique et Téléphonique, reflects the political view of this sector. Standardization appears as the coordination of nations in a specific technical sphere. It is a core activity within the international telecommunications regime. The principle of national representation in the CCITT/ITU-T, which concedes to every nation—be it the United States or Luxembourg—a single vote, is one basis of this regime, which can be designated an international negotiation regime (Scharpf 1993b). Members of national delegations engage in negotiations which are institutionally structured by cooperation norms, by the expectation of open information exchange, and by the exigency of consensus if not unanimity. Where (as is typical for international regimes) no effective policing or sanctioning power exists, the attainment of a broad consensus is the appropriate way to achieve compliance with the results of negotiations (Haggard and Simmons 1987; Young 1992). The technical nature of matters in the CCITT/ITU-T may, in general, make it easier to attain this consensus. By no means, however, does it guarantee this result.

Negotiating standards among representatives of nations opens the door to political considerations which are not directed at technical questions of compatibility but which tend to regard a national standard as a strategic element in the global competition among nations. This suggests that relative-gains orientations prevail among the relevant players—that is, the actors not only want to improve their position; they also want to gain more than the others. Grieco (1990) encountered such orientations in his analysis of the GATT Tokyo round of talks on nontariff barriers to trade (including technical standards). Although the removal of barriers to trade was generally seen as a means of increasing economic wealth in all countries, in the negotiations the potential positive-sum game was transformed into a zero-sum game when the desire to receive a relatively higher benefit than the others determined the strategies of the actors involved. From Grieco's findings we can conclude that the more standards become a matter of national prestige or an element of a national industrial policy strategy, the more salient are relative-gains orientations. When nations regard standards as a means of gaining competitive advantages, antagonistic negotiations will tend to be the consequence.

The standardization of videotex provides an example of this. Complete national systems were submitted to the CCITT for approval as standards, and the central focus was on standards for terminals. During the 1970s and the early 1980s, in Europe especially, the systems were developed in a close symbiosis of political and economic actors in the hierarchically structured constellation of network operators and manufacturers, including, in this case, some "outsiders"—such as the producers of TV sets, who traditionally had not supplied the PTTs. The inclusion of these outsiders very much restricted the delegations' leeway for compromise with others. In contrast to the traditional "court suppliers," the outsiders' cooperation in the national projects was based not on enduring business relations with the PTTs (which would imply some degree of dependence) but on more specific contracts and mutual agreements. Thus, as we have shown in detail for the United Kingdom, some countries had practically lost all flexibility at the international level. Videotex, a large project, had reached such a high degree of institutionalization and commitment on the part of the initiators in these countries[1] that a disapproval of its central specifications as standards by the CCITT would have been perceived as a considerable loss of national technological reputation and of export chances for the national industry. In the case of France, political sensitivity even motivated an unusual move in the Plenary Assembly, the "political" level of the CCITT, in 1984. France tried to counteract a compromise achieved by the responsible working party of Study Group VIII, in which French delegates had participated and had not objected to the undertaking. The compromise included the North American NAPLPS standard, the Japanese CAPTAIN standard, and the European CEPT standard as options. CEPT itself comprised various options, one of them being the French Antiope. Thus, Antiope was part of the recommendation (T.101), though not as directly visible as in the old recommendation (S.100). The mere threat of a symbolic disadvantage mobilized French resistance, but by this stage the majority of countries was unwilling to concede and ruled in favor of the new recommendation (although, as an exception to the rule, unanimity of adoption had to be sacrificed).

International coordination of technology-related matters does not always involve politics, and when it does relative-gains concerns do not necessarily prevail (Ancarani 1995). These concerns must be treated as a variable, directing scholarly endeavors to specify the historical context and the systematic conditions under which they emerge (Jervis 1988). Moreover, as Mastanduno (1991) has shown, relative-gains concerns can be present without affecting policy. Mastanduno's case in point is the US government's economic policy toward Japan in the area of high-definition television (HDTV). Since 1985 it has been observed that there is a general tendency on the part of US political actors to seek relative gains in international economic relations, especially vis-à-vis Japan. Despite this general tendency, the fragmentation of the policy process, with different agencies within the executive branch pursuing different institutional missions, has prevented the United States from setting up a coherent industrial policy outside the military sector. In the matter of HDTV, the agencies committed to the principles of international free trade and the nonintervention of the state in the economy successfully blocked far-reaching measures in favor of a national initiative (Mastanduno 1991, pp. 101–112; McKnight and Neuman 1995).[2]

The lack of a coherent national position can indeed hamper political intervention in the standardization process even when relative gains are at stake (OTA 1992, pp. 15–18). Not only may political agencies hold divergent opinions, but often competing national companies are unwilling to compromise for the sake of an abstract increase of national welfare. Some marginal evidence is provided by the case of message handling standardization. Parallel to the activities in the CCITT, the US National Bureau of Standards (NBS) was involved in standardization. Contributions commissioned by the NBS were intended for submission as US proposals to the CCITT once they had been adopted internally. Since it transpired that the initiative only mobilized opposition from some of those who were simultaneously active in the CCITT work, the NBS retreated from MHS standardization.

At the technical working level of the CCITT, the actors tend to blame politics when difficulties in reaching a consensus arise. Where the economic motives of rival manufacturers' agents or even the rather idiosyncratic tactics of individual delegates may have caused the trouble, the suspicion that some hidden or even conspiratorial political strategy is at work easily emerges. We found an example of this in the conflict over the Group 3 digital standard for facsimile. Opponents to this proposal, which was seen as threatening the existing Group 4 standard, speculated that the US Department of State, backed by the Department of Defense, was pushing the standard—with help from a consultant to AT&T—in order to put the Japanese manufacturers in a state of uncertainty. Proponents suspected that Japan would defend the Group 4 standard because the Group 3 standard was regarded as a threat to Japanese manufacturers' monopoly in the market for facsimile machines.

Two conclusions about the role of politics in international standardization can be drawn. First, since the dissolution of the hierarchical national order in telecommunications, in which either a postal, telephone, and telegraph administration or a recognized private operating agency clearly dominated, defining unambiguous national positions has become much more difficult.[3] Only when complete technical systems of collectively perceived great technical and economic significance are at stake is it possible to politicize standardization. Second, once achieved, a national position is not very flexible, because intended deviations from a pre-set stance necessitate time-consuming internal renegotiation. Therefore, in a two-level game, with a homogeneous national position and/or a strong PTT, the national delegations in the CCITT have considerable flexibility for international agreement. The more heterogeneous the national actors, and/or the weaker the position of the PTT, the narrower is the scope for the international agreement. As flexibility decreases, political intervention confines itself to monitoring standardization and to exercising what, in a different institutional context, has been called "negative coordination" (Scharpf 1994; Mayntz and Scharpf 1975). If a proposed standard is regarded as interfering with a national interest, the affected country tries to delay or to completely block its adoption. If blocking an opposed proposal turns out to be impossible, the country will at least try to get its own solution adopted as one option of the standard. This sometimes results in CCITT recommendations comprising several incompatible options in a single standard. For an organization whose institutional goal is parsimony and internal consistency of standards options, this can well be seen as a functional variant of obstruction. Terminating a conflict through the adoption of incompatible options, however, keeps the organization viable. Otherwise controversies would be perpetuated ad infinitum, restraining the actors from turning to new standardization tasks.

Politics remains blind to the technicality of standards. It is activated when standardization (most likely of a comprehensive technical system, such as videotex, not of isolated components), has gained considerable national political significance. As these situations are rare, though always spectacular, politics typically remains in a monitoring role. Occasionally, if at the technical working level of the CCITT no consensual solution can be found, politics helps to achieve an agreement precisely because it does not consider technical details. This facilitates the trading of concessions among the several standards to be decided upon, much as in other political package deals. When politics matters, some smaller countries support certain other countries even though the constellation of interests within their delegation would imply a neutral position.

Standardization as a Special Case of Interorganizational Coordination

With the gradual transformation of the old PTTs and monopolistic RPOAs into "ordinary" business organizations competing with others at both the national and the international level, the role of politics in international standardization has changed and the need to coordinate technology-related activities has increased. The partial convergence of telecommunications and data-processing technology and the concomitant multiplication of the number of actors able to offer telecommunications technology and services have contributed even more to the increased need of coordination, especially since IBM has lost its technological hegemony in data processing.

This development suggests that the need for coordination has shifted from the coordination of nations to the coordination of (mostly business) organizations (Genschel 1995). From an institutional perspective, the emergence of international standardization organizations of a more "private" character (e.g., ECMA, SPAG, EWOS), as well as several reforms of the CCITT and the ITU, can be attributed to this shift. However, the preeminent position of the intergovernmental ITU, with its ITU-T active in telecommunications standardization, indicates that the legitimacy and the normative significance of standards endowed through the political entrenchment of the standard-setting process are still important to the business organizations, not least because they increase the likelihood of implementation and compliance. As we have shown, this was one reason why in the case of X.400 the initial promoters of a standard for message handling chose the CCITT as the appropriate organization to issue the standard. CCITT/ITU-T recommendations are likely to be adopted by national standards organizations, and this often grants them a status similar to that of a legal norm.

In addition, many of the actors (network operators, service providers, manufacturers, major users) still have a national basis. Even most of the large multinational corporations have territorial roots (Porter 1990; Hollingsworth and Streeck 1994). Hence, we encounter a considerable similarity of interest between "home-based" multinationals with a clear national identity and "their" nation states. Where one manufacturer is the "national champion," as has been typical of telecommunications, that firm can count on the government's support, because strengthening the international position of this manufacturer lies in the national interest. Thus, an intergovernmental political basis of international standardization is well appreciated by a number of business organizations.

This, however, does not affect the need of the organizations to coordinate their technology-related activities with other organizations. Strategic alliances, joint ventures, horizontal and vertical integration (takeovers), and many other bilateral and multilateral measures of concerted action serve this coordination interest. So do technical standards, which leave considerably more leeway to the actors they address than do other modes of concerted action. Interface standards, in particular, demand only that conformity is secured at the interconnection points of subsystems or components. How the components are internally structured is not determined by the standards; it is a matter of individual choice. Standards allow for variety, which they constrain only when the total system's performance is at stake. Concepts of systems, including notions of adequate functionality, are not given, of course, but are themselves the results of constructive imagination, negotiated design, mix-and-match experimentation, or pure evolutionary accident.

Although international political coordination of telecommunications is blind to the technical details that are to be contained in a standard, for business organizations the question of producing and safeguarding effective functionality by means of technical standards may go as far as to affect their competitive position in the market. Thus, these corporate actors are compelled to involve themselves in the technical details. Moreover, technical scrutiny interacts with cost considerations, adding complexity to the standardization process. For manufacturers, operators, and users, a common aversion to the incompatibility of a system's components can generally be assumed. Without compatibility, components are useless and systems do not work. With compatibility, however, not only interoperation of complementary components but also competition of functionally equivalent and thus substitutive components is possible. This built-in tension shapes interorganizational coordination through standard setting. From this perspective, the standardization game oscillates between zero-sum conflicts and pure coordination, typically displaying a battle-of-the-sexes payoff structure.

Facsimile standardization provides many instances of conflicts of the battle-of-the-sexes type. Once the public telephone networks of most countries had been opened to facsimile traffic, compatibility was regarded as a central issue by all terminal manufacturers. This primarily required adapting the technology to the opportunities and constraints of the telephone network. Thus, all the standards and capabilities of the analog network optimized for the telephone service had to be accepted as given, and the de facto quality of transmission could not be changed by those designing and producing terminals intended for connection to that network. Accordingly, the network operators left most of the playing field to the terminal manufacturers so that they could coordinate the technical features of the facsimile machines internally. The different brands had started with diverging properties that rendered interoperation impossible, but the machines did not differ in all their components. Hence, in order to achieve compatibility, some components would have to be modified, whereas others could remain unchanged. For the actors, the theoretical decomposability of facsimile machines into distinct functional components—only loosely coupled with other components—kept the negotiations on standards from turning into an all-or-nothing situation. Moreover, controversies could not easily be politicized, because the (national) network operators usually played no decisive role and the diverging preferences of manufacturers did not coincide with national frontiers. American manufacturers in particular often had no common position. Only Japan generally relied on an explicitly defined national position, even at times when the international significance of Japanese manufacturers was low. In general, compatibility was seen as a necessary condition for the takeoff of facsimile technology. The potential benefits of cooperation aiming at compatibility clearly overshadowed relative-gains considerations. This was evident from the actors' tendency to use quasi-scientific methods of testing and experimentation when a decision on competing standards proposals had to be taken. Here the criteria of choice referred mainly to technical efficiency.

It was most apparent that different countries were involved and that facsimile standardization in the CCITT was not completely similar to comparable activities within a country's national standards organization when the specific characteristics of national telephone networks had to be taken into account. Most of these characteristics were regarded as lying outside the study group's responsibility, being a matter of adequate national adaptation to and implementation of the basic standards recommended by the CCITT. Others, however, such as the nationally different levels of transmission quality, stimulated additional efforts to respond to them by providing standardized optional features of facsimile machines. Standardization of an error-correction mode (ECM) for Group 3 machines, for example, evoked a certain amount of controversy in which organizational positions were partly congruent with national positions. Proposed by British subsidiaries of Japanese manufacturers, ECM was supported by firms and national delegations from countries with low-quality transmission networks and from countries with high-quality service concepts. In North America, where the networks were rather more reliable, opinions were split. Japanese manufacturers in Japan had already developed internally coordinated proprietary methods of error correction and therefore, unlike their British subsidiaries, saw no need for an international standard. Owing to the congruence of organizational and national positions, political lobbying became possible, and the British managed to convince the Japanese to surrender their opposition to an international ECM standard.

In message handling (X.400) we discovered parallels with facsimile standardization. The need for organizational coordination with respect to message handling was strongly felt in North America at a time when communication from computer to computer was unfamiliar in virtually all other parts of the world. This need found an early expression in the pre-standardization work of the International Federation of Information Processing, an organization in which membership was available to individuals and to organizations (including universities) but not to countries. Most computer manufacturers offered proprietary message-handling systems (MHS) that were incompatible with the products of other vendors in North America. None of these existing systems was used as a basis for the new MHS, which was supposed to become integrated into the OSI frame of reference. The intention was not so much to replace the existing systems with X.400 as to offer a new super system that would make communication between proprietary systems possible and which might render these systems obsolete only at a later stage. Thus, in many instances the work on standards in the CCITT resembled a joint development project which did not directly jeopardize the commercial interests of organizations running proprietary systems. As with facsimile, the interests of PTTs and other network operators were mainly concentrated on the ability of MHS to generate traffic for their packet-switched data networks. Hence, they were not involved in the details of standardization work early on, but they were eager to prevent an organization from designing a system of message handling that, by treating private and public networks as equivalent, would allow their networks to be bypassed. The initial distinction between administration and private management domains, and the reservation of all transborder electronic mail traffic for the administration management domains, which at the same time embodied a transparent technical solution to the problem of addressing and routing, served the network operators' economic interests.

Developing standards for worldwide computer communication was a very ambitious and complex undertaking with a long-term perspective. No single organization had the solutions to all problems ready to hand. The prevailing spirit of cooperation left little room for relative-gains considerations, and even the absolute return to be gained could not be properly calculated. All the participating organizations expected to reap considerable benefits from MHS. Features of pure coordination or even joint problem-solving characterized the X.400 standardization process.

Our case studies show that the coordination of organizations through standards characteristically combines technical and economic aspects, which can sometimes be distinguished only analytically. The criterion of technical efficiency is often employed when a choice between options for a standard has to be made. We found, as did Weiss and Sirbu (1990), that in this sense the standards that are clearly technically inferior are not accepted. But it is not always possible to determine efficiency unequivocally. A standard that would be optimal in the technical environment of one organization would be technically suboptimal in another. In addition, we find that some organizations place much more emphasis on technical perfection than others, which prefer less perfect, short-term solutions. This tends to be associated with a firm's "nationality." German and Japanese manufacturers, for instance, often value technical perfection more highly than American firms (Hollingsworth and Streeck 1994). Often the best solution technically is not the cheapest one, and, moreover, the costs of a standard's implementation vary from organization to organization, depending on earlier investments and sunk costs.

The battle-of-the-sexes type of conflict in organizational coordination through standards makes it difficult for actors to reach an agreement. The standard game-theoretical characterization of the "battle" shows that in a game with two players two outcomes, each favoring a different player, are equally likely. When each side tries to maximize its individual benefit, the two sides will be unable to reach a stable consensus (Knight 1992, pp. 48–53). Thus, the specific interests, resources, and strategic options of each player, on the one hand, and the institutional provisions for reaching an agreement, on the other, are salient variables generating actor-process dynamics which eventually explain the output of the standardization process.

11

Coordination of Standardization

The central purpose of standardization organizations is to provide an arena in which standards can be worked out by those interested. Collectively uncoordinated standardization via markets may result in an underprovision of standards and/or in the emergence of suboptimal standards, particularly in network technologies such as telecommunications. We have shown that international standardization organizations, especially the CCITT, have developed a specific set of formal and informal rules that shape negotiation and decision making. These rules change, but they usually have to be taken as given when work on a standard is in progress. It would be inadequate to simply regard these rules as constraints. Rather, they coordinate and channel collective action, excluding some options but opening up others. The consensus principle, for example, levels the playing field by strengthening the positions of less powerful actors relative to those of actors with great resources.

We will now come back to our general model of the standardization process. The graphic condensation of the model in figure 4.6 categorized aspects and variables that generally have to be taken into consideration if processes of committee standardization are to be compared. In a given institutional context and in a given period of time, not many of them will vary. Thus, at best it is possible to assess their influence on standards in general, but not to causally attribute to the variables specific features of a single standard. The "structural aspects" in figure 4.6 comprise elements and properties of the institutional framework, the constellation of actors, and the technological foundation. They can be regarded as influencing a standard indirectly by framing the process of standardization. Within the CCITT, these factors have assumed a certain form that is pertinent to telecommunications and not to shipbuilding, for instance. As the CCITT issues telecommunications standards, the technological foundation of its work relates to dominant designs of telecommunications networks, to the stock of knowledge in telecommunications, to specific technical problems in telecommunications, and so on. The set of actors includes those "interested" in telecommunications, and the actors' motives, perceptions, etc. are oriented to this technology. Also, the institutional framework has been designed for the purpose of issuing international voluntary standards for global telecommunications networks so as to contribute to the international and interorganizational coordination of this technology's development. From this point of view our focus is highly selective, and only in this context do we substantiate the role of the structural aspects in the standardization process. In each negotiation and decision-making process on a specific telecommunications standard, certain structural elements combine to frame the process. In chapter 1 we pointed out that our understanding of a frame is rather similar to Bijker's (1995a) concept of a "technological frame."

Hence, we begin with a look at the institutional and organizational framework, and at how it affects standardization. Then we turn to the technological foundation that constitutes the cognitive technical frame of the actors' negotiations and decisions. We then illustrate the actor-process dynamics unfolding in the standardization processes as a result of the interaction of "interested" actors, framed by the structure of the actor set as well as by institutional and (cognitive) technical factors. Then, by means of a short synopsis of the factors that differ significantly among videotex, X.400, and facsimile, we summarize the results of our three case studies before finally coming to the general conclusion of this book.

Institutional and Organizational Framework

The principles and rules governing organized standardization—as our case studies have shown—do not guarantee organizational efficiency, nor do they guarantee a smooth and ultimately successful working and decision-making process. Nevertheless, they influence the output. Alongside the differences between market and committee standards, we encounter variations between technical rules issued by the CCITT and those adopted by the ISO with respect to a specific functionality that has to be realized. The case of message-handling systems provides certain indicators. The ISO's MOTIS was not compatible with the CCITT's 1984 version of X.400; even the organizational design of the whole system differed. The CCITT regarded the Administration Management Domain (ADMD) as the appropriate functional unit for concentrating traffic and handling transborder exchange, and thus confined the Private Management Domain (PRMD) to the subordinate role of serving only the purposes of internal communication. The ISO, however, rejected this hierarchical design and treated ADMD and PRMD as functionally equivalent. This conflicting view was the consequence of the two standardization organizations' having different areas of responsibility, membership rules, working routines, voting procedures, and so on, although both placed great emphasis on the benefits of compatibility.

Rules and principles influence the standardization process through framing effects. The concept of frame used in the psychology of choice indicates that variations in the perception of an "identical" issue are controlled by the formulation of the problem, by norms and habits, and by the personal characteristics of the decision maker (Tversky and Kahneman 1985, 1988; Lindenberg 1993). Changing perceptions can lead to changing preferences and choices, though without affecting the basic individual characteristics of the actors. If we take the circle of actors and their personal characteristics as given, it is mainly the formulation of the problem and the organizational norms that cause a differentia specifica of standards development, be it between standardization processes outside and inside standardization organizations or be it, as in our case, between different standardization organizations.

We will readdress some of the norms and rules of the CCITT below in our discussion of actor-process dynamics. In the present context of framing effects, we want to draw attention to an observation that refers to the informal and formal rules governing controversy and to the concomitantly prevailing type of reasoning in the CCITT. Problems of technical coordination are discussed and evaluated from a number of angles. At the international level, at least four perspectives are frequently encountered: the political, the scientific, the economic, and the technical. The differentiation of perspectives is not merely an analytical exercise[1]; it points to the means available to an organization to officially privilege, or at least tolerate, or sometimes even exclude a specific type of reasoning in internal negotiations. In this sense an organization puts up "boundaries of irrelevance" (Scharpf 1991, p. 297) that separate types of reasoning from one another and eliminate some of them from the arena, thus reducing the complexity of the discourse. In this respect we notice again an important difference from studies where the generation of technology does not take place in an institutionalized context.

The CCITT, as an international organization set up to provide technical standards, is almost exclusively confined to the technical appraisal of problems, though there is the theoretical possibility for it to mobilize politics as a last resort through its institutional structure. Work begins with a technical study question, which is supposed to be processed in a *technical frame*. This frame is institutionally established at the meso level of a society—the level of organizations and associations. In our case in point, the frame is provided by the CCITT. Thus, the frame is not a feature of individual actors; rather, it "structures the interaction among the actors" (Bijker 1995a, p. 123).[2]

The CCITT's system of formal and informal rules leaves little room for argumentation from perspectives other than the technical one. The work is strongly based on the common search for the best technical solution with the classical technical orientation: "to make things function or improve efficiency." This is also regarded as the most appropriate way of reaching consensus on a technical standard. Accordingly, in most cases, technical experts, especially engineers, are delegated to the CCITT by the organizations participating. The dominance of these experts in the working groups supports and reinforces this technical perspective, although the experts are no longer exclusively drawn from the telecommunications field; nowadays some are from the field of computing. Hence, technical reasoning, formally institutionalized as the most legitimate way of interacting in the working hierarchy below the Plenary Assembly, has remained rather stable. Engineering of consensus by engineers through technical optimization appears to be the CCITT's prevailing frame of action.[3] The significance of the technical frame, though a little less predominant, can be observed in many standardization organizations at the national, regional, and international levels.[4]

Although in most cases the institutionalized technical frame is likely to be compatible with the technical experts' professional perception of the problems, viewing the problems from other perspectives is by no means excluded. Indeed, as we have already mentioned, a political perspective is formally institutionalized at the level of the CCITT's Plenary Assembly. Although in times of international liberalization and deregulation the nationally based political approach to the coordination of technical development appears anachronistic, we have found that political reasoning has occasionally facilitated standardization (though usually it has hampered it).

At the level of the technical groups, political reasoning is unwanted, although its general institutional legitimacy means that it cannot be banned. However, existing conflicts can be labeled "political" and thus attributed to sources external to the working level. Technical work, if it has not stopped completely because of these outside pressures, may then continue. In any case, the search for the best technical solution need not be questioned in the conflict.

In contrast to the political and technical perspectives, a scientific perspective is not formally accommodated in the institutional framework. Informally it is not categorically excluded either, probably because of the close relationship of the scientific and the technical fields (Aitken 1978; Schmidt and Werle 1993). Still, the only occasion where scientific reasoning is totally unproblematic is where it merely supplements and assists technical argumentation. The standardization of Group 2 and Group 3 facsimile provides many instances of this. One example: for the modulation procedure to be employed in Group 2, different methods were available, and the manufacturers had diverging preferences. In order to facilitate an agreement, a commission was set up to study evaluation criteria and test conditions for the selection of an appropriate procedure. Building on the commission's conclusions, science-based tests of the modulation procedures were carried out, and a consensus was reached on the basis of the test results.

But science-based testing as a neutral yardstick is not always the final option for establishing criteria for deciding between alternatives. When different criteria for testing are preferred or test methods are not universally accepted, other means of reaching a consensus must be found. In addition, not all kinds of scientific reasoning are equally welcome. People from universities or from the PTT research centers in particular are often regarded as "too theoretical," promoting solutions that are pure and scientific but totally unworkable for the practitioners.[5]

The institutional emphasis on technical reasoning, which permits a political view of standardization problems only at the top level of the CCITT and which makes use of a scientific approach only as an instrument for facilitating technology-related decision making, may appear unworldly and far removed from the "real" (i.e., economic) problems of technical coordination. In fact, the coordination among the organizations with regard to their involvement in the construction, operation, and use of large telecommunications systems always implies the need to deal with investment decisions and other financial commitments. We have argued above that technical and economic factors interact and that sometimes they cannot even be clearly distinguished. However, despite its real-world significance, the economic perspective lacks viability in formal and informal argumentation in the CCITT. Even informally it is not possible to evaluate the costs of different technical options. Figures are not mentioned in the CCITT; moreover, we have gained the impression that the technical experts had considerable difficulty estimating the costs of different standard alternatives with any degree of accuracy. On this point we agree with Cowhey (1990, p. 183) that engineers delegated to the CCITT "were not hired to think about politics or economies of the market." However, we have reservations about Cowhey's conclusion that "for years economists accepted the premise of engineers because these engineers had the only cost data." It appears, rather, that the engineers' comprehension of costs is diffuse, and they tend to treat the technically most efficient solution as the best economic one.

As long as technical coordination in telecommunications was mostly a matter of the coordination of nations and more or less an exclusive business of national PTTs and of accredited RPOAs (such as AT&T), the potentially different economic implications of standards approved in the CCITT either were rarely considered, remained unknown, or were neglected because they did not affect the position of the PTTs. The prevailing tendency was to regard the agreed-upon technical standard as a plausible, if not the best, economic solution. This view has since changed, and now many manufacturers, service providers, major users, and network operators (including some of the "liberalized" PTTs) are especially reluctant to recommend a standard that is impracticable and more expensive to realize than an existing technical alternative. Neither would they agree to a recommendation that is too sophisticated to be implemented within a reasonable time span—such recommendations are often called "too theoretical." However, officially introducing an economic perspective into the standardization process in the CCITT continues to be difficult. Outsiders who are not directly involved in the work of the technical groups tend to face great difficulties in understanding "the subtle aspects of the technical issues" (Cowhey and Aronson 1991, p. 310). Insiders often lack the expertise needed to calculate the commercial repercussions of different standard alternatives for their "home" organization, especially when the alternatives do not differ radically. But even in the case of clearly distinguishable alternatives, openly economic reasoning is not officially tolerated. It is almost unthinkable that during a regular meeting of a study group or its subgroups delegates proposing a specific standard would offer side payments to opponents in order to compensate for the economic disadvantage they could incur from the standard.

Whenever economic gains and losses can be calculated—correctly or not—by the participants in the standardization committees, they exert an intricate influence on the negotiation process, because the affected interests cannot be revealed officially. To be introduced into the discourse, the economic preferences must be "translated" into legitimate reasons. Sometimes political reasoning helps to defend economic interests, such as when economic stakes can be translated into national political interests. Once again the distinction between Administration Management Domains and Private Management Domains in message handling provides a case in point. The PTTs' interest in concentrating as much traffic as possible in their public networks initially drove them to insist on this distinction, which implied that the prerogative of handling transnational traffic through the ADMD would rest with the management domains of the PTTs and the RPOAs. Arguments relating to political sovereignty were introduced to prevent private organizations from extending their management domains across the borders without being subject to any political control. For a while Germany (i.e., the delegates of the Bundespost) even announced it would refuse to accept any electronic-mail traffic in its territory that had originated with foreign PRMDs, thus declaring ADMDs to be the only welcome communication partners.

In this case there were also technical arguments in favor of the PTTs' position. Where there is only one management domain per country as the official clearing agency for incoming and outgoing traffic, international addressing and routing of electronic messages is less complicated, because part of the technical coordination problem is left for internal hierarchical solution by the national PTT. In the CCITT context it could easily be argued that this procedure was technically more efficient than having several of these domains operating as equivalent agencies in every country. There was no doubt that, in general, only a PTT's ADMD could be positioned in such a privileged place.

In facsimile standardization we found many instances of economic views being translated into technical reasons. Firms fought for a specific standard based on their proprietary practice because they had already invested money in it and had produced several thousand machines. However, they did not refer to this economic dimension; rather, they praised the technical superiority of their solution in the standardization negotiations. Convinced that they would gain from a common standard, they accepted that science-based experimental methods should be used to select the technically most efficient standard.

This official neglect to take into account cost estimates and other means of economically assessing standards can be regarded as an official "veil of ignorance" (Rawls 1972). It leaves the commercial implications of different standard variants collectively unexplored, thus often making it easier to reach a consensus. If, on the other hand, contradicting economic preferences have been successfully translated into competing technical proposals but no best technical solution has been found, the different proposals can be adopted as options of one standard. In this case, economic factors are accepted in determining what is the most appropriate standard. But these factors, the rules of the market, are seen as lying outside the CCITT's area of responsibility. The negotiators agree to "let the market decide." This phrase was used by some delegates at the end of the controversy over the Group 3 and the Group 4 standard in facsimile, when no consensus could be reached. In such a situation, the adoption of few different options of a standard may appear as the second-best solution, worse than approving a single standard but better than having no CCITT recommendation at all. This cannot be generalized, however. In the case of videotex, for instance, the multitude of incompatible options resulted in the stabilization of the different national standards, none of which could gain significance at the international level.

It is not always possible to translate an economic reason into a technical reason or an alternative technical proposal. If it cannot be translated, a strong economic interest related to a specific standard finds expression in strategic maneuvers aimed at delaying technical work and blocking decisions on formal procedural grounds without going into technical detail. This type of obstructive behavior is generally attributed to hidden economic motives, but sometimes personal idiosyncrasies are involved. For example, during the X.400 standardization the de facto alliance of IBM and Siemens against a message-store standardization proposal submitted by academic participants from the United Kingdom was attributed to the two firms' vested economic interests in an existing document-filing-and-retrieval standard that did not form part of X.400 but could be used in its context. In fact, IBM and Siemens did not see the necessity of adding message-store features to X.400, and they were hesitant to load too much into a standard because they wanted to comply with the principle of parsimony. The Siemens delegate attacked the message-store proposal because he was involved in the work on the existing document-filing-and-retrieval standard in the European Computer Manufacturers Association (a private standardization body); however, after this delegate retired from CCITT work it became clear that his strategy was prompted more by his personal commitment than by the economic business interest of Siemens.

These examples demonstrate the ambiguous effects of the CCITT's institutional frame. Emphasizing technical reasoning and delegitimizing economic argumentation has certainly proved to be helpful in reducing the complexity of many negotiations and routinizing the work of the technical experts. In many ways, economic argumentation is more prone to result in a distributive conflict than technical reasoning is. Technical reasoning appeals to "higher" technical values as a common unifying interest, whereas the economic perspective is regarded as based solely on self-interested behavior. However, the institutionalized premise that technical concerns generally deserve precedence over economic or other concerns underrates the implications of many contentious decisions. In addition, simply banning the specific intricacies of economic interests from the agenda can keep some "technical" conflicts from being resolved by keeping the driving economic motives hidden.

Cognitive Technical Framework

We have argued that technical reasoning is officially the most appropriate way to deal with standardization problems in the CCITT, and that it can also be regarded as the best means of reaching consensus within the constraints of that international organization. At the same time, technical reasoning expresses the professional interests of the technical experts in international standardization in protecting their domain against other professions.[6] These professional interests are linked to some cognitive elements of the technical frame that are relevant to the work of the CCITT.

The differentiation of the technical from the economic perspective on standards is closely related to the distinction between the knowledge aspect and the product aspect of standards that we drew in chapter 2. It can be regarded as an aspect of the societal institutionalization of an autonomous or self-referential technical subsystem vis-à-vis other functional subsystems (e.g., the economy, science, the political sphere) (Luhmann 1984). Rikard Stankiewicz has recently developed a three-stage model of the process of differentiation and autonomization of the technical subsystem from other societal subsystems. The cognitive (or, in Stankiewicz's terms, "sociocognitive") elements of technology play the most crucial role in this process. "Coherent fields of generic technological knowledge [emerge] which are increasingly autonomous to the concrete [technical] systems." (Stankiewicz 1992, p. 34) At the first stage, technology can be described as a "knowledge pool" providing solutions for highly specific "local" problems. The second stage is already a "systemic" one. Accordingly, the cognitive model captures the complexity and the interconnectedness of technology and its components. The knowledge elements of technology, "the engineering traditions or, as they are also called, 'paradigms' constitute [already] differentiated but [still] basically subordinate components of the dominant concrete systems" (ibid.). Only at the third stage does Stankiewicz suggest that knowledge gains complete autonomy from any concrete technical system.

With the aid of Stankiewicz's model, telecommunications can be perceived as being in a transition from the second to the third stage of autonomization. Some actors still look at this technology with a very concrete technical system in mind; others hold a more abstract view. The transition is a source of conflict. The dominant concrete system which has guided the majority of actors' perceptions in the CCITT is the public telephone system. Thus, standardization processes in the CCITT have typically centered on the telephone system, which has gained the status of a materialized technological paradigm (Schmidt and Werle 1994). Electromechanical switches and analog connection-oriented (point-to-point) voice-transmission technology were originally core components of the dominant design. Around this technical configuration a rather small and stable set of actors evolved, including PTTs and RPOAs as network operators and up to three large manufacturers of telecommunications technology per country. They, together with other actors, constituted the organizational community that held the relevant technical knowledge concerning telephone service.

Technical changes interacting with changes in the organizational community initiated a functional diversification and transformation of the telephone system, which had a considerable impact on the stock of relevant technological knowledge (Schmidt and Werle 1994).[7] In particular, the digitization of switching and transmission has made it possible to base the technical processes of telecommunication on technical components similar to those used in computing and to rely on knowledge about coding, modulation, multiplexing, switching, and the structuring of networks, most of which has been accumulated more in the context of computing than in the context of telecommunications.[8] In parallel to this, knowledge has become more abstract.

These developments have left their mark on the cognitive frame of the CCITT. The institutional concentration on a technical discourse could not avoid an attendant increase in complexity and dissent, which was directly related to new technical opportunities, new technological knowledge, and new actors in the telecommunications domain, the boundaries of which are now becoming blurred. The technical changes have been welcomed by the actors traditionally involved in CCITT work, and the perceived need for coordination that has resulted has been actively taken up and approached (initially in the area of telematic services).[9]

The cases we have studied belong to the category of telematic services, although facsimile was invented long before the digital revolution gained momentum. The Group 3 machines are the first generation of terminals that work on a digital basis. However, modems are needed to convert the digital signals for transmission over the telephone network, large parts of which are still analog. They provide a gateway from the digital to the analog world. For Group 3 the architectural constraints of the telephone network always had to be considered. The standardization of an error-correction mode, for example, was a reaction to deficiencies of the analog mode of transmission in general and to the comparatively low quality requirements of voice transmission in specific. Smaller transmission errors can be easily compensated for by communicating human beings, who have the cognitive ability to fill gaps (as, for example, when a syllable or even a whole word has been lost on its way from the sender to the receiver). Through the use of a facsimile machine, the same error may be conceived as noise, rendering it impossible to interpret a message correctly. Thus, features capable of correcting transmission errors must be added.

The almost universal availability of telephone connections, however, has always been an incentive to use the network for other types of communication. Facsimile has taken off in this network, and incremental improvements of the Group 3 machines have made them more and more attractive. The system architecture of the telephone network, historically developed in a piecemeal manner rather than designed on the basis of a central plan, has remained the cognitive frame for Group 3 facsimile. This does not mean that standardization of Group 3 occurred without conflict. We have sketched the controversies that surrounded resolution, coding, and error correction. Against the background of a common understanding of the technical opportunities and constraints of facsimile in the telephone network, however, it was possible to agree explicitly on criteria or to use tacit common criteria facilitating consensual choice.

In contrast to Group 3, Group 4 facsimile was initially designed for public data networks, but was seen in the context of plans or projections to migrate to an integrated digital network such as ISDN or another type largely conforming to the OSI frame of reference. Such a network did not exist when standardization began. Thus, the standardization of Group 4 had an ex ante character both with respect to the network in which it should be operated and with respect to the terminals. With no sunk costs involved and no proprietary solutions existing, the negotiation process was rather smooth, and a complex standard was adopted that was open to interworking with other services and to "mixed-mode" operation. The standardization of Group 4 was a highly theoretical exercise oriented toward a future, multifunctional, integrated digital network. Because of this completely different cognitive reference, Group 4 was not compatible with Group 3.

The recent controversy over these two groups demonstrates that not every "pure" technical conflict can be resolved easily. It was the submission of a proposal for a Group 3 digital standard that provoked the conflict. The proponents of such a standard suggested adopting an improved version of the successful Group 3 standard for use in digital networks (more specifically, in the "old," now digitized, telephone network). The opponents insisted on regarding the existing Group 4 as the single solution to be adopted for digital networks. They, however, offered a more streamlined version of the Group 4 standard (called UDI, which stands for "unrestricted digital information") as a compromise. However, being based on the architectural principles of the future integrated digital network and relying on corresponding protocols, UDI was not compatible with Group 3 either.

Because the two camps started from different basic architectural concepts (a rather principled approach oriented toward OSI and

ISDN versus a pragmatic approach relying on the telephone network), not even a consensus on common criteria for the evaluation of the competing proposals could be achieved. The CCITT's principle of parsimony of standards options failed, and in the end both Group 3 digital and UDI—though they were not compatible—were approved as options to the Group 3 base standard. In addition, the old Group 4 standard continued to exist.

Videotex was linked to different basic architectural orientations from its inception, combining knowledge from the fields of computing, consumer electronics, and telecommunications. Yet this heterogeneity was mostly absorbed at the national level, where it tended to make the attainment of a common position difficult. The highly politicized conflict in the CCITT did not refer to the cognitive heterogeneity so much as it concentrated on easy-to-grasp technical differences in the various national proposals. Thus, the potential conflict arising from the different types of knowledge remained latent.

Choices are structured by technical orientations concerning the architecture of systems or networks. But less explicit technical traditions linked with a specific organizational or corporate culture are also sources of technical orientations that influence preferences (Dierkes 1991). "Not invented here" is just such a principle. Apart from its confrontational aspect, it points to the relevance of specific cognitive efforts in structuring technical systems in a coherent way within defined circles of actors. Alternative technical options from a different background are neither considered nor replaced with endogenous solutions, because they do not directly match the prevailing cognitive structure. AT&T's resistance to accepting one of the interactive videotex standards submitted to the CCITT and its late effort at developing a proposal of its own were due in part to this orientation. Thus, technical traditions and cultures of companies, but sometimes also those of whole countries, lend substance to such principles as "not invented here."

One example of a national technical tradition is the joint shaping of Germany's telecommunications system by the Bundespost and the leading German manufacturer of telecommunications installations, Siemens. Siemens and the Bundespost share a distinct technical tradition that dates back to the time when the lack of competition left considerable scope for following technical predilections that were not severely restrained by economic considerations (Werle 1990). Apart from their complementary economic interests, their common technical knowledge and their similar architectural concepts of telecommunications systems provided the basis for an alliance at the international level. Thus, little internal coordination was needed for the delegates from Siemens and from the Bundespost to jointly and coherently attack the proposed Group 3 digital standard for facsimile.

Controversies rooted in diverging technical orientations are difficult to handle. When basic architectural orientations, dominant designs, or technological paradigms collide, the CCITT is invariably incapable of achieving consensus, especially by means of a pure technical discourse, because the cognitive technical frame is too heterogeneous. The technical conflict is of the zero-sum type, and tolerating instead of excluding economic reasoning in this case would be more likely to strengthen the position of those whose cognitive model is more closely linked to a concrete technical system. These actors can more easily demonstrate the economic utility of what they propose, whereas those with "autonomous" cognitive models (in Stankiewicz's sense) would appear "too theoretical" though technically "elegant." Switching from technical to economic reasoning therefore presents an opportunity to reach an agreement without reconciling the underlying divergence of architectural orientations. Thus, a "consensus of interests" regarding business could perhaps be achieved, notwithstanding the persistence of "orientational dissent" concerning the technical discourse (Schimank 1992).

Differentiated technical traditions of countries and companies within a basic architectural orientation are wont to produce battle-of-the-sexes conflicts. Their transformation into pure-coordination problems always appears likely when medium-term or long-term perspectives on systems development constitute the collective effort's central focus, as in the case of message handling. These perspectives leave room for gradual modifications and perhaps for the convergence of differentiated (though not too divergent) cognitive traditions. However, under these conditions and those of short-term planning, consideration of both economic and technical reasons does not generally support conflict resolution; instead, it causes additional complexity and creates new sources of dissent.

Actor-Process Dynamics

Actor-process dynamics relate to the concrete standardization procedures—the processes of negotiation and decision making that are governed by formal and informal rules, framed through institutionalized patterns and types of legitimate reasoning, structured by individual and corporate knowledge as well as interest positions, and distinctively shaped by very personal characteristics and interaction orientations on the part of the individuals participating. This enumeration of the factors that must be considered when actor-process dynamics are involved already indicates that the output is overdetermined, and that it is very difficult (if not impossible) to figure out which factors have been most significant. Nevertheless, the feeling that standardization processes (like other collective interaction processes) create their own history, which feeds back into these processes and gives them a direction that is from the start difficult to project and to control, has been widespread, not only among the participants but also among the observers. Since a sufficiently complex theory of autodynamic social processes is still missing (Mayntz and Nedelmann 1987), we can, without claiming general validity, directly settle on exemplifying how actor-process dynamics affect standardization processes and their output.

The videotex case provides an example of how technical reasoning at the CCITT's working level can end up as a ritual without consequences. The small window of opportunity for compromise and agreement that existed in videotex's early days was closed very soon, and afterward the political interests of the contending parties were almost exclusively directed at gaining a competitive edge, either by eliminating other systems or at least by getting their own system accepted as a viable alternative. At the working level, in a competitive atmosphere the discussions often repeated well-known arguments instead of introducing new perspectives. Little effort was invested in possible compromises, although participants conceded in interviews that the technical differences between systems like the French and the British were not insurmountable. Simply terminating the debate was impossible. Normally the study group to which a study question has been addressed must deal with the question and reach a collective conclusion at the end of the four-year study period. With videotex, the successive approval of the different incompatible national proposals failed to offer a way out of the conflict. On the contrary, it kept the fire smoldering, and even cautious steps toward a partial compromise taken before 1984 failed. Within the tension-stressed CCITT arena nobody could find a way out of the conflict. Accordingly, compromises were sought outside the CCITT, in the CEPT or in the WWUVS initiative. However, only developments external to the CCITT—the economic failure of many national videotex systems and especially the rapid diffusion of personal computers and online services—caused the conflict to subside as interests in videotex standardization generally waned.

Compared to videotex, the standardization of message handling was a harmonious and creative process. The rather small group within Study Group VII that initiated the work on X.400 in the CCITT conceived of developing a system that would integrate rather than replace proprietary message handling with all the zeal of a mission. Nobody was excluded from participation, but the complexity and the time-consuming nature of the work—more joint development than negotiation—prevented many from becoming involved. Even Study Group I, which dealt with the service aspects of telecommunications systems, initially refrained from considering message handling. And later, when it did become involved, Study Group I never really influenced the standardization of message handling. New members entering the small Special Rapporteur Group were socialized into the long-term perspective and the specific mission of the undertaking. Although the coordinator of the group, the Special Rapporteur, changed in every study period, continuity was secured by choosing someone who had already proved his commitment by actively contributing to the work. This reinforced an informal rule of the CCITT according to which only those who devote time and energy to furthering the common goal are entitled to determine the direction and the results of the work.

The in-group atmosphere prepared the ground for creative work with a tendency toward principled technical perfectionism that was not curbed by a clearly focused built-in stop rule. Institutionally, perfectionism is confined only by the cycle of study periods. At the beginning of a four-year period there is ample time to study problems and assess their scope without having to worry about finalizing decisions prematurely; however, toward the end of the period the danger of a delay for another four years creates pressure to come to an agreement, which encourages compromises on perfectionism. As long as the confines of a technical discourse remain untouched, however, work can be highly detailed and even esoteric during a study period.

Hence, standardization of X.400 was never severely disturbed from inside the CCITT; the trouble was caused by parallel work on message handling in the International Standardization Organization in the mid 1980s. The resulting jurisdictional conflict was resolved through an informal agreement in favor of cooperation, settled by the two international organizations. Even while cooperating, the CCITT members claimed a prerogative because they had already officially adopted a version of the X.400 standard in 1984. Unwilling to sacrifice the time and effort they had invested and to begin the work on standardization anew, they insisted on safeguarding backward compatibility even though the 1984 version was not entirely harmonious with the OSI frame of reference and, from this point of view, was inferior to the proposal made by some of the ISO members. The position of the CCITT members was based on their organization's strong degree of self-commitment. Once a standard has been recommended, it is not supposed to be changed or replaced unless compatibility is guaranteed. This stability is considered necessary to protect the members' investments in standardized telecommunications technology and the long-term confidence in the CCITT, which in turn is an important incentive for implementation of the nonobligatory CCITT recommendations.

In the CCITT context, most of the work on X.400 was based on trust and familiarity among the participants. It was partly induced by the overall uncertainty connected to future-oriented work in a complex and rapidly changing environment. Besides a general confidence, few concrete indications for the direction of the development of the "office of the future" (Uhlig et al. 1979) and for the role of electronic mail existed, and the criteria for choice between alternative options were diffuse. To a large extent, the trust and familiarity were institutionally generated. The four-year study period is connected to the expectation that participants do not resign between the beginning and the end. During the period, involving at times rather frequent meetings of a study group's subgroups, whose international venue regularly alternates, personal contacts grow and can be cultivated. At the same time, the organizations sending the participants are literally kept at a distance. The delegates find themselves caught between the demands of the work on joint standards and allegiance to their delegating organization. The organizations face severe difficulties monitoring their delegates' work. The more complex and the more similar to research and development the standardization work is, the less it can be influenced from outside and the more leeway the participants enjoy. In the complex X.400 standardization process, participants could react flexibly; this stabilized an atmosphere of trust and familiarity and, in turn, facilitated a convergence of perceptions and other cognitive factors. Whereas in the work on videotex early tensions escalated (as a result of cardinally divergent positions rooted in major political and economic commitments at national level), and this reduced the technical debates to a ritualistic exercise, in message handling within the CCITT an early collective mission orientation was reinforced by an atmosphere of trust and continuous cooperation.

In facsimile standardization as well, a cooperative (though rather routinized) climate prevailed, at least until recently. Some problems, however, date back to an internal reorganization of the CCITT. The dissolution of Study Group XIV in 1980 and the transfer of facsimile standardization to Study Group VIII (also responsible for teletex and videotex) spoiled the working process somewhat, although facsimile was not directly affected by the videotex conflict. The two study groups had differentiated technical traditions. Thus, a former Study Group XIV member argued, they would have never developed the Group 4 facsimile standard, which was produced in Study Group VIII. Yet other points may be more remarkable. Lacking for a long time the ex ante character of X.400 standardization, facsimile was more directly related to the economic interests of the manufacturers. As long as the differing technical preferences of the participants were anchored in the same architectural frame, standardization was a cooperative undertaking, and science-based methods were used whenever they could be to decide on competing proposals.

It was acknowledged, moreover, that—in order to keep the delegating organizations from interfering in the negotiation process and to keep the cooperative spirit alive—concessions to the active individual participants were necessary so that, once in a while, everybody could return home with something that legitimized his or her work. This collective perception is typical of the CCITT, but it is not always possible to distribute concessions so generously. The decomposability or the degree of coupling of a technical system is a crucial variable. Though affected by given physical constraints, the variable is to a considerable extent shaped by earlier choices. In the case of facsimile, the coupling of components was sufficiently loose, and hence there was enough flexibility to distribute concessions. Some concessions are not even very "expensive," as was exemplified by the case of error correction in facsimile Group 3. Here the possibility of recommending standards including several options was used to evade an unnecessary conflict and, at the same time, grant a concession to the delegates of the USSR. The USSR quite forcefully promoted an error-limiting mode (ELM) as an alternative to the error-correction mode (ECM) supported by most of the participants. As it was considered unlikely that ELM would be implemented outside the USSR and its periphery, the majority could easily adopt ELM as an option for error correction alongside ECM.

Recent developments in facsimile have demonstrated that—as might have been concluded from X.400—ex ante standardization does not guarantee smooth or low-conflict working procedures. On the contrary, in this instance two competing standards rooted in different system architectures caused a conflict that ended in a deadlock. The routinized working procedures did not indicate a way out of the impasse. Moreover, the conflict damaged the spirit of cooperation and the basis of mutual trust. Strategic moves on the part of the contending parties were not interpreted as resulting from strict company mandates or as a legitimate step toward gaining an advantage; they were interpreted as misuses of the rules through opportunism, cheating, and polishing of individual profiles—misuses that eventually might lead to breaches of confidence.

On both sides, influential actors who had built up reputations of trustworthiness over the years were involved. This is normally an assurance that correct information is being given. Owing to their long-term membership in the study group after the obligatory years of "quiet participation in the back row," they knew how to play the standardization game in the CCITT. For instance, two of the actors pushing the Group 3 digital standard managed to find themselves with additional responsibility for facsimile as members of Study Group VIII and Study Group I. Thus, questions directed at Study Group I concerning service aspects of the new standards were drafted by these actors in Study Group VIII and then answered by them on behalf of Study Group I. Similarly, the official submission of a newly created simplified version of the Group 4 standard (UDI, which was intended as a compromise option for Group 3 in place of the Group 3 digital) can be seen as a strategic move on the part of highly experienced CCITT members. If the proposal had been accepted, the position of the Group 4 standard would have been strengthened and the opportunity of the Group 3 standard to migrate into digital networks would have been abolished. The experts on the other side realized this and insisted on their version of Group 3 digital.

With the consensus principle—a principle nobody wanted to violate—serving as the informal guide for decision making, the controversy had to end in a deadlock. A unanimous agreement on one of the two proposals, linked to clearly diverging architectural orientations, was unlikely, although both proposals were of an ex ante nature. The only way out of this conflict was to adopt both proposals as a rather empty compromise. Thus, UDI provides an example of innovative technical knowledge genuinely produced by members of the committee during the standardization process. The value of this compromise lies in closing a contentious process of standardization and keeping the organization viable. At the same time, however, it means that the CCITT loosened its hold on the development of technical systems. The CCITT can only hope that one of the options offered in the recommendation will be chosen and that there is no outside option that benefits from the CCITT's inability to offer a single solution. Such an outside option might also supply compatibility, but it would imperil the future of the CCITT.

12

Sources and Effects of Variation in a Stable Framework

Our analysis has concentrated on the institutionally embedded definition and choice of standards. We selected the CCITT as the central organizational focus, but we did not ignore other significant organizations and mechanisms that provide technical compatibility through standardization. However, as a rule, we analyzed their roles only when they influenced the CCITT process. Through concentrating on the CCITT, we have been able to scrutinize more closely the standardization cases we chose to study. The processes of negotiation and decision making and their outcomes differed from case to case, although institutional and organizational factors either remained quite constant or varied in a way that affected the cases similarly. This demonstrates that institutions and organizations exert only moderate pressure for uniformity of processes and output, especially at the international level. Sanctioning power is low and scarcely employed. Hence, some room is left for the actors to disregard the rules or to acquiesce only when positive utility is promised. At any rate, organizational selectivity and specificity as far as membership rules, decision rules, working procedures, etc. are concerned have to be taken into consideration when the output of standardization processes is assessed. They channel attention, set incentives and disincentives, allocate legitimacy, and structure the playing field. But above all, the framework they provide reduces complexity and thus facilitates decision making, thus providing an inducement to comply with the expectations and rules.

Salient factors framing the work in the CCITT are the consensus principle operating at the working level and the institutional emphasis on technical reasoning (which simultaneously confines economic arguments to a hidden discourse). Majority voting (instead of consensus) and political reasoning (instead of technical reasoning) are tolerated at the upper level of the CCITT's working hierarchy. PTTs and RPOAs still constitute a very significant fraction of the membership, but many manufacturers of telecommunications equipment and an increasing number of service providers, as well as firms in the computer and consumer electronics industries, actively participate in or monitor the work of the CCITT. Users of telecommunications equipment and services, and other groups with interests relating to telecommunications, are represented poorly or not at all. This fact also frames standardization. Other factors relevant to the CCITT in general have been mentioned in the preceding chapters. Many of them have remained stable or have changed only gradually.

With regard to the three standardization cases, we have extracted several factors that shape both the process and the output of standardization. They account for much of the variation in the output. Table 12.1 presents a synopsis of the factors that proved to be crucial in this respect. It also includes three important features of the dependent variables: the number of options, the consistency of options, and the architectural entrenchment of the standards. Such a summary condenses information and complexity to an almost intolerable limit, but it does illustrate that the CCITT, which from the outside appears so homogeneous, manifests remarkable internal diversity and strain. In contrast to X.400 and videotex, facsimile standardization must be regarded as a twofold process, dedicated to analog telephone networks on the one hand and to digital telecommunications networks on the other. This is why we provided two columns for facsimile in the table. By facsimile for digital networks we chiefly refer to the present situation and the recent history of this technology, and we mainly have the controversy over Group 3 digital and Group 4 in mind. Early standardization of facsimile for digital networks was in many respects similar to standardization of X.400. It resulted in the Group 4 standard, which later was challenged by Group 3 digital.

The factors characterizing the process of negotiation and decision making on standards within the general framework of the CCITT are grouped into two categories. We call them actor-related and technology-related factors, although actors have always provided the focal point of our observation of the standardization process. In the general model of the standardization process, condensed in figure 4.6, the actor-related and technology-related factors can be located on the side of the structural aspects in the boxes "technological foundation" and "actors." They relate to the "decision-making process," in which they

Table 12.1
Factors shaping the processes and the results of standardization in the CCITT.

			Facsimile for type of network	
	Videotex	X.400	Analog	Digital
Actor-related factors				
Homogeneity of actor constellation	High[a]	High/medium	High	Medium
Cooperative-interaction orientation	Low	High	High	Medium
Type of standardization game	Battle/ Zero sum	Pure coordination	Coordination/ Battle	Battle/ Zero sum
Political salience	High	Low	Low	Low
Economic salience	High	Medium	Low	Medium
Sunk costs	High	Medium[b]	Medium	Low
Term of orientation	Short/ medium	Long	Medium	Medium/ long
Technology-related factors				
Reliance on "old" telephone network	Medium	Low	High	Medium/ low
Systemness	High	High	Low	Low
Decomposability	Low	High	High	High/ medium
Coherence of cognitive technical frame	Medium[a]	High	High	Low
Complexity of technical problems	High	High	Low	Medium
Output: Features of standards				
Number of options included	Several	Few	Few	Few
Compatibility and consistency of options	No	Yes	Yes	No
Architectural entrenchment	Medium	High	High	High

a. At CCITT level only.
b. In proprietary systems.

interact with elements of the "institutional framework" as the most stable factors in our study. This *general model* of the standardization process must be adequately specified to explain the output of specific standardization processes. Our discussion of the institutional and organizational frames, the cognitive technical framework, and the actor-process dynamics in our standardization cases refers to specificities of the processes. In our cases, some of the actor-related and some of the technology-related factors can be regarded as the main sources of variance in a basically stable institutional and structural framework. This does not mean, however, that the stable elements of the framework were irrelevant. Without their framing effects (e.g., leveling the playing field) the repercussions of the other factors would have been different. Needless to say, we cannot generalize the results of our studies of three specific cases. However, some tentative general conclusions concerning the effects of the factors listed in table 12.1 can be drawn.

The factors we call actor-related refer to the actor constellation, the interaction orientation, the actors' payoffs from the standardization game, their perception of the political and economic salience of standards, the amount of money and other assets already invested in a technology, and the time horizon or term of orientation the actors have in relation to the different standards. In our context these factors are necessarily linked to technology; however, they approach the technology from different, more general angles, not grounded in but rather detached from specific features of technology. Their roots can be identified in areas of nontechnical knowledge (such as economic knowledge), in management principles, in political goals and values, and in organizations other than the CCITT. Thus, as we see it, the actors carry these "external" factors into the CCITT working procedures, where they are framed (i.e., processed and influenced) by the organizational rules and other institutional elements.

With the category of technology-related factors, as we have said, we do not want to suggest that these factors have some kind of absolute "objective" quality. The whole standardization process is, of course, shaped by actors insofar as technical features per se do not affect the output. Rather, the effects of technical features are contingent on how the features are perceived and evaluated by the actors, individually and collectively, and which of them are introduced into the standardization process. However, the technical features do not represent some kind of purely individually constructed, somewhat artificial or voluntaristic reality. The perception of technical features does not vary by chance; it is decisively structured, though not determined, by the physical reality of technology, by engineering science, and (in the present case) by the CCITT, which also channels the view of this reality. In short, the technology-related factors must be considered from this point of view, and in this sense they can be treated as a separate set of variables influencing standardization. These factors include the reliance of the technology to be standardized on the telephone network (instead of OSI, for instance), the systematic character ("systemness") and the decomposability of the technology in the actors' perception, the complexity of the standardization problem, and the coherence of the actors' cognitive technical framework.

The first actor-related factor in table 12.1 relates to the homogeneity of actor constellations. Though the constellation of actors in the CCITT has changed greatly in recent years with the involvement of more and more actors from computer manufacturers, service providers, and other non-PTT organizations, we did not encounter much heterogeneity in our cases. By itself, the degree of homogeneity does not have a direct impact on the work; rather, it becomes decisive in combination with other actor-related factors. For instance, it seems as if, ceteris paribus, a cooperative interaction orientation may be more difficult to establish if participating actors tend to be heterogeneous, whereas a homogeneous constellation of participants (e.g., only computer manufacturers or only PTTs) favors a cooperative orientation and a problem-solving attitude.

That the cooperative interaction orientation was practically missing in the case of videotex, although the constellation of actors was homogeneous, must be regarded as a result of the payoff structure of the standardization game. The videotex game was transformed from an already-tough battle of the sexes to a zero-sum game very early. The expected payoffs are normally closely linked to the political and economic salience of the standardization issue. However, in videotex both the political salience and the economic salience were high, and in this case a zero-sum game was played with a competitive interaction orientation.

Sunk costs generally reduce willingness of actors to compromise on a standard when diverging proposals have to be decided upon. The danger of losing invested assets because they cannot be adapted to other standards is always present when technical development runs parallel in different firms without being coordinated ex ante. Zero-sum conflicts are likely to result unless long-term orientations guide the actors' strategies. A long-term perspective often reduces the weight of current investments and thus opens up new possibilities for cooperation and compromise. In the case of X.400, despite some existing sunk costs, the collective long-term perspective helped to create a completely new set of standards. It was possible to interact in a very cooperative way, treating the standardization process as a pure-coordination game from which all participants would profit in the same manner.

The successful standardization of X.400 and facsimile for analog networks suggests that a relationship exists between cooperative interaction orientations and specific technology-related factors. Two of the technology-related factors, a high degree of decomposability and a coherent cognitive technical framework, appear to facilitate a cooperative interaction orientation. A coherent cognitive technical framework allows a similar evaluation of technical alternatives, while a high degree of decomposability provides for incremental and therefore "fair" decision making in which participants can alternate in realizing their different interests. On the other hand, actors with a cooperative orientation favor decomposable technical solutions because they facilitate compromise.

Although high degrees of both "systemness" and technical-problem complexity complicate a standardization process, X.400 demonstrates that difficult technical problems may also be treated as a pure-coordination game and need not arouse conflict as they did with videotex. The coherent cognitive technical framework within which standardization of message handling proceeded was one crucial factor. One element of this framework was the comprehensive technical architecture of OSI. A comparable effect facilitating standardization can be discovered with respect to the telephone network. When (as in the case of early facsimile) standardization relies on common knowledge of the requirements of the old telephone network, criteria for the evaluation of technical proposals can be derived from this knowledge and easily agreed upon. When (as in facsimile for digital networks) a coherent cognitive framework is missing and the old telephone network cannot serve as a collective technical guidepost either, standardization tends to be contentious.

The actor-related and technology-related factors influence standardization within the general framework of the CCITT, with its consensus rule, its principle of parsimony of standards options, and all the other rules that facilitate successful standardization. In table 12.1, instead of referring to all the detailed features of the videotex, X.400, and facsimile standards described in our case studies, we have summarized the output of the standardization processes—the dependent variable—in a way that makes it possible to consider success and failure from an institutional perspective.

Most relevant from the CCITT's point of view are the number of options included in a standard and the consistency of these options. If too many options are available for the design of the numerous components of a modern telecommunications system, it may very soon reach a state of opaqueness and confusion, especially when the standards are not embedded in an all-embracing architectural frame of reference.[1] The situation will be even worse if the options are inconsistent or even incompatible. Thus, parsimony and consistency of standards and standard options are salient institutionalized goals of organized standardization in general and of the CCITT in particular. These goals can be used to measure the success or the failure of standardization.

According to these criteria, only the standardization of message handling was a success, although some inconsistencies between the 1984 and the 1988 version of that standard have been noticed. The success of facsimile standardization is confined to the original Group 3 standard and its predecessors for the analog telephone network, whereas for digital networks two incompatible options have been adopted. Even the invention of a new option during the controversy surrounding facsimile standards for digital networks did not help to overcome the conflict but instead eventually reinforced incompatibility. However, since facsimile will not significantly influence network evolution but will rather be determined by this evolution, the consequences of unsuccessful standardization may not be very serious. In contrast, the standardization of videotex (which was designed as a virtually new telecommunications system) clearly failed. Not only two but several options were issued by the CCITT, none of them compatible with any of the others. This failure had highly detrimental repercussions for the development of videotex at the international level.

The cases we have studied relate to the new technical opportunities in telecommunications based on digitization. The coordinated development of an infrastructure that would make use of these opportunities under the control of the CCITT was the explicit goal of this organization, and this goal was to be achieved through standardizing telematic services. Although videotex failed and facsimile has been stumbling, the X.400 process can be considered a success, at least as far as standardization is concerned. Ironically, in terms of diffusion it has not been doing well. (In this sense it is similar to videotex, which did much worse at the standardization stage.) Electronic mail within the Internet, the great competitor in the market, which is based on standards developed outside the CCITT, clearly has taken the lead. However, the question of implementation of standards lies outside the framework of the CCITT.

Moreover, the CCITT's rules and procedures apparently do not provide a framework that excludes failures of the standardization process such as those we have encountered. In facsimile standardization for digital networks especially, restricting argumentation to technical reasoning as the only legitimate type of reasoning actually reduced the zone of negotiation in a way that rendered a compromise impossible. Opening the floor to economic reasoning might have been useful, at least in this case, and might very well prove useful in other cases where routinized standard setting is likely to end in a deadlock. Thus, more flexibility and more variability of the still rather stable institutional framework may be the appropriate and necessary answer of the CCITT to the challenges resulting from the new actor-related and technology-related sources of variation in standardization.

Conclusion

A large technical system, particularly a telecommunications system, is built up from a multitude of technical components. Such systems cross national borders and develop into global networks which are neither controlled nor even coordinated by any single authority. Compatibility is the crucial requirement that has to be fulfilled by the various components; otherwise, there will be segmentation instead of integration, and malfunctioning instead of smooth interoperation of components. Even in a world with divergent and separate telecommunications systems, compatibility remains an essential requirement. Standardization is one possible way to achieve compatibility.

Compatibility standards are rules specifying relational properties of technical artifacts. They are needed when these artifacts are meant to interoperate with others. Compliance with these rules affirms compatibility and hence makes it possible to build up comprehensive technical systems. Seen from this perspective, standardization is an essential aspect of technical development. As elements of the stock of technical knowledge, standards represent the "state of the art" in both a factual and a normative sense. They help to control technical change without preventing it.

Our study of standard setting by committees shows that by no means do the participants simply execute what an invisible "techno-logic" prescribes. Standardization is a process of choosing among several alternatives or options, none of which is necessarily the "best." That technology or technical development is socially shaped has been demonstrated time and again, especially by those who take social constructivist approaches. It would be misleading, however, to infer from this that technical knowledge, architectures of technical systems, and the existing material technology are of only minor importance in standardization. In an institutional context we find that, although technology, economy, politics, and science all affect the work of standardization, they do not form a "seamless web" (Hughes 1986).

In many cases of institutionally embedded, organized standardization, a technical perspective is systematically privileged. In the Comité Consultatif International Télégraphique et Téléphonique, the central organizational focus of our analysis, technical reasoning is in fact the only explicitly legitimate type of reasoning operating at the working level. Economic argumentation is officially banned and must, if possible, be translated into technical reasoning in order to influence the course of standardization.

The technical perspective does not guarantee broad support in committee standardization, even if the consensus being striven for has traditionally been reached without much effort. Digitization has made possible a convergence of basic elements of telecommunications and data-processing technology. This development, together with the liberalization of telecommunications in many countries, has transformed negotiations on standards from a pure-coordination game into a "battle of the sexes." The actors share an aversion to incompatibility, but they prefer different standards on technical grounds and on economic or other grounds. Each standard carries the promise of a positive payoff for each player, but the benefits are distributed unequally and therefore it is difficult to reach an agreement. Formal and informal rules of the CCITT, including the privileging of technical reasoning, are meant to prepare the way for cooperation. This sometimes helps to transform the controversy into a pure-coordination game. In other cases this transformation fails and the game turns into a zero-sum game, in which the winner could take all and therefore no agreement is reached.

We found instances where, during the processes of negotiation (which alternated between bargaining and joint development), novel technical proposals were worked out that could be explained only in terms of the actor-process dynamics operating in the CCITT. Thus, the CCITT and other standardization organizations are not simply clearing agencies that select a technical solution from a given pool. Rather, they create their own solutions, which later, as elements of the organizations' history and tradition, shape the work on related standards.

In looking for the seams in the web of factors influencing and framing standardization, we also noticed the social selectivity of the standards organizations. Not every business firm, public administration, user organization, or other interested group participates in standardization. Some (nowadays, relatively few) are unwelcome or are not formally admitted; others cannot afford to participate, especially at the international level. In the CCITT, public network operators (postal, telephone, and telegraph administrations) and regulated private network operators (recognized private operating agencies) still play prominent roles. Engineers find access to standardization organizations easier than business managers. In addition, the CCITT is embedded in a context of diplomacy—a state of affairs more pronounced than with other organizations. Though political actors rarely interfere directly with standard setting, they monitor the work, standing ready to protect "national interests" from harm.

Thus, international technical coordination through standardization in the telecommunications industry embodies a peculiar mix of actors and interests—a mix influenced by technical compatibility requirements and by organizational as well as national coordination interests. The rules and working procedures of the CCITT can be regarded as an attempt to conciliate or neutralize various perspectives and interest positions on the common bases of technical reasoning and consensual choice. Whether this provides an adequate institutional basis for coping with the social and technical complexity and the heterogeneity of telecommunications remains an open question. Our study of the standardization of facsimile, interactive videotex, and message handling shows that to simply suggest an institutional mismatch would be to miss the point. On the other hand, some weaknesses rooted in an overemphasis on the technical perspective and in a cognitive bias toward the traditional demands of the telephone network are obvious. They cannot be overcome by merely speeding up standardization work or by simply changing the division of labor within the International Telecommunication Union; rather, they require extension and augmentation of the ITU's cognitive frame. They also require that the ITU develop a critical distance from its own tradition.

Notes

Introduction

1. Electronic mail is currently a growing service, but its diffusion has been based largely on standards competing with X.400, mainly on the Internet protocols.

Chapter 1

1. For a critical assessment of the implications of this "radical" postulate in the analysis of science, see Collins and Yearley 1992. See also Collins 1995.

2. The specific concept of the technological trajectory in economics "is linked with (a) the development of specific infrastructures; (b) system scale economies; (c) complementary technologies; and (d) particular technical standards that positively feed upon specific patterns of innovation" (Dosi 1988, p. 1146).

3. From an institutionalist point of view it is not surprising that Sørensen and Levold also challenge the assumption of a high degree of similarity between science and technology. In our argument, however, this point is only marginal.

4. A first empirical analysis of the development of technical systems, which illustrates the perspective of this approach, was published in 1995 by Schneider and Mayntz. Unfortunately, it is available only in the German language.

Chapter 2

1. For a summary see pp. 53–58 of Chandler 1990.

2. In order to survive the challenge, Edison's company merged with Thomson-Houston in 1892 to become the General Electric Company (Hirsh 1989, p. 17).

3. In 1899, after a stock swap with American Bell, AT&T became the parent company of the Bell System.

4. In its functions, the NTEA is comparable to the American National Standards Institute Standards Committee for Telecommunications ("ANSI T1"), which was set up after the divestiture of AT&T, when hierarchical coordination was no longer considered feasible. (See chapter 3 below; see also Hawkins 1992.)

5. For more recent examples of the significance of quality standards for compatibility purposes in telephone networks and other systems technologies, see Rosenberg 1992.

6. For a general discussion of the advantages and disadvantages of hierarchical coordination that is applicable to our case in point, see Scharpf 1993a.

7. Henderson and Clark (1990, p. 12) characterize the replacement of analog with digital components in telephone handsets as a modular innovation that confines changes to the "core concepts" of components without changing the system's architecture.

8. Anderson and Tushman (1990, p. 628) acknowledge the need to explore the social dynamics of standardization more specifically in future research.

9. This also applies to the concepts of "technological guideposts" (Sahal 1981), "inducement mechanisms and focusing devices" (Rosenberg 1976), and "meta models" and "meta designs" (Disco et al. 1992; Rip 1992).

10. Discontinuities include competing product and process designs, rivalry between alternative technological regimes, and enhancement of certain competencies and destruction of others. After a period of substantial uncertainty, the competition of different designs and regimes in the era of ferment is eventually closed by the emergence of a dominant design (Tushman and Rosenkopf 1992).

11. For a more theoretical treatment of the underlying structure of social functional differentiation, see Luhmann 1984.

12. OSI defines interconnection as a set of hierarchically ordered functions. Each of the seven layers uses the service of the lower layer and provides a service to the upper layer. Service protocols are agreed conventions on what functions the services provide.

13. In Thompson's (1967, p. 56) organizational perspective, standardization "involves the establishment of routines and rules which constrain action of each unit into paths consistent with those taken by others in the interdependent relationship."

Chapter 3

1. For a list of the IEC's technical committees and subcommittees relevant or related to telecommunication, see and compare Macpherson 1990 (pp. 115–116) and Wallenstein 1990 (p. 86).

2. ANSI T1 and TTC are virtually national organizations (representing the United States and Japan, respectively) that have regional significance (Hawkins 1992). Foreign members are admitted. The membership structure and voting procedure of T1 and TTC are company-based, whereas ETSI's top-level decisions are taken by nation-based weighted voting. At the working level, a company-based consensus orientation prevails in all organizations (Besen 1990a,b; Lifchus 1985). The only direct regional counterpart to ISO/IEC is the European CEN/CENELEC.

3. For other examples, especially of groups promoting standardization in conformance with the Open Systems Interconnection reference model, see Genschel 1995; Jerman-Blazic 1988; Dankbaar and van Tulder 1989.

4. For a more comprehensive analysis (which, however, neglects the professional dimension), see Genschel 1995. Genschel also provides a detailed description of the multitude of coordination agreements, bodies, and procedures that render the peaceful coexistence of international standardization organizations possible. See also Egyedi 1996.

5. The concept of organizational fields suggests that a change of organizations belonging to the same line of business leads to isomorphism but not necessarily to enhanced efficiency. The forces that trigger isomorphism are coercion (by governments, legal regulations, etc.), imitation, and normative pressures (stemming especially from professionalization) (DiMaggio and Powell 1991).

6. Egyedi 1996, pp. 111–120; David and Greenstein 1990; Cargill 1988; Robinson 1986; Cerni 1984; Jones 1979; Verman 1973, pp. 150–188.

7. The literature on compatibility standards (see Branscomb and Kahin 1995; Gabel 1991, pp. 1–15, 174–176; Reddy 1990; Besen and Saloner 1989, pp. 178–186; Berg 1989b; Kindleberger 1983) has not devoted much attention to this aspect. The institutionalization of standardization in international organizations seems to have gained so much momentum that underprovision is not perceived to be a problem.

8. This does not preclude considerations of institutional and regulatory settings which might be created to overcome the problems of market failure (Blankart and Knieps 1993a).

9. Similar developments could be observed in Germany (Lundgreen 1986; Voelzkow et al. 1987). A systematic comparative analysis of the history of standardization organizations remains to be written, however.

10. Our analysis of the reality of standard setting, however, shows that internally problems arising from the institutional compounding of political, commercial, and technical perspectives have always been present.

11. This view has been questioned by Liebowitz and Margolis (1990a, 1994).

12. The competition between Betamax and VHS is often taken as an example of the possibility that market standardization can pick the wrong horse, as

Betamax was supposedly the superior technology (Arthur 1990, p. 80). This possible market outcome notwithstanding, we avoid making any reference to the supposed inferiority of VHS.

Chapter 4

1. In chapter 1 we introduced a conceptual distinction between organizations and institutions. We regard institutions as systems of abstract rules, whereas organizations embody a specific selection, combination, operationalization, and implementation of these rules with a higher normative obligation based on formal and informal sanctioning mechanisms. As institutions are to a large extent consciously designed, organizations are not only influenced by them but can also enter into what Mayntz calls "constitutional negotiations" in order to change the institutional arrangement ("constitution") of a sector (Mayntz 1993a, p. 13).

2. "Public evil" is Hirschman's (1970, pp. 98–105) term for such a good with negative individual effects. As long as no individual commitments in favor of a specific solution have evolved (e.g. investments "sunk" in another solution), no single coordinative standard will be perceived as a "public evil"; any standard is better than none at all.

3. The underrepresentation of users has recently turned into a political issue. On the potential role and institutional integration of users in standardization, see Foray 1992; Foray 1994; Tamarin 1988; Dankbaar and van Tulder 1992; Voelzkow and Eichener 1992; Voelzkow 1995.

4. We have already quoted some relevant studies. Other notable works are Fudenberg et al. 1983; Farrell and Saloner 1986; Swann 1987; Farrell and Shapiro 1988; Cowan 1992; Katz and Shapiro 1992. For a comprehensive review, see David and Greenstein 1990 and Besen and Farrell 1994.

5. Outlining an agreement made between the US delegation and the French delegation, Sirbu and Zwimpfer (1985) also disclose several factors at work other than the technical or the economic. Although the general US position was to reject a standard that had obvious technical flaws, the US delegates agreed to accept the 1976 version of X.25 since France offered cooperation on other unrelated issues at the Plenary Assembly (ibid., p. 40).

6. As was mentioned above, DEC, Xerox, IBM, and Honeywell were also ECMA members because they had computer manufacturing facilities in Europe.

7. From this perspective, the relations resemble political "issue networks" (Heclo 1978), which tend to lack a powerful core and which display quite variable degrees of mutual commitment, rendering tradeoffs as well as coalition changes easily possible (see also Mayntz 1993b).

8. Practitioners of standardization sometimes hint at the nature of these dynamics by offering good advice on ways to overcome blockage, waste of

time, or personal frustration (Mooney 1986; Daughtrey et al. 1986; Buckley 1986; Verman 1973; Cerni 1984, pp. 190–196).

9. Katz and Shapiro (1985, p. 425) propose this definition: "If two brands of hardware can use the same software, then the hardware brands are said to be compatible." For broader discussions of technical interrelatedness, complementarity, and compatibility, see Schmidt and Werle 1992 and Genschel 1995.

10. See Colman 1982 for a good, nonmathematical introduction to game theory. See Ryll 1989 for a discussion of the scope and the problems of the application of game theory in social research.

11. The story behind this game relates to a couple wanting to spend the evening out. Each prefers going out together to being alone; however, while he prefers boxing, she wants to go to the theatre. If neither readily gives in (which is facilitated if a routine exists for dealing with such distributive issues—e.g., he simply always gives in, or she invites him for dinner if he goes to the theatre), both may end up going out alone, even though this is worse than spending the evening together at the partner's favored event. (Colman 1982, p. 97)

12. The story behind this game is well known. Two prisoners, held in separate cells, are suspected of a serious crime. The prosecutor tells each prisoner that if neither of them confesses they are both to be released (as there are no other clues). If both confess, each will receive a moderate sentence. However, if only one confesses, he will get a very low sentence, while the other one denying the deed will receive the maximum penalty. The dilemma is that both would be better off not confessing, but without communication and binding agreements between them not confessing is a risky strategy: it could result in the maximum penalty should the other one confess. Both will therefore confess, and thus they will realize neither the common nor the individually optimum outcome. Nevertheless, under these specific conditions, rationally acting individuals will pursue this "dominant" strategy, because it "leaves each player better off no matter what the opponent does" (Tsebelis 1990, p. 62).

13. A recent game-theoretical approach, which explicitly uses negotiations in the International Telecommunication Union on standards for international telegraphy at the end of the nineteenth century as an example, combines "distributional" with "informational" problems (Morrow 1994). The presence of "informational" problems (e.g., uncertainty about payoffs, or lack of common knowledge generally) "creates a mutual interest in sharing information" (ibid., p. 400). In the game's first round, the actors select the game to be played—which, in the case of two players, is either a battle of the sexes or one of two variants of an "assurance game."

14. Weiss and Sirbu's (1990) statistical analysis of factors that influence the choice of technologies in standards committees can be regarded as an exception, although their multivariate statistical design, based on questionnaire data, necessarily neglects the dynamics of the negotiation process in committees and reduces the choice to a single binary "adopt or not adopt" decision.

15. They are "corporate actors," usually with a corporate constitution, with their own resources and concomitant interests (Coleman 1990, pp. 325–370, 421–450; Schneider and Werle 1990). However, their actor characteristics are not only internally generated but also externally attributed. A stream of organizational actions is interpreted by others as corresponding to a consistent set of collective interests (compare Scharpf 1991).

16. On international negotiation, see Kremenyuk 1991.

17. Luhmann (1983b), in his treatise on the sociology of law, differentiates two types of expectation: the cognitive and the normative. The first type of expectation, one is willing to change whenever frequent deviance is experienced. The second type, however, is stabilized "contrafactually"; i.e., even if observed behavior deviates from expected behavior, one holds onto the expectation. In Luhmann's terms, market standards definitely rely on cognitive expectations, whereas committee standards initially embody normative expectations. Unlike (usually national) legal rules, however, "our" international standards lack any policing agency that could enforce compliance. Thus, from the point of view of the addressees of standards, the normative nature of the expectations is rather weak.

Chapter 5

1. The concept of telematic services was adopted by the seventh Plenary Assembly of the CCITT in 1980 (Wallenstein 1990, p. 22).

Chapter 6

1. The informality of national coordination in the CCITT contrasts with the situation in the ISO, whose members are the national standardization organizations (e.g., DIN, BSI, AFNOR, ANSI).

2. The nomenclature of the standards is likely to cause confusion. The CCITT's teletex is an advanced version of the well-known telex service. Teletext is the CCIR's system of broadcasting pages of text via the TV. What the CCITT calls (interactive) videotex is indeed the interactive version of the CCIR's teletext. In Germany, Teletext is called Videotext.

3. An extraordinary Plenipotentiary Conference in Geneva in December 1992 resulted in many changes. As we have already mentioned in chapter 3, the CCITT as such does not exist any more; it was replaced by the ITU-T, which operates quite similarly to the former CCITT yet which relies heavily on postal ballots to accelerate the approval of standards. This procedural element, first introduced in the 1989–1992 study period, was slightly modified and reinforced in 1993 by the first World Telecommunication Standardization Conference (the successor of the Plenary Assembly). For our analysis of the three

standardization processes, however, it is sufficient to discuss the structure of the old CCITT.

4. See pp. 226–236 of Schelling 1984 for a detailed example of how a decision on which of the two parts of a problem to decide first can greatly bias the outcome.

Chapter 7

1. Such conflict, which is quite unique to the CCITT, is more akin to the controversies in the CCIR surrounding the standardization of color TV in the 1960s (Crane 1979) or to the ongoing problems experienced with the standardization of high-definition television (Clarke 1979; Nickelson 1990).

2. The name "videotex" was originally used on a provisional basis for this "public network-based interactive information retrieval system." The questions were as follows. SG I: "What new operational Recommendations are required and what additions or modifications to existing Recommendations are needed for the provision of these services internationally, specifically: 1. Definition of the international system; 2. Definition of the international services to be provided; 3. Operational procedures. Taking account of: a) communication via general switched telephone networks and/or public data networks; b) interworking with other services as appropriate, e.g. telex and teletex; c) broadcasting teletext services for general reception; d) retrieval of information from remote data banks; e) interactive communication with data processing services; f) message store and forward facilities; g) procedures for entering information into the data banks used in the international service." SG VIII: "What new technical Recommendations are required and what additions or modifications to existing Recommendations are needed for the provisions of these services internationally when taking account in particular, of: [point a) to g) above]." Annex I to this question contained a nonexhaustive "list of important points"—for example: "There could well be a requirement to identify the respective interests of CCITT and other international standard organizations, e.g. ISO. 1. Display characteristics; 2. Character repertoire; 3. Subset of the character repertoire used for commands from the terminal; 4. Line control procedures; 5. End-to-end control procedures." (CCITT Circular 98/Dir. 24.5.1978 [erroneously dated as 24.3.1978]).

3. Viewdata was later to be called Prestel.

4. The CCETT was created in 1972 as a joint research body of the Centre National d'Études des Télécommunications and the Office de Radiodiffusion et Télévision Française.

5. Antiope stands for "acquisition numérique et télévisualisation d'images organisées d'écriture."

6. Compte Rendu de la Reunion Tripartite Bundespost, GPO, CCETT les 26 et 27 Janvier 1978.

7. CAPTAIN stands for "character and pattern telephone access information network."

8. At the same conference, CBS presented a common teletext system together with France and Canada. It was later endorsed by the CCIR as one of three options, along with a British and a Japanese system (O'Brien 1982, p. 24).

9. The Swedish version of Prestel differed from the original in the type of graphics used. It did not enjoy as much support for investment and implementation as the British and French standards.

10. On conditions for videotex in the US, see Sullivan and Noll 1982. For a reference emphasizing standards, see Lipinski et al. 1981.

11. The agreement included the following basics: 1. Concept of WWUVS; 2. Definition: Superset of functionalities coded in a compatible way in the international service, based on the existing systems; 3. Use of PPDU (Presentation Protocol Data Unit) to identify blocks of functionalities to make transcoding easier and to allow a step-by-step approach to WWUVS; 4. The PPDU structure should permit the development of new functionalities which must be coded in a compatible way; 5. Formalization of the best working structure to implement points 1 to 4. (Lazak 1984, p. 197)

12. Twice in France and once each in the US, Canada, and Japan.

13. Modern online services and the Internet present serious challenges to videotex, which cannot be technically upgraded and adapted to them (Schneider and Werle 1996).

14. For an overview of the concepts of the various systems and related technical and standards issues, see Gecsei 1983 and Ray-Barman 1985.

15. A code designates which combination of binary digits represents which alphabetical, numerical, or other type of character, including special symbols and graphical information. The code is used to process and transmit information at high speed. A decoding procedure translates the information back into its original form. The number of signs a code comprises depends on the number of bits used for coding. In principle, the lower the number of bits required, the lower the costs of coding and decoding are but the earlier the capacity limits are reached. This is problematic for low-bit codes because codes generally also include several control codes reserved for special functions.

16. For details see Zimmermann 1979 and Clarke 1979.

17. Teletext is based on the finding that the frame flyback periods to TV picture transmission may carry a limited amount of data.

18. It is said that Antiope was first demonstrated at a fair in Moscow. A Russian engineer asked whether Antiope could be used as an interactive service, and

when this was affirmed he demanded to see a demonstration. (Giraud 1991, pp. 68–69)

19. It was only after an unexpected demonstration before the Canadian Deputy Minister that the decision went in favor of this domestic development (Parkhill 1981, 1985).

20. See Binder 1985 (pp. 61–149) and Hancock and Balko 1983. For a list of organizations involved in Canadian videotex—notably Norpak, Communications Research Centre, and Bell Northern Research—see p. 557 of Bown et al. 1980.

21. See p. 224 of Tydeman et al. 1982 for a list of some of the organizations that were involved in videotex and teletext, among them American Express, Citibank, CBS, GTE, IBM, GEC, Texas Instruments, and Western Electric. See also p. 86 of Mosco 1982.

22. A Prestel International service, however, was commercialized after a field trial in the second half of 1981. Via international calls, Prestel databases could be accessed from several countries. Nevertheless, this "de facto standardization" strategy ran up against significant regulatory problems (Ford 1981; Cooke 1981).

23. The CEPT, with its more homogeneous and smaller circle of participants, is less formal; hence, political aspects can be discussed more openly in that body.

24. Sometimes a national mandate could be converted into a legitimate technical argument.

25. Although by 1980 it was already assumed that business applications would dominate private use, this had no real repercussions for terminal standardization; the TV set and its capabilities continued to be the reference point.

26. Siemens first had a multistandard software platform to interconnect different national systems in 1985 ("Der Gleisanschluß zum internationalen Rechnerverbund," *Blick durch die Wirtschaft* 14 (1985), no. 6: 5). Once economic expectations had dwindled, the new technological possibilities were at least able to facilitate the standardization work that ensued.

Chapter 8

1. For the history of facsimile, see Wenger 1989, McConnell et al. 1989, and McConnell et al. 1992.

2. In Germany, however, use of the PSTN for facsimile transmission was permitted in the early 1940s, and again in 1964 (Bohm et al. 1979, pp. 173, 178).

3. On parallels between the facsimile sector and the computing sector in this period, see Brock 1975 and Brock 1986.

4. The decision to divide facsimile standards into groups based on the transmission time for a typical A4 page (Group 1 = 6 minutes) was taken in the 1973–1976 study period (recommendation T.0). The distinction is introduced here, out of chronological order, for ease of presentation. For a short overview of the different facsimile groups see Kroemer 1987.

5. Recommendation T.1 is for phototelegraphy. "Facsimiles are classified into either *photographic* facsimile, in which the original copy is reproduced faithfully with graded tonal densities, or *document* facsimile, in which the original copy is reproduced primarily with black and white." (Kobayashi 1985, p. 27) Specialized facsimile equipment for the press, the police, or meteorologists do not concern us here, only office equipment.

6. Significant expectations were based on this work, as is evident from the international colloquia organized by the CCITT in 1979 and 1980.

7. For example, multiple addressing, deferred delivery, and standards conversion facilities in the network.

8. In 1987, market shares per company were as follows: Ricoh 21%, Matsushita 20.5%, Canon 13%, NEC 13%, Toshiba 13%; others 19.5% (McConnell et al. 1989, p. 177). Since the start of Japanese dominance in production, OEM channels have played a significant role in overseas marketing, often via firms that used to produce their own machines.

9. With regard to the mechanical aspects of the terminal, it has been found that fewer movable parts reduce the susceptibility to malfunctioning. For that reason, this criterion was instrumental in orienting the direction of development.

10. Group 1 and Group 2 have a resolution of 3.85 pel/mm vertically; Group 3 has 8 pel/mm horizontally and 3.85 pel/mm vertically (7.7 as an option); Group 4 has 200, 240, 300, or 400 pel/inch horizontally and vertically.

11. A digital network (such as the ISDN) that integrates a multitude of services, including voice, data, and pictures, does not technically differentiate types of information (Nussbaum and Noller 1986).

12. Graphic Sciences, founded by two defectors from Xerox in 1967, was acquired by Burroughs in 1975. Half of the approximately 50,000 facsimile machines in use in the United States in 1972 had been installed by Xerox; most of these were Telecopier 400 machines. Japan had about 60,000 machines in operation in 1972 (Scherer 1992, pp. 97–102).

13. Graphic Sciences, 3M, Muirhead, Plessey, Rank Xerox, Rapifax, and the Xerox Corporation (COM XIV (3E), 1977/80).

14. With this conflict we leave the written document base and turn to our interview material. Many interviewees who had been active in facsimile standardization in the 1970s made no reference to the conflicts mentioned above

but tended to regard the ECM case as the first minor turbulence and the Group 3 digital conflict as the only major one.

15. There was no consensus among those interviewed as to when the proposal was first made. Though in most interviews 1988 was mentioned, some informants believed that the idea had been around earlier, or that it had originated not in SG VIII but in SG I.

Chapter 9

1. Bell Northern Research is a joint R&D organization owned by Bell Canada (30%) and Northern Telecom (70%). In the late 1970s about 30% of its research funds were provided by Bell Canada, but Northern Telecom now provides 90%. Bell Canada and Northern Telecom have another connection: BCE Inc., the owner of Bell Canada, holds 52% of the shares in Northern Telecom.

2. Germany with Siemens and Sweden with Ericsson (Schenke et al. 1981; Henricson 1984).

3. The organizations that participated in the meetings, organized by country, were as follows. Canada: BNR (2 delegates), CTCA/TCTS (1), Teleglobe (1). France: PTT (2), CNET (2). Germany: Siemens (1). Japan: CIAJ (1), KDD (5), NTT (1), OIAJ (1). Netherlands: PTT (1). Spain: CTNE (1). Sweden: Televerket (2). United Kingdom: GPO (2), Cable & Wireless (2), Plessey (1). United States: AT&T (1), BBN (1), DEC (1), GTE/CNS (1), GTE (1), GTE/Telenet (1), NBS (1), Xerox (1). (Source: COM VII (10E: 5), 1981/84) The meetings took place in France, the United States, and the Netherlands.

4. For an overview of the work of various organizations (IFIP, NBS, CCITT, ECMA, ISO) involved in MHS standardization, see Wilson 1984. On the ECMA, the ISO, and the CCITT, see Willmott 1989. On procedural differences among the organizations and differences in technical specifications for MHS, see Campbell-Grant 1985.

5. After JTC1 had been set up by ISO and IEC to deal with information technology standardization (see chapter 3 above), MHS standardization took place in its WG4 of Sub-Committee 18.

6. On this point there is some inconsistency in our material. It was often maintained in interviews that joint meetings had been impossible, owing to reservations on the part of ISO and the CCITT, with the result that, until recently, they had to be "co-located." In general, it seems that the administrative leaders of the CCITT and the ISO were not involved much in the decisions on interorganizational coordination.

7. Of all the national bodies that commented on the draft international standard (DIS) for MOTIS, the one from the UK was the most active. However, it focused on technical perfection and the appropriate ISO point of view rather than on compatibility with X.400.

8. Differences between the CCITT and ISO versions were published as an annex to the standards. See also Manros 1989.

9. In 1990, the following companies planned to support X.400 or had implemented it: Apollo, AT&T, Bell Canada, British Telecom/Tymnet, Bull, Concurrent, Convergent, Digital, Data General, DBP Telekom, Ferranti, France Telecom, GEC, Hewlett-Packard, IBM, ICL, Indisy, Logica, MCI, Motorola, NCR, Network Systems, Norsk Data, Novell, Olivetti, Philips, Plessey, Prime, Pyramid, Retix, Sema, Sprint International, Sun Microsystems, Tandem, Torch, Unysis, Wang, Wollongong, and 3Com.

10. A new window of opportunity has been opened for X.400 in the context of the Defense Message System (DMS) initiative of the US Department of Defense. A new military messaging system based on the 1988 X.400 recommendation and on the X.500 series of directory recommendations adopted by the ITU-T in 1993 is expected to provide messaging with business-quality reliability, robustness, and security. It is projected that by the year 2000 some 2 million users will be connected to this network, which does not exclude Internet mailing components but which has X.400-based technology at its core. Such a high-quality messaging system is likely to attract corporate users, who tend to regard e-mail via the Internet as insecure.

11. P2 was mistaken in the 1984 version as a protocol; it then turned out to be in conflict with the appropriate definition of protocols in the OSI context. However, the specifications kept the name P2 in X.420.

12. Höller (1993, p. 144, n.148) assumes that the former generation of point-to-point transmission systems, in which terminals are always ready to receive, was taken as a model.

13. On the conflict between the national regulatory situation and the ADMD/PRMD distinction, see Kirstein 1984. The distinction between ADMD and PRMD may have also been introduced by SG I had that group been involved from the beginning, the reason being not so much the routing of messages as the need for an organizational distinction to facilitate the definition of service requirements, accounting rules, and service tariffs. The allocation of this responsibility to public operators and ADMDs was a very natural decision in this context.

14. In 1988, only ISO/IEC/DIS 10021–5 (the message store) was voted on for the first time. All the other parts had been distributed before: 10021 parts 1–3 as ISO/IEC/DIS 8505, parts 4 and 6 as DIS 8883, and part 7 as DIS 9065.

15. For instance, MOTIS protocols were likely to stop the whole operation of X.400 implementations.

16. Ironically, the technical compromises accepted in the CCITT to maintain backward compatibility—as opposed to ISO's position of revising earlier mistakes in favor of the superior, long-term technical solution—failed in the end to achieve compatibility without involving renewed adaptations.

17. Consequently, as early as 1986, KDD of Japan introduced a text-oriented electronic mail service based on the 1984 X.400 standard. Two years later this was upgraded to provide multimedia communication, including Group 3 and Group 4 facsimile as well as telex and teletex intercommunication (Miki 1991).

Chapter 10

1. On the process of institutionalization of techno-economic projects, see Lundin 1986.

2. However, in May of 1989 the United States retracted an earlier endorsement "in international standards-setting bodies of Japan's HDTV production standard, due to pressure from the US electronics industry and the EC" (Mastanduno 1991, p. 106). This change of direction in standardization was not accompanied by an independent alternative proposal, and therefore it was a "cheap" means of causing trouble for the Japanese. For the set of relevant actors and the general context of US telecommunications policy, see Dutton 1992, Pitt and Morgan 1992, and Geller 1995.

3. Such positions, however, remain easier to achieve in "state-centered" systems such as that of France than in, say, the United States.

Chapter 11

1. Other approaches—albeit from different theoretical traditions—distinguishing "modes of reasoning" (Hagendijk 1990), "modes of behavior" (Aitken 1985, p. 17), "dimensions of rationality" (Hård 1994), or "types of discourse" (Bora and Döbert 1993) have demonstrated the usefulness of this strategy. For further theoretical references, see Schmidt and Werle 1993.

2. Bijker (1995a, p. 123) argues that technological frames provide the goals, the thoughts, and the tools for action. They guide thinking and action. A technological frame offers both the key problems and the related problem-solving strategies. Bijker, however, does not introduce the concept in order to distinguish the technological frame from other, nontechnological, frames. Rather, he wants to point to internal differentiations in the image of technology: different notions of a technology can exist in parallel and can compete as different technological frames.

3. Thus, as far as standardization is concerned, we disagree with Cowhey's claim (1990, p. 176) that "the conventional view of the telecommunications regime as primarily a technocratic exercise in technical collaboration [is] wrong"; the conventional view is still adequate in our opinion. This fact notwithstanding, a critical assessment of the ITU may well come to the conclusion that this intergovernmental organization is the core of a rather inefficient international telecommunications regime. Our research does not provide an empirical basis for appraisal of this conclusion.

4. The Internet Engineering Task Force (IETF) also emphasizes the value of a technical and engineering perspective if standardization problems are at issue. (See chapter 3 above.)

5. Commercially unsuccessful standards also are often labeled "too theoretical," just as unsuccessful standardization processes are often labeled "too political."

6. For a general discussion of the role of "jurisdictional conflicts" between professions engaged in the process of establishing and consolidating professional domains from the perspective of the theory of professionalization, see Abbott 1988.

7. Rosenkopf and Tushman (1994) have used the concept of "coevolution" to decipher the interplay of technological and organizational community development.

8. On network structures, a former generic domain of telecommunication knowledge, see Huber 1987.

9. The other (more network-oriented than services-oriented) standardization project of the CCITT is the Integrated Digital Services Network (ISDN), the origins of which date back to the mid 1970s (Rutkowski 1985).

Chapter 12

1. However, the embeddedness of the standards in a more comprehensive system of standards (architectural entrenchment) is generally high or at least medium (videotex) in our cases. This follows from the fact that standardization was either based on a coherent cognitive technical frame (as in the cases of X.400 and facsimile for analog networks) or aimed, as a result, at the integration of the "new" facsimile service into the "old" (but gradually digitized) telephone network.

References

Abbate, Janet. 1994a. The Internet challenge: Conflict and compromise in computer networking. In *Changing Large Technical Systems,* ed. J. Summerton. Westview.

Abbate, Janet. 1994b. From ARPANET To Internet: A History Of ARPA-Sponsored Computer Networks, 1966–1988. Dissertation, University of Pennsylvania.

Abbott, Andrew. 1988. *The System of Professions: An Essay on the Division of Expert Labor.* University of Chicago Press.

Abernathy, William J., and Kim B. Clark. 1988. Innovation: Mapping the winds of creative destruction. In *Readings in the Management of Innovation,* ed. M. Tushman and W. Moore. Ballinger.

Adams, Walter, and James W. Brock. 1982. Integrated monopoly and market power: System selling, compatibility standards, and market control. *Quarterly Review of Economics and Business* 22, no. 4: 29–42.

Aitken, Hugh G. 1978. Science, technology and economics: The invention of radio as a case study. In *The Dynamics of Science and Technology,* ed. W. Krohn et al. Reidel.

Aitken, Hugh G. 1985. *The Continuous Wave: Technology and American Radio, 1900–1932.* Princeton University Press.

Akrich, Madeleine. 1992. Beyond social construction and things in the innovation process. In *New Technology at the Outset,* ed. M. Dierkes and U. Hoffmann. Campus.

Alchian, Armen A. 1984. Specificity, specialization, and coalitions. *Journal of Institutional and Theoretical Economics* 140: 34–49.

Allen, David. 1988. New telecommunications services: Network externalities and critical mass. *Telecommunications Policy* 12: 257–271.

Amos, S. W. 1978. A History of Ceefax. In *Ceefax: Its History and the Record of Its Development by BBC Research Department.* BBC Engineering.

Ancarani, Vittorio. 1995. Globalizing the world: Science and technology in international relations. In *Handbook of Science and Technology Studies,* ed. S. Jasanoff et al. Sage.

Anderson, Philip, and Michael L. Tushman. 1990. Technological discontinuities and dominant designs: A cyclical model of technological change. *Administrative Science Quarterly* 35: 604–633.

Antonelli, Cristiano. 1992. The economic theory of information networks. In *The Economics of Information Networks,* ed. C. Antonelli. North-Holland.

Aronson, Jonathan D., and Peter F. Cowhey. 1988. *When Countries Talk: International Trade in Telecommunications Services.* Ballinger.

Arthur, Brian W. 1989. Competing technologies, increasing returns, and lock-in by historical events. *Economic Journal* 99: 116–131.

Arthur, Brian W. 1990. Positive feedbacks in the economy. *Scientific American,* February: 80–85.

Barke, Richard P. 1993. Managing technological change in federal communications policy: The role of industry advisory groups. In *Policy Change and Learning,* ed. P. Sabatier and H. Jenkins-Smith. Westview.

Barnett, William P. 1990. The organizational ecology of a technological system. *Administrative Science Quarterly* 35: 31–60.

Barnett, William P., and Glenn R. Carroll. 1987. Competition and mutualism among early telephone companies. *Administrative Science Quarterly* 32: 400–421.

Baumol, William J., John C. Panzar, and Robert D. Willig. 1982. *Contestable Markets and the Theory of Industry Structure.* Harcourt.

Berg, Sanford V. 1989a. The production of compatibility: Technical standards as collective goods. *Kyklos* 42: 361–383.

Berg, Sanford V. 1989b. Technical standards as public goods: Demand incentives for cooperative behavior. *Public Finance Quarterly* 17: 29–54.

Besen, Stanley M. 1990a. The Economic Dimension of Standards in Information Technology: Telecommunications and Information Technology Standard-Setting in Japan. Mimeo, OECD, Paris (published in part in OECD 1991).

Besen, Stanley M. 1990b. The European Telecommunications Standards Institute: A preliminary analysis. *Telecommunications Policy* 14: 521–530.

Besen, Stanley M., and Joseph Farrell. 1994. Choosing how to compete: Strategies and tactics in standardization. *Journal of Economic Perspectives* 8: 117–131.

Besen, Stanley M., and Leland L. Johnson. 1986. *Compatibility Standards, Competition, and Innovation in the Broadcasting Industry.* Rand Corporation.

Besen, Stanley M., and Garth Saloner. 1989. The economics of telecommunications standards. In *Changing the Rules,* ed. R. Crandall and K. Flamm. Brookings Institution.

Bickers, Kenneth N. 1991. Transformations in the governance of the american telecommunications industry. In *Governance of the American Economy,* ed. J. Campbell et al. Cambridge University Press.

Bijker, Wiebe E. 1993. Do not despair: There is life after constructivism. *Science, Technology & Human Values* 18, no. 1: 113–138.

Bijker, Wiebe E. 1995a. *Of Bicycles, Bakelites, and Bulbs: Toward a Theory of Sociotechnical Change.* MIT Press.

Bijker, Wiebe E. 1995b. Sociohistorical technology studies. In *Handbook of Science and Technology Studies,* ed. S. Jasanoff et al. Sage.

Binder, Michael. 1985. *Videotex and Teletext: New Online Resources for Libraries.* Jai.

Blankart, Charles B., and Günter Knieps. 1993a. State and standards. *Public Choice* 77: 39–52.

Blankart, Charles B., and Günter Knieps. 1993b. Network evolution. In *On the Theory and Policy of Systemic Change,* ed. H.-J. Wagener. Physica.

BMFT (Bundesministerium für Forschung und Technologie). 1977. *Entwicklungslinien der technischen Kommunikation. Stand und Entwicklungstendenzen der Faksimiletechnik. Forschungsbericht.*

Bohm, Jürgen, Roswitha Wolf, Hartmut Nitsch, and Friedrich Bardua. 1979. Der Telefaxdienst der Deutschen Bundespost. In *Jahrbuch der Deutschen Bundespost* 29, ed. K. Gscheidle and D. Elias. Verlag für Wissenschaft und Leben.

Bora, Alfons, and Rainer Döbert. 1993. Konkurrierende Rationalitäten: Politischer und technisch-wissenschaftlicher Diskurs im Rahmen einer Technologiefolgenabschätzung von genetisch erzeugter Herbizidresistenz in Kulturpflanzen. *Soziale Welt* 44: 75–97.

Borenstein, Nathaniel S. 1991. Why do people prefer fax to email? In *Message Handling Systems and Application Layer Communication Protocols,* ed. P. Schicker and E. Stefferud. North-Holland.

Bouwman, Harry, Mads Christoffersen, and Tomas Ohlin. 1992a. Videotex in a broader perspective: From failure to future medium? In *Relaunching Videotex,* ed. H. Bouwman and M. Christoffersen. Kluwer.

Bouwman, Harry, Mads Christoffersen, and Tomas Ohlin. 1992b. Videotex: Is there a life after death? In *Relaunching Videotex,* ed. H. Bouwman and M. Christoffersen. Kluwer.

Bowden, Gary. 1995. Coming of age in STS. Some methodological musings. In *Handbook of Science and Technology Studies,* ed. S. Jasanoff et al. Sage.

Bowers, Albert W., and Edward B. Connell. 1985. A checklist of communications protocol functions organized using the open system interconnection seven-layer reference model. In *Computer Communications,* ed. W. Stallings. IEEE Computer Society Press.

Bown, Herbert, C. Douglas O'Brien, and W. Sawchuk. 1980. Telidon technology development in Canada. In *Proceedings of Viewdata '80.* Online Publications.

Bradach, Jeffrey L., and Robert G. Eccles. 1989. Price, authority, and trust: From ideal types to plural forms. *Annual Review of Sociology* 15: 97–118.

Bradner, Scott O., and Allison Mankin, eds. 1996. *IPng—Internet Protocol Next Generation.* Addison-Wesley

Branscomb, Lewis M., and Brian Kahin. 1995. Standards processes and objectives for the national information infrastructure. In *Standards Policy for Information Infrastructure,* ed. B. Kahin and J. Abbate. MIT Press.

Braunstein, Yale M., and Lawrence J. White. 1985. Setting technical compatibility standards: An economic analysis. *The Antitrust Bulletin* 30: 337–355.

Brock, Gerald W. 1975. Competition, standards and self-regulation in the computer industry. In *Regulating the Product,* ed. R. Caves and M. Roberts. Ballinger.

Brock, Gerald W. 1986. The computer industry. In *The Structure of American Industry,* ed. W. Adams. Macmillan.

Broomfield, J. 1981. Videotex: A regulator's nightmare. In *Proceedings of Viewdata '81.* Online Publications.

Bruce, Robert R. 1981. US legal and regulatory issues relating to teletext and videotex: A layman's guide to a lawyer's "no-man's land." In *Proceedings of Viewdata '81.* Online Publications.

Buchanan, James M., and Gordon Tullock. 1971. *The Calculus of Consent: Logical Foundations of Constitutional Democracy.* University of Michigan Press.

Buckley, Fletcher J. 1986. An overview of the IEEE Computer Society standards process. In *Computer Standards Conference 1986. Proceedings.* Computer Society Press.

Burns, Tom R. 1985. Actors, rule systems, social action: Rule system analysis for the social sciences. In *Man, Decisions, Society,* ed. T. Burns et al. Gordon and Breach.

Burns, Tom R., and Thomas Dietz. 1992. Technology, sociotechnical systems, technological development: An evolutionary perspective. In *New Technology at the Outset,* ed. M. Dierkes and U. Hoffmann. Campus.

Callon, Michel. 1987. Society in the making: The study of technology as a tool for sociological analysis. In *The Social Construction of Technological Systems,* ed. W. Bijker et al. MIT Press.

Campbell-Grant, I. R. 1985. Open systems application layer standards for text and office systems. In *Networks and Electronic Office Systems.* Institution of Electronic and Radio Engineers.

Cargill, Carl F. 1988. A modest proposal for business based standards. In *Computer Standards Evolution.* IEEE.

Cargill, Carl F. 1989. *Information Technology Standardization: Theory, Process, and Organizations.* Digital Press.

Carlson, W. Bernard. 1992. Artifacts and frames of meaning: Thomas A. Edison, his managers, and the cultural construction of motion pictures. In *Shaping Technology/Building Society,* ed. W. Bijker et al. MIT Press.

Carlton, Dennis W., and J. Mark Clamer. 1983. The need for coordination among firms, with special reference to network industries. *University of Chicago Law Review* 50: 446–465.

Cerni, Dorothy M. 1982. The CCITT: Organization, US Participation, and Studies Toward the ISDN. NTIA Report, US Department of Commerce, Boulder.

Cerni, Dorothy M. 1984. Standards in Process: Foundations and Profiles of ISDN and OSI Studies. Institute for Telecommunication Sciences, US Department of Commerce, Boulder.

Cerni, Dorothy M., and E. M. Gray. 1983. International Telecommunication Standards: Issues and Implications for the '80s. NTIA Report, US Department of Commerce, Boulder.

Chandler, Alfred D. Jr. 1977. *The Visible Hand: The Managerial Revolution in American Business.* Harvard University Press.

Chandler, Alfred D. Jr. 1990. *Scale and Scope: The Dynamic of Industrial Capitalism.* Harvard University Press.

Chapuis, Robert J., and Amos E. Joel Jr. 1990. *Electronics, Computers and Telephone Switching: A Hundred Years of Telephone Switching. Volume 2: 1960–1984.* North-Holland.

Childs, G. H. L. 1982. Chances for a worldwide videotex standard. In *Bildschirmtext Kongress 1982.* Diebold.

Chisholm, Donald. 1989. *Coordination without Hierarchy.* University of California Press.

Church, Jeffrey, and Neil Gandal. 1992. Network effects, software provision and standardization. *Journal of Industrial Economics* 40: 85–103.

Clark, Kim B. 1985. The Interaction of Design Hierarchies and Market Concepts in Technological Evolution. *Research Policy* 14: 235–251.

Clark, Richard. 1981. International technical standards. In *Viewdata in Action,* ed. R. Winsbury. McGraw-Hill.

Clarke, K. E. 1979. International Standards for Videotex Codes. *Proceedings of the Institution of Electrical Engineers* 126: 1355–1361.

Coase, Ronald H. 1937. The nature of the firm. *Economica* N. S. 4: 386–405.

Codding, George A. 1977. The United States and the ITU in a changing world. *Telecommunication Journal* 44: 231–235.

Codding, George A. 1984. Politicization of the International Telecommunication Union: Nairobi and after. In *Policy Research in Telecommunications,* ed. V. Mosco. Ablex.

Codding, George A. 1991a. Introduction: Reorganizing the ITU. *Telecommunications Policy* 15: 267–269.

Codding, George A. 1991b. Evolution of the ITU. *Telecommunications Policy* 15: 271–285.

Codding, George A., and Dan Gallegos. 1991. The ITU's 'federal' structure. *Telecommunications Policy* 15: 351–363.

Codding, George A., and Anthony M. Rutkowski. 1982. *The International Telecommunication Union in a Changing World.* Artech House.

Cole, Robert, Steve Kille, and Wah Ming Lee. 1985. Naming, Addressing and Directories in Open Systems. Technical report, Department of Computer Science, University College, London.

Coleman, James S. 1990. *Foundations of Social Theory.* Harvard University Press.

Collins, H. M. 1995. Science studies and machine intelligence. In *Handbook of Science and Technology Studies,* ed. S. Jasanoff et al. Sage.

Collins, H. M., and Steven Yearly. 1992. Epistemological chicken. In *Science as Practice and Culture,* ed. A. Pickering. University of Chicago Press.

Collins, Randall. 1986. *Weberian Sociological Theory.* Cambridge University Press.

Colman, Andrew M. 1982. *Game Theory and Experimental Games: The Study of Strategic Interaction.* Pergamon.

Comer, Douglas E. 1991. *Internetworking With TCP/IP,* volume I: *Principles, Protocols, and Architecture,* second edition. Prentice-Hall.

Constant, Edward W. II. 1984. Communities and hierarchies: structure in the practice of science and technology. In *The Nature of Technological Knowledge,* ed. R. Laudan. Reidel.

Cooke, T. M. 1981. The Prestel service in international markets. In *Proceedings of Viewdata '81.* Online Publications.

Cool, Karel, and H. Landis Gabel. 1992. Industry restructuring through alliances: 'Open systems' and the European mainframe computer industry. In *European Industrial Restructuring in the 1990s,* ed. K. Cool et al. Macmillan.

Cowan, Robin. 1992. High technology and the economics of standardization. In *New Technology at the Outset,* ed. M. Dierkes and U. Hoffmann. Campus.

Cowhey, Peter F. 1990. The international telecommunications regime: The political roots of regimes for high technology. *International Organization* 44: 169–199.

Cowhey, Peter, and Jonathan D. Aronson. 1991. The ITU in transition. *Telecommunications Policy* 15: 298–310.

Craigie, Jim. 1988–89. ISO 10021–X. 400(88): A tutorial for those familiar with X. 400(84). *Computer Networks and ISDN Systems* 16: 153–160.

Crane, Rhonda J. 1979. *The Politics of International Standards.* Ablex.

Crane, Rhonda. 1992. TV technology and government policy. In *The Telecommunications Revolution. Past, Present and Future,* ed. H. Sapolsky et al. Routledge.

Cunningham, Ian. 1983. Message-handling systems and protocols. *Proceedings of the IEEE* 71, Special Issue: 1425–1430.

Cunningham, Ian, D. Delestre, I. Kerr, T. Myer, Y. Sekido, D. Touillet, C. Ware, and N. West. 1982. Emerging protocols for global message exchange. In *Proceedings of COMPCON '82.* IEEE.

Cyert, Richard M., and James G. March. 1963. *A Behavioral Theory of the Firm.* Prentice-Hall.

Dankbaar, Ben, and Rob van Tulder. 1989. The Construction of an Open Standard: Process and Implications of Specifying the Manufacturing Automation Protocol (MAP). Working document, Netherlands Organization for Technology Assessment, The Hague.

Dankbaar, Ben, and Rob van Tulder. 1992. The influence of users in standardization: The case of MAP. In *New Technology at the Outset,* ed. M. Dierkes and U. Hoffmann. Campus.

Daughtrey, Taz, Roger Fujii, and Dolores Wallace. 1986. Case history: Development of a software engineering standard. In *Computer Standards Conference 1986.* Computer Society Press.

David, Paul A. 1985. Clio and the economics of QWERTY. *American Economic Review* 75: 332–337.

David, Paul A. 1987. Some new standards for the economics of standardization in the information age. In *Economic Policy and Technological Performance,* ed. P. Dasgupta and P. Stoneman. Cambridge University Press.

David, Paul A. 1992. Heroes, herds and hysteresis in technological history: Thomas Edison and "the battle of the systems" reconsidered. *Industrial and Corporate Change* 1: 129–180.

David, Paul A., and Julie Ann Bunn. 1988. The economics of gateway technologies and network evolution: Lessons from electricity supply history. *Information Economics and Policy* 3: 165–202.

David, Paul A., and Shane Greenstein. 1990. The economics of compatibility standards: An introduction to recent research. *Economics of Innovation and New Technology* 1: 3–41.

Day, John, and Hubert Zimmermann. 1983. The OSI Reference Model. *Proceedings of the IEEE* 71: 1334–1340.

Dekker, W. 1984. Systems for sale. In *Links for the Future,* ed. P. Dewilde and C. May. North-Holland.

Denning, Peter J. 1990. The ARPANET after twenty years. In *Computers Under Attack,* ed. P. Denning. Addison-Wesley.

des Jardins, Richard. 1983. Afterword: Evolving towards OSI. *Proceedings of the IEEE* 71: 1446–1448.

Dierkes, Meinolf. 1991. Vom Technology Assessment zum "Leitbild Assessment": Zur Rolle von Organisationskultur und professionellen Leitbildern in der Technikgenese. In *Reichweite und Potential der Technikfolgenabschätzung,* ed. K. Kornwachs. Poeschel.

DiMaggio, Paul J., and Walter W. Powell. 1991. The iron cage revisited: Institutional isomorphism and collective rationality in organizational fields. In *The New Institutionalism in Organizational Analysis,* ed. W. Powell and P. DiMaggio. University of Chicago Press.

Disco, Cornelis, Arie Rip, and Bahrend van der Meulen. 1992. Technical innovation and the universities: Divisions of labour in cosmopolitan technical regimes. *Social Science Information* 31: 465–507.

Dosi, Giovanni. 1982. Technological paradigms and technological trajectories. *Research Policy* 11: 147–162.

Dosi, Giovanni. 1988. Sources, procedures, and microeconomic effects of innovation. *Journal of Economic Literature* 26: 1120–1171.

Dosi, Giovanni, and Luigi Orsenigo. 1988. Coordination and transformation: An overview of structures, behaviours and change in evolutionary environments. In *Technical Change and Economic Theory,* ed. G. Dosi et al. Pinter.

Drake, William J. 1988. WATTC-88. Restructuring the international telecommunication regulations. *Telecommunications Policy* 12: 217–233.

Drake, William J. 1993. The Internet religious war. *Telecommunications Policy* 17: 643–648.

Drake, William J. 1994. The transformation of international telecommunications standardization. In *Telecommunications in Transition,* ed. C. Steinfield et al. Sage.

Drake, William J., ed. 1995. *The New Information Infrastructure: Strategies for US Policy.* Twentieth Century Fund.

Dunphy, Ed. 1991. *The UNIX Industry: Evolution, Concepts, Architecture, Applications, and Standards.* QED Information Sciences.

Dutton, William H. 1992. The ecology of games in telecommunications policy. In *The Telecommunications Revolution,* ed. H. Sapolsky et al. Routledge.

Dybvig, Philip H., and Chester S. Spatt. 1983. Adoption externalities as public goods. *Journal of Public Economics* 20: 231–247.

Economides, Nicholas. 1989. Desirability of compatibility in the absence of network externalities. *American Economic Review* 79: 1165–1181.

Economides, Nicholas, and Steven C. Salop. 1992. Competition and integration among complements, and network market structure. *Journal of Industrial Economics* 40: 105–123.

Edelman, Peter B. 1988. Japanese product standards as non-tariff trade barriers: When regulatory policy becomes a trade issue. *Stanford Journal of International Law* 24: 389–446.

Edwards, Paul N. 1996. *The Closed World: Computers and the Politics of Discourse in Cold War America.* MIT Press.

Egan, Michelle. 1991. "Associative Regulation" in the European Community: The Case of Technical Standards. Mimeo, West European Studies Program, University Center for International Studies, University of Pittsburgh.

Egyedi, Tineke. 1996. *Shaping Standardization: A Study of Standards Processes and Standards Policies in the Field of Telematic Services.* Delft University Press.

Eichener, Volker, Rolf G. Heinze, and Helmut Voelzkow. 1991. Von staatlicher Technikfolgenabschätzung zu gesellschaftlicher Techniksteuerung. *Aus Politik und Zeitgeschichte* B 43: 1–14.

Eicher, Lawrence D. 1990. Building global consensus for information technology standardization. In *An Analysis of the Information Technology Standardization Process,* ed. J. Berg and H. Schumny. North-Holland.

Farrell, Joseph, and Garth Saloner. 1985. Standardization, compatibility, and innovation. *Rand Journal of Economics* 16: 70–83.

Farrell, Joseph, and Garth Saloner. 1986. Installed base and compatibility: Innovation, product preannouncements, and predation. *American Economic Review* 76: 940–955.

Farrell, Joseph, and Garth Saloner. 1988. Coordination through committees and markets. *Rand Journal of Economics* 19: 235–252.

Farrell, Joseph, and Garth Saloner. 1992. Converters, compatibility, and the control of interfaces. *Journal of Industrial Economics* 40: 9–35.

Farrell, Joseph, and Carl Shapiro. 1988. Dynamic competition with switching costs. *Rand Journal of Economics* 19: 123–137.

Fedida, Sam. 1977. Viewdata display characteristics and future enhancement. In *Eurocon '77*. European Conference on Electrotechnics.

Fischer, Claude S. 1987. The revolution in rural telephony: 1900–1920. *Journal of Social History* 21: 5–26.

Fischer, Claude S. 1992. *America Calling: A Social History of the Telephone to 1940*. University of California Press.

Foray, Dominique. 1990. Exploitation des externalités de réseau versus évolution des normes: Les formes d'organisation face au dilemme de l'efficacité dans le domaine des technologies de réseau. *Revue D'Economie Industrielle* 51: 113–140.

Foray, Dominique. 1992. The Economic Dimension of Standards: The Role of Users in Information Technology Standardisation. Mimeo, Organization for Economic Cooperation and Development, Paris.

Foray, Dominique. 1994. Users, standards and the economics of coalitions and committees. *Information Economics and Policy* 6: 269–293.

Ford, Michael L. 1979. The search for international standards. *Intermedia* 7: 48–50.

Ford, Michael L. 1981. Legal and administrative problems for the interconnection of national videotex services. In *Videotex 81: International Conference and Exhibition*. Online.

Freiburghaus, K. 1983. The new international telematics services and their standardization. *Telecommunication Journal* 50: 561–565.

Fudenberg, Drew, Richard Gilbert, Joseph Stiglitz, and Jean Tirole. 1983. Preemption, leapfrogging and competition in patent races. *European Economic Review* 22: 3–31.

Gabel, H. Landis. 1987. Open standards in the European computer industry: The case of X/Open. In *Product Standardization and Competitive Strategy*, ed. H. Landis Gabel. North-Holland.

Gabel, H. Landis. 1991. *Competitive Strategies for Product Standards*. McGraw-Hill.

Gabel, Jürgen, Wolfgang Heidrich, and Manfred Worlitzer. 1984. Der CEPT-Standard als Grundlage des Bildschirmtext-Dienstes. *Nachrichtentechnische Zeitung* 37: 214–220.

Galambos, Louis. 1988. Looking for the boundaries of technological determinism: A brief history of the US telephone system. In *The Development of Large Technical Systems*, ed. R. Mayntz and T. Hughes. Campus.

Garnet, Robert W. 1985. *The Telephone Enterprise: The Evolution of the Bell System's Horizontal Structure, 1876–1909*. Johns Hopkins University Press.

Gecsei, Jan. 1983. *The Architecture of Videotex Systems*. Prentice-Hall.

Geller, Henry. 1995. Reforming the US telecommunications and information process. In *The New Information Infrastructure,* ed. W. Drake. Twentieth Century Fund.

Genschel, Philipp. 1995. *Standards in der Informationstechnik. Institutioneller Wandel in der internationalen Standardisierung*. Campus.

Genschel, Philipp. 1997. The dynamics of inertia: Institutional persistence and change in telecommunications and health care. *Governance* 10: 43–66.

Genschel, Philipp, and Raymund Werle. 1993. From national hierarchies to international standardization: Historical and modal changes in the governance of telecommunications. *Journal of Public Policy* 13: 203–225.

Gibson, Richard B. 1995. The global standards process: A balance of the old and the new. In *Standards Policy for Information Infrastructure,* ed. B. Kahin and J. Abbate. MIT Press.

Gilbert, Richard J. 1992. Symposium on compatibility: Incentives and market structure. *Journal of Industrial Economics* 40: 1–8.

Gingras, Yves. 1995. Following scientists through society? Yes, but at arm's length! In *Scientific Practice: Theories and Stories of Doing Physics,* ed. J. Buchwald. University of Chicago Press.

Giraud, Alain. 1991. The technical genesis. In *European Telematics,* ed. J. Jouët et al. North-Holland.

Graham, Nicholas. 1986. Contribution to a political economy of mass communication. In *Media, Culture and Society,* ed. R. Collins et al. Sage.

Grallert, Hans-Joachim, and Bernard Hammer. 1979. Faksimileübertragung: Eine Übersicht. *Nachrichten-Elektronik* 33, no. 4: 112–117.

Granovetter, Marc. 1978. Threshold models of collective behavior. *American Journal of Sociology* 83: 1420–1443.

Grieco, Joseph M. 1990. *Cooperation among Nations: Europe, America, and Non-Tariff Barriers to Trade*. Cornell University Press.

Grindley, Peter. 1990. Standards and the open systems revolution in the computer industry. In *An Analysis of the Information Technology Standardization Process,* ed. J. Berg and H. Schumny. North-Holland.

Grindley, Peter. 1995. *Standards Strategy and Policy: Cases and Stories*. Oxford University Press.

Grindley, Peter, and Ronnie McBryde. 1990. Standards strategy for personal computers. In *An Analysis of the Information Technology Standardization Process,* ed. J. Berg and H. Schumny. Elsevier.

Grindley, Peter, and Saadet Toker. 1994. Establishing standards for Telepoint: Problems of fragmentation and commitment. In *Global Telecommunications Strategies and Technological Changes,* ed. G. Pogorel. Elsevier.

Hagedoorn, John. 1993. Strategic technology alliances and modes of cooperation in high-technology industries. In *The Embedded Firm,* ed. G. Grabher. Routledge.

Hagendijk, Rob. 1990. Structuration theory, constructivism, and scientific change. In *Theories of Science in Society,* ed. S. Cozzens and T. Gieryn. Indiana University Press.

Haggard, Stephan, and Beth A. Simmons. 1987. Theories of international regimes. *International Organization* 41: 491–517.

Hall, Peter A., and Rosemary C. R. Taylor. 1996. Political Science and the Three New Institutionalisms. Discussion paper 96/6, Max-Planck-Institut für Gesellschaftsforschung.

Hancock, K. E., and C. L. Balko. 1983. *A Study of the Implementation of Doc's Role in Information Technology Standardization.* Ottawa: Department of Communications.

Hård, Mikael. 1994. *Machines Are Frozen Spirit: The Scientification of Refrigeration and Brewing in the 19th Century.* Campus.

Hart, Jeffrey A., Robert R. Reed, and Francois Bar. 1992. The building of the Internet: Implications for the future of broadband networks. *Telecommunications Policy* 16: 666–689.

Hawkins, Richard W. 1992. The doctrine of regionalism: A new dimension for international standardization in telecommunications. *Telecommunications Policy* 16: 339–353.

Hawkins, Richard W. 1995. Standards-making as technological diplomacy: Assessing objectives and methodologies in standards institutions. In *Standards, Innovation and Competitiveness,* ed. R. Hawkins et al. Elgar.

Heclo, Hugh. 1978. Issue networks and the executive establishment. In *The New American Political System,* ed. S. Beer et al. American Enterprise Institute for Public Policy Research.

Hemenway, David. 1975. *Industrywide Voluntary Product Standards.* Ballinger.

Henderson, Rebecca M. 1992. Technological change and the management of architectural knowledge. In *Transforming Organizations,* ed. T. Kochan and M. Useem. Oxford University Press.

Henderson, Rebecca M., and Kim B. Clark. 1990. Architectural innovation: The reconfiguration of existing product technologies and the failure of established firms. *Administrative Science Quarterly* 35: 9–30.

Henricson, A. 1984. The non-voice services of Swedish Telecom. In *Links for the Future,* ed. P. Dewilde and C. May. North-Holland.

Hergert, Michael. 1987. Technical standards and competition in the microcomputer industry. In *Product Standardization and Competitive Strategy,* ed. H. Landis Gabel. Elsevier.

Heys, F. A. 1981. International videotex standards: How will they affect the user? In *Proceedings of Viewdata '81.* Online Publications.

Hirschman, Albert O. 1970. *Exit, Voice, and Loyality: Responses to Decline in Firms, Organizations, and States.* Harvard University Press.

Hirsh, Richard F. 1989. *Technology and Transformation in the American Electric Utility Industry.* Cambridge University Press.

Hohn, Hans-Willy, and Volker Schneider. 1991. Path-dependency and critical mass in the development of research and technology: A focused comparison. *Science and Public Policy* 18: 111–122.

Höller, Heinzpeter. 1993. *Kommunikationssysteme: Normung und soziale Akzeptanz.* Vieweg.

Hollingsworth, J. Rogers. 1991. The logic of coordinating American manufacturing sectors. In *Governance of the American Economy,* ed. J. Campbell et al. Cambridge University Press.

Hollingsworth, J. Rogers, and Wolfgang Streeck. 1994. Countries and sectors: Concluding remarks on performance, convergence, and competitiveness. In *Governing Capitalist Economies,* ed. J. Hollingsworth et al. Oxford University Press.

Huber, Peter W. 1987. *The Geodesic Network: Report on Competition in the Telephone Industry.* Government Printing Office.

Hughes, Thomas P. 1983. *Networks of Power: Electrification in Western Society, 1880–1930.* Johns Hopkins University Press.

Hughes, Thomas P. 1986. The seamless web: Technology, science, etcetera, etcetera. *Social Studies of Science* 16: 281–292.

Hughes, Thomas P. 1987. The evolution of large technological systems. In *The Social Construction of Technological Systems,* ed. W. Bijker et al. MIT Press.

Hummel, E. 1980. The role of the CCITT in the introduction of new telecommunication services. *Telecommunication Journal* 47: 26–27.

Hunter, Troy, and A. Harry Robinson. 1980. International digital facsimile coding standards. *Proceedings of the IEEE* 68: 854–867.

Irmer, Theodor. 1990. CCITT after the IXth Plenary Assembly: A Review, and a Preview, on Its Standardization Activities. Resolution 18, Ad Hoc Group Report R1, CCITT, Geneva.

ITU. 1990. Final Acts of the Plenipotentiary Conference: Nice 1989. Constitution and Convention of the International Telecommunication Union, Optional Protocols, Decisions, Resolutions, Recommendations and Opinions.

Jerman-Blazic, Borka. 1988. Implementation of policies for standards development and application. In *Information Technology for Organisational Systems,* ed. H.-J. Bullinger et al. North-Holland.

Jervis, Robert. 1988. Realism, game theory, and cooperation. *World Politics* 40: 317–349.

Jones, W. T. 1979. Standards for telecommunication. The future role of CCITT. *Telecommunication Journal* 46: 730–736.

Kashiwagi, T., and H. Taniike. 1981. An evaluation of the information provided and videotex society in the future. In *Proceedings of Viewdata '81.* Online Publications.

Katz, Michael L., and Carl Shapiro. 1985. Network externalities, competition, and compatibility. *American Economic Review* 75: 424–440.

Katz, Michael L., and Carl Shapiro. 1986. Technology adoption in the presence of network externalities. *Journal of Political Economy* 94: 822–841.

Katz, Michael L., and Carl Shapiro. 1992. Product introduction with network exernalities. *Journal of Industrial Economics* 40: 55–83.

Keck, Otto. 1993. The new institutionalism and the relative-gains-debate. In *International Relations and Pan-Europe,* ed. F. Pfetsch. LIT.

Kelley, Harold H., and John W. Thibaut. 1978. *Interpersonal Relations: A Theory of Interdependence.* Wiley.

Kerr, Ian H. 1981. Interconnection of electronic mail systems: A proposal on naming, addressing and routing. In *Computer Message System,* ed. R. Uhlig. North-Holland.

Kille, Steve. 1983. Survey of Standardisation Work on Message Handling Systems. Final Report for Project. Department of Computer Science, University College London.

Kindleberger, Charles P. 1983. Standards as public, collective and private goods. *Kyklos* 36: 377–396.

Kirstein, Peter T. 1984. The interconnection of message handling systems in the light of current CCITT recommendations. In *Computer-Based Message Services,* ed. H. Smith. North-Holland.

Knight, Jack. 1992. *Institutions and Social Conflict.* Cambridge University Press.

Knorr, Henning. 1993. *Ökonomische Probleme von Kompatibilitätsstandards.* Nomos.

Kobayashi, Kazuo. 1985. Advances in facsimile art. *IEEE Communications Magazine* 23, no. 2: 27–35.

Konangi, Vijaya K., and C. R. Dhas. 1985. An introduction to network architectures. In *Computer Communications,* ed. W. Stallings. IEEE Computer Society Press.

Krasner, Stephen D. 1991. Global communications and national power: Life on the Pareto frontier. *World Politics* 43: 336–366.

Kremenyuk, Victor A., ed. 1991. *International Negotiation. Analysis, Approaches, Issues.* Jossey-Bass.

Kroemer, Frithjof. 1987. Telefax im internationalen Boom. *Nachrichten—Elektronik und Telematik* 41: 266–273.

Kubicek, Herbert, and Peter Seeger. 1992. The negotiation of data standards: A comparative analysis of EAN and EFT/POS systems. In *New Technology at the Outset,* ed. M. Dierkes and U. Hoffmann. Campus.

La Porte, Todd R., ed. 1991. *Social Responses to Large Technical Systems: Control or Anticipation.* Kluwer.

Labarrère, C. 1985. *L'Europe des Postes et des Télécommunications.* Masson.

Langlois, Richard N. 1986. Rationality, institutions, and explanation. In *Economics as a Process,* ed. R. Langlois. Cambridge University Press.

Langlois, Richard N., and Paul L. Robertson. 1992. Networks and innovation in a modular system: Lessons from the microcomputer and stereo component industries. *Research Policy* 21: 297–313.

Larratt, Richard. 1981. Market factors. In *The Telidon Book,* ed. D. Godfrey and E. Chang. Porcépic.

Latour, Bruno. 1987. *Science in Action: How to Follow Scientists and Engineers through Society.* Harvard University Press.

Latour, Bruno. 1992. Where are the missing masses? The sociology of a few mundane artifacts. In *Shaping Technology/Building Society,* ed. W. Bijker et al. MIT Press.

Laudan, Rachel. 1984. Cognitive change in technology and science. In *The Nature of Technological Knowledge,* ed. R. Laudan. Reidel.

Law, Carl Edgar. 1989. X. 400 and OSI Electronic Messaging into the 1990s. IBC Technical Services, London.

Law, John, and Michel Callon. 1988. Engineering and sociology in a military aircraft project: A network analysis of technological change. *Social Problems* 35: 284–297.

Law, John, and Michel Callon. 1992. The Life and Death of an Aircraft: A Network Analysis of Technical Change. In *Shaping Technology/Building Society,* ed. W. Bijker et al. MIT Press.

Lazak, Dieter. 1984. *Bildschirmtext. Technische Leistung und wirtschaftliche Anwendung neuer Kommunikationstechnik.* CW-Publikationen.

Lehr, William. 1995. Compatibility standards and interoperability: lessons from the Internet. In *Standards Policy for Information Infrastructure,* ed. B. Kahin and J. Abbate. MIT Press.

Leibenstein, Harvey. 1984. On the economics of conventions and institutions: An explanatory essay. *Journal of Institutional and Theoretical Economics* 140: 74–86.

Liebowitz, S. J., and Stephen E. Margolis. 1990a. The fable of the keys. *Journal of Law & Economics* 23: 1–25.

Liebowitz, S. J., and Stephen E. Margolis. 1990b. The Economics of Standards. Paper, University of Texas, Dallas.

Liebowitz, S. J., and Stephen E. Margolis. 1994. Network externality: An uncommon tragedy. *Journal of Economic Perspectives* 8: 133–150.

Lifchus, Ian M. 1985. Standards Committee T1: Telecommunications. *IEEE Communications Magazine* 23, no. 1: 34–37.

Lindenberg, Siegwart. 1993. Framing, empirical evidence, and applications. In *Jahrbuch für Neue Politische Ökonomie. Band 12: Neue Politische Ökonomie von Normen und Institutionen,* ed. P. Herder-Dorneich et al. Mohr.

Lipinski, Hubert, John Tydeman, and Laurence Zwimpfer. 1981. *Teletext and Videotex Standards for the United States: Report of a Policy Workshop.* Institute for the Future.

Lorange, Peter, and Johan Roos. 1992. *Strategic Alliances: Formation, Implementation, and Evolution.* Blackwell.

Luhmann, Niklas. 1983a. *Legitimation durch Verfahren.* Suhrkamp.

Luhmann, Niklas. 1983b. *Rechtssoziologie.* Westdeutscher Verlag.

Luhmann, Niklas. 1984. *Soziale Systeme.* Suhrkamp.

Lundgreen, Peter. 1986. *Standardization, Testing, Regulation: Studies in the History of the Science-Based Regulatory State.* Kleine.

Lundin, Rolf A. 1986. Organizational economy: The politics of unanimity and suppressed competition. In *Organizing Industrial Development,* ed. R. Wolff. De Gruyter.

Maciejewski, Paul G. 1989. Internationale Standardisierung elektronischer Kommunikation. *Computer Magazin* 18, no. 3: 49–52.

MacKenzie, Donald. 1990. *Inventing Accuracy: A Historical Sociology of Nuclear Missile Guidance*. MIT Press.

MacKenzie, Donald, and Wajcman, Judy, eds. 1985. *The Social Shaping of Technology*. Open University Press.

Macpherson, Andrew. 1990. *International Telecommunication Standards Organizations*. Artech House.

Madden, John C. 1979. *Videotex in Canada*. Minister of Supply and Services (Canada).

Majone, Giandomenico. 1992. Market integration and regulation: Europe after 1992. *Metroeconomica* 43: 131–156.

Malamud, Carl. 1993. *Exploring the Internet: A Technical Travelogue*. Prentice-Hall.

Manros, Carl-Uno. 1989. *The X. 400 Blue Book Companion. CCITT X. 400 MHS 1988. ISO/IEC MOTIS. Message Oriented Text Interchange System*. Dotesios.

Mansell, Robin. 1993. *The New Telecommunications: A Political Economy of Network Evolution*. Sage.

Mansell, Robin, and Richard Hawkins. 1992. Old roads and new signposts: Trade policy objectives in telecommunication standards. In *Telecommunication*, ed. F. Klaver and P. Slaa. International Organization Standards Press.

Mao, Y. M., and E. Hummel. 1981. Recommendation, Standard or Instruction? *Telecommunication Journal* 48: 747–748.

March, James G., and Johan P. Olsen. 1989. *Rediscovering Institutions: The Organizational Basis of Politics*. Free Press.

Marti, Bernard. 1979. Videotex developments in France. *Computer Communications* 2: 60–64.

Marti, Bernard. 1982. Efforts toward a European standard. In *Bildschirmtext Kongress 1982*. Diebold.

Marti, Bernard, and C. Schwartz. 1980. Videotex standardization. In *Videotex, Viewdata, Teletext*. Online.

Marti, Bernard, Hervé Layec, Christiane Schwartz, and Francis Thabard. 1990. *Télématique. Techniques, Normes, Services*. Dunod.

Marwell, Gerald, and Pamela Oliver. 1993. *The Critical Mass in Collective Action*. Cambridge University Press.

Mastanduno, Michael. 1991. Do relative gains matter? *International Security* 16: 73–113.

Matutes, Carmen, and Pierre Regibeau. 1988. "Mix and match": Product compatibility without network externalities. *Rand Journal of Economics* 19: 221–234.

Matutes, Carmen, and Pierre Regibeau. 1992. Compatibility and bundling of complementary goods in a duopoly. *Journal of Industrial Economics* 40: 37–54.

Mayntz, Renate. 1993a. Modernization and the logic of interorganizational networks. In *Societal Change Between Market and Organization,* ed. J. Child et al. Avebury.

Mayntz, Renate. 1993b. Networks, issues, and games: Multiorganizational interactions in the restructuring of a national research system. In *Games in Hierarchies and Networks,* ed. F. Scharpf. Campus.

Mayntz, Renate. 1993c. Große technische Systeme und ihre gesellschaftstheoretische Bedeutung. *Kölner Zeitschrift für Soziologie und Sozialpsychologie* 45: 97–108.

Mayntz, Renate, and Thomas P. Hughes, eds. 1988. *The Development of Large Technical Systems.* Campus.

Mayntz, Renate, and Birgitta Nedelmann. 1987. Eigendynamische soziale Prozesse. *Kölner Zeitschrift für Soziologie und Sozialpsychologie* 39: 648–668.

Mayntz, Renate, and Fritz W. Scharpf. 1975. *Policy-Making in the German Federal Bureaucracy.* Elsevier.

Mayntz, Renate, and Fritz W. Scharpf. 1995. Der Ansatz des akteurzentrierten Institutionalismus. In *Gesellschaftliche Selbstregelung und politische Steuerung,* ed. R. Mayntz and F. Scharpf. Campus.

Mayntz, Renate, and Volker Schneider. 1988. The dynamics of system development in a comparative perspective: Interactive videotex in Germany, France and Britain. In *The Development of Large Technical Systems,* ed. R. Mayntz and T. Hughes. Campus.

Mazda, Fraidoon. 1992. Standardizing on standards. *Telecommunications* 26: 40–50.

McConnell, Kenneth R., Dennis Bodson, and Richard Schaphorst. 1989. *FAX: Digital Facsimile Technology and Applications.* Artech House.

McConnell, Kenneth R., Dennis Bodson, and Richard Schaphorst. 1992. *FAX: Digital Facsimile,* second edition. Artech House.

McKnight, Lee, and W. Russell Neuman. 1995. Technology policy and the national information infrastructure. In *The New Information Infrastructure,* ed. W. Drake. Twentieth Century Fund.

Miki, Toshiaki. 1991. KDD's facsimile interworking and X. 400 interconnection. In *Message Handling Systems and Application Layer Communication Protocols,* ed. P. Schicker and E. Stefferud. North-Holland.

Misa, Thomas J. 1992. Controversy and closure in technological change: Constructing "steel." In *Shaping Technology/Building Society,* ed. W. Bijker et al. MIT Press.

Misa, Thomas J. 1994. Retrieving sociotechnical change from technological determinism. In *Does Technology Drive History?* ed. M. Smith and L. Marx. MIT Press.

Mooney, James D. 1986. Software interface standards: Lessons from the MOSI project. In *Computer Standards Conference 1986.* Computer Society Press.

Morganti, G. 1982. Videotex standards: User needs and relevant fulfilment possibilities. In *Proceedings of Viewdata '82.* Online Publications.

Morrow, James D. 1994. Modeling the forms of international cooperation: Distribution versus information. *International Organization* 48: 387–423.

Mosco, Vincent. 1982. *Pushbutton Fantasies: Critical Perspectives on Videotex and Information Technology.* Ablex.

Müller, Jörg, and Paul Kuhn. 1988. *Information Technology Standardization in Japan.* Siemens (Tokyo).

Nash, John F. 1953. Two-person cooperative games. *Econometrica* 21: 128–140.

National Research Council (US). 1995. *Standards, Conformity Assessment, and Trade.* National Academy Press.

Nelson, Richard R., and Nathan Rosenberg. 1993. Technical innovation and national systems. In *National Innovation Systems,* ed. R. Nelson. Oxford University Press.

Nelson, Richard R., and Sidney G. Winter. 1977. In search of useful theory of innovation. *Research Policy* 6: 36–76.

Nelson, Richard R., and Sidney G. Winter. 1982. *An Evolutionary Theory of Economic Change.* Harvard University Press.

Nickelson, R. L. 1990. HDTV standards: Understanding the issues. *Telecommunication Journal* 57: 302–312.

Noam, Eli. 1992. *Telecommunications in Europe.* Oxford University Press.

Noble, David F. 1977. *America by Design: Science, Technology, and the Rise of Corporate Capitalism.* Knopf.

Nora, Simon, and Alain Minc. 1978. *L'informatisation de la société. Rapport à M. le Président de la République.* La Documentation française.

Norberg, Arthur L., and Judy E. O'Neill. 1996. *Transforming Computer Technology: Information Processing for the Pentagon, 1962–1986.* Johns Hopkins University Press.

North, Douglass C. 1990. *Institutions, Institutional Change and Economic Performance.* Cambridge University Press.

North, Douglass C. 1993. Institutional change: A framework of analysis. In *Institutional Change,* ed. S.-E. Sjöstrand. Sharpe.

Nussbaum, Eric, and Walter R. Noller. 1986. Integrated network architectures: Alternatives and ISDN. *IEEE Communications Magazine* 24, no. 3: 8–12.

O'Brien, C. Douglas. 1982. Videotex standards shake down. *Canadian Electronics Engineering,* June: 24–29.

O'Brien, C. Douglas, and Herbert G. Bown. 1983. A perspective on the development of videotex in North America. *IEEE Journal on Selected Areas in Communications* 1: 260–266.

OECD. 1991. *Information Technology Standards: The Economic Dimension.* Organization for Economic Cooperation and Development.

OECD. 1993. *Economic and Trade Issues in the Computerised Database Market.* Organization for Economic Cooperation and Development.

Olshan, Marc A. 1993. Standards-making organizations and the rationalization of American life. *Sociological Quarterly* 34: 319–335.

Olson, Mancur. 1971. *The Logic of Collective Action: Public Goods and the Theory of Groups.* Harvard University Press.

Ostrom, Elinor. 1986. A method of institutional analysis. In *Guidance, Control, and Evaluation in the Public Sector,* ed. F.-X. Kaufmann et al. De Gruyter.

OTA (US Office of Technology Assessment). 1992. *Global Standards: Building Blocks for the Future.* Government Printing Office.

Parkhill, Douglas F. 1981. The Telidon story. Presentation at Rotary Club, Toronto.

Parkhill, Douglas F. 1985. The Evolution of Telidon: Background Paper for Remarks. Prepared for Videotex Canada Meeting, Toronto, March 4.

Perabo, Richard. 1981. Technik der Fernkopierer. *Nachrichten-Elektronik* 35, no. 10: 401–403.

Perabo, Richard. 1982. Technik der Fernkopierer. *Nachrichten-Elektronik* 36, no. 2: 49–53.

Perrow, Charles. 1984. *Normal Accidents: Living with High-Risk Technologies.* Basic Books.

Pfeiffer, Günter. 1989. *Kompatibilität und Markt.* Nomos.

Pinch, Trevor J., and Wiebe E. Bijker. 1984. The social construction of facts and artefacts: or how the sociology of science and the sociology of technology might benefit each other. *Social Studies of Science* 14: 399–441.

Pitt, Douglas C., and Kevin Morgan. 1992. Viewing divestment from afar. In *The Telecommunications Revolution,* ed. H. Sapolsky et al. Routledge.

Porter, Michael E. 1990. *The Competitive Advantage of Nations.* Free Press.

Putnam, Robert D. 1988. Diplomacy and domestic politics: The logic of two-level games. *International Organization* 42: 427–460.

Quarterman, John S. 1990. *The Matrix: Computer Networks and Conferencing Systems Worldwide.* Digital Press.

Radack, Shirley M. 1994. The federal government and information technology standards: Building the national information infrastructure. *Government Information Quarterly* 11, no. 4: 373–385.

Raiffa, Howard. 1982. *The Art and Science of Negotiation.* Harvard University Press.

Rallapalli, Krishna. 1981. Fax: What is in it for LSI? In *International Conference on Communications 1981,* volume 2. IEEE Service Center.

Rammert, Werner. 1992. From mechanical engineering to information engineering: Phenomenology and social roots of an emerging type of technology. In *New Technology at the Outset,* ed. M. Dierkes and U. Hoffmann. Campus.

Rankine, L. John. 1990. Information technology standards: Can the challenges be met? In *An Analysis of the Information Technology Standardization Process,* ed. J. Berg and H. Schumny. North-Holland.

Rawls, John. 1972. *A Theory of Justice.* Harvard University Press.

Ray-Barman, Arun. 1985. *L'ère du Videotex. Principes et applications.* Éditests.

Reddy, N. Mohan. 1990. Product self-regulation: A paradox of technology policy. *Technological Forecasting and Social Change* 38: 49–63.

Renaud, Jean-Luc. 1990. The role of the International Telecommunication Union: Conflict, resolution and the industrialized countries. In *The Political Economy of Communications,* ed. K. Dyson and P. Humphreys. Routledge,.

Rip, Arie. 1992. Science and technology as dancing partners. In *Technological Development and Science in the Industrial Age,* ed. P. Kroes and M. Bakker. Kluwer.

Robertson, Paul L., and Richard N. Langlois. 1992. Modularity, innovation, and the firm: The case of audio components. In *Entrepreneurship, Technological Innovation, and Economic Growth,* ed. F. Scherer and M. Perlman. University of Michigan Press.

Robinson, Gary S. 1986. Accredited Standards Committee for Information Processing Systems, X3. In *Computer Standards Conference 1986: Proceedings.* Computer Society Press.

Rogers, Everett M. 1995. Diffusion of innovations: Modifications of a model for telecommunications. In *Die Diffusion von Innovationen in der Telekommunikation,* ed. M. Stoetzer and A. Mahler. Springer.

Rogers, Juan D. 1996. Implementation of a National Information Infrastructure: Science and the Building of Society. Dissertation, University of Virginia.

Rohlfs, Jeffrey. 1974. A theory of interdependent demand for a communications service. *Bell Journal of Economics* 5: 16–37.

Rosario, Marta, and Susanne K. Schmidt. 1991. Standardization in the European community: The example of ICT. In *Technology and the Future of Europe,* ed. C. Freeman et al. Pinter.

Rosen, Barry N., Steven P. Schnaars, and David Shani. 1988. A comparison of approaches for setting standards for technological products. *Journal of Product and Innovation Management* 5: 129–139.

Rosen, Paul. 1993. The social construction of mountain bikes: Technology and postmodernity in the cycle industry. *Social Studies of Science* 23: 479–513.

Rosenberg, Nathan. 1976. *Perspectives on Technology.* Cambridge University Press.

Rosenberg, Nathan. 1992. Science and technology in the twentieth century. In *Technology and Enterprise in a Historical Perspective,* ed. G. Dosi et al. Oxford University Press.

Rosenkopf, Lori, and Michael L. Tushman. 1994. The coevolution of technology and organization. In *Evolutionary Dynamics of Organizations,* ed. J. Baum and J. Singh. Oxford University Press.

Russell, Stewart. 1986. The social construction of artefacts: A response to Pinch and Bijker. *Social Studies of Science* 16: 331–346.

Rutkowski, Anthony M. 1985. *Integrated Services Digital Networks.* Artech House.

Rutkowski, Anthony M. 1991. The ITU at the cusp of change. *Telecommunications Policy* 15: 286–297.

Rutkowski, Anthony M. 1995. Today's cooperative competitive standards environment and the Internet standards-making model. In *Standards Policy for Information Infrastructure,* ed. B. Kahin and J. Abbate. MIT Press.

Ryll, Andreas. 1989. Die Spieltheorie als Instrument der Gesellschaftsforschung. Discussion Paper 89/10, Max-Planck-Institut für Gesellschaftsforschung.

Sabel, Charles F. 1993. Constitutional ordering in historical context. In *Games in Hierarchies and Networks,* ed. F. Scharpf. Campus.

Sahal, Devendra. 1981. *Patterns of Technological Innovation.* Addison-Wesley.

Salsbury, Stephen. 1988. The emergence of an early large-scale technical system: The American railroad network. In *The Development of Large Technical Systems,* ed. R. Mayntz and T. Hughes. Campus.

Saunders, Stephen, and Peter Heywood. 1992. X. 400's last windows of opportunity. *Data Communications,* May 21: 73–76.

Savage, James G. 1989. *The Politics of International Telecommunications Regulation.* Westview.

Scharpf, Fritz W. 1989. Decision rules, decision styles and policy choices. *Journal of Theoretical Politics* 1: 149–176.

Scharpf, Fritz W. 1991. Games real actors could play: The challenge of complexity. *Journal of Theoretical Politics* 3: 277–304.

Scharpf, Fritz W. 1992. Koordination durch Verhandlungssysteme: Analytische Konzepte und institutionelle Lösungen. In *Horizontale Politikverflechtung,* ed. A. Benz et al. Campus.

Scharpf, Fritz W. 1993a. Coordination in hierarchies and networks. In *Games in Hierarchies and Networks,* ed. F. Scharpf. Campus.

Scharpf, Fritz W. 1993b. Versuch über Demokratie im verhandelnden Staat. In *Verhandlungsdemokratie, Interessenvermittlung, Regierbarkeit,* ed. R. Czada and M. Schmidt. Westdeutscher Verlag.

Scharpf, Fritz W. 1994. Games real actors could play: Positive and negative coordination in embedded negotiations. *Journal of Theoretical Politics* 6: 27–53.

Scharpf, Fritz W. 1997. *Games Real Actors Play: Actor-Centered Institutionalism in Policy Research.* Westview.

Schelling, Thomas C. 1960. *The Strategy of Conflict.* Oxford University Press.

Schelling, Thomas C. 1984. *Choice and Consequence.* Harvard University Press.

Schenke, Klaus, Rolf Rüggeberg, and Jens Otto. 1981. Teletex, ein neuer internationaler Fernmeldedienst für die Textkommunikation. *Jahrbuch der Deutschen Bundespost* 32: 277–349.

Scherer, Frederic M. 1992. *International High-Technology Competition.* Harvard University Press.

Scheuerer, Johann. 1990. X. 400: Internationaler Standard für E-Mail-Systeme. In *Das Dach der TelematikWelt* 11: 29–31.

Schimank, Uwe. 1992. Spezifische Interessenkonsense trotz generellem Orientierungsdissens. Ein Integrationsmechanismus polyzentrischer Gesellschaften. In *Kommunikation und Konsens in modernen Gesellschaften,* ed. H.-J. Giegel. Suhrkamp.

Schlesinger, Leonard A., Davis Dyer, Thomas N. Clough, and Diane Landau. 1987. *Chronicles of Corporate Change: Management Lessons from AT&T and Its Offspring.* Lexington Books.

Schmidt, A. C. 1983. New developments in high-speed digital facsimile and graphics. In *International Conference on Communications 1983,* volume 2. IEEE Service Center.

Schmidt, Susanne K. 1991. Taking the long road to liberalization. *Telecommunications Policy* 15: 209–222.

Schmidt, Susanne K. 1993. Coordinating Complementarities: The Case of Institutionalized Standardization. Paper presented at research seminar on Institutional Change and Network Evolution, Stockholm.

Schmidt, Susanne K. 1996. Privatizing the federal postal and telecommunications services. In *A New German Public Sector?* ed. A. Benz and K. Goetz. Dartmouth.

Schmidt, Susanne K., and Raymund Werle. 1992. The development of compatibility standards in telecommunications: Conceptual framework and theoretical perspective. In *New Technology at the Outset,* ed. M. Dierkes and U. Hoffmann. Campus.

Schmidt, Susanne K., and Raymund Werle. 1993. Technical Controversy in International Standardization. Discussion paper 93/5, Max-Planck-Institut für Gesellschaftsforschung.

Schmidt, Susanne K., and Raymund Werle. 1994. Koordination und Evolution: Technische Standards im Prozeß der Entwicklung technischer Systeme. In *Technik und Gesellschaft, Jahrbuch 7,* ed. W. Rammert and G. Bechmann. Campus.

Schneider, Volker. 1989. *Technikentwicklung zwischen Politik und Markt: Der Fall Bildschirmtext.* Campus.

Schneider, Volker. 1991. The governance of large technical systems: the case of telecommunications. In *Social Responses to Large Technical Systems,* ed. T. La Porte. Kluwer.

Schneider, Volker. 1992. Kooperative Akteure und vernetzte Artefakte: Überlegungen zu den Formen sozialer Organisation großtechnischer Systeme. In *Technik und Gesellschaft, Jahrbuch 6,* ed. G. Bechmann and W. Rammert. Campus.

Schneider, Volker. 1993. Networks and games in large technical systems: The case of Videotex. In *Games in Hierarchies and Networks,* ed. F. Scharpf. Campus.

Schneider, Volker, Jean-Marie Charon, Ian Miles, Graham Thomas, and Thierry Vedel. 1991. The dynamics of videotex development in Britain, France and Germany: A cross-national comparison. *European Journal of Communication* 6: 187–212.

Schneider, Volker, and Renate Mayntz. 1995. Akteurzentrierter Institutionalismus in der Technikforschung. Fragestellungen und Erklärungsansätze. In *Technik und Gesellschaft. Jahrbuch 8,* ed. J. Halfmann. Campus.

Schneider, Volker, and Raymund Werle. 1990. International regime or corporate actor? The European Community in telecommunications policy. In *The Political Economy of Communications,* ed. K. Dyson and P. Humphreys. Routledge.

Schneider, Volker, and Raymund Werle. 1996. Co-evolution and development constraints: The development of large technical systems in evolutionary perspective. In *Proceedings of the Conference "Management and New Technology."* COTEC.

Sebenius, James K. 1991. Negotiation analysis. In *International Negotiation,* ed. V. Kremenyuk. Jossey-Bass.

Shibui, Michiro. 1984. The commercial CAPTAIN system and the CAPTAIN PLPS. In *Links for the Future,* ed. P. Dewilde and C. May. North-Holland.

Siebe, Wilfried. 1991. Game theory. In *International Negotiation,* ed. V. Kremenyuk. Jossey-Bass.

Simon, Herbert A. 1962. The architecture of complexity. *Proceedings of the American Philosophical Society* 106: 467–482.

Sirbu, Marvin A. 1989. Telecommunication Standards, Innovation and Industry Structure. Paper, University of Pittsburgh.

Sirbu, Marvin A., and Deborah L. Estrin. 1989. Standards. In *International Encyclopedia of Communications 4,* ed. E. Barnouw et al. Oxford University Press.

Sirbu, Marvin A., and Kent Hughes. 1986. Standardization of Local Area Networks. Paper presented at 14th Annual Telecommunications Policy Research Conference, Airlie, Virginia.

Sirbu, Marvin A., and Laurence E. Zwimpfer. 1985. Standards setting for computer communication: The case of X. 25. *IEEE Communications Magazine* 23, no. 3: 35–45.

Smit, Wim A. 1995. Science, technology, and the military: Relations in transition. In *Handbook of Science and Technology Studies,* ed. S. Jasanoff et al. Sage.

Smith, George D. 1985. *The Anatomy of Business Strategy: Bell, Western Electric, and the Origin of the American Telephone Industry.* Johns Hopkins University Press.

Snidal, Duncan. 1985. Coordination versus prisoners' dilemma: Implications for international cooperation and regimes. *American Political Science Review* 79: 923–942.

Snidal, Duncan. 1991. Relative gains and the pattern of international cooperation. *American Political Science Review* 85: 702–726.

Solomon, Jonathan. 1984. The future role of international telecommunications institutions. *Telecommunications Policy* 8: 213–221.

Sørensen, Knut H., and Nora Levold. 1992. Tacit networks, heterogeneous engineers, and embodied technology. *Science, Technology & Human Values* 17, no. 1: 13–35.

Spring, Michael B., Christal Grisham, Jon O'Donnell, Ingjerd Skogseid, Andrew Snow, George Tarr, and Peihan Wang. 1995. Improving the standardization process: Working with bulldogs und turtles. In *Standards Policy for Information Infrastructure,* ed. B. Kahin and J. Abbate. MIT Press.

Stalk, George, and Thomas M. Hout. 1990. *Competing Against Time: How Time-Based Competition Is Reshaping Global Markets.* Free Press.

Stallings, William. 1985. IEEE Project 802: Setting standards for local-area networks. In *Computer Communications,* ed. W. Stallings. IEEE Computer Society Press.

Stankiewicz, Rikard. 1992. Technology as an Autonomous Sociocognitive System. In *Dynamics of Science-Based Innovation,* ed. H. Grupp. Springer.

Staudinger, Wilhelm. 1985. Evolution of telematic terminal equipment in an OSI & ISDN Environment, as seen from the viewpoint of the CCITT Study Group VIII. In *The New World of the Information Society,* ed. J. Bennett and T. Pearcey. North-Holland.

Stein, Arthur A. 1982. Coordination and collaboration: Regimes in an anarchic world. *International Organization* 36: 299–324.

Stein, Arthur A. 1990. *Why Nations Cooperate: Circumstance and Choice in International Relations.* Cornell University Press.

Streeck, Wolfgang, and Philippe Schmitter, eds. 1985. *Private Interest Government: Beyond Market and State.* Sage.

Stukenbröker, B. 1980. Wettkampf der Systeme. In *Bildschirmtextkongress 1980.* Diebold.

Sullivan, Dennis J., and A. Michael Noll. 1982. A Bell System View of Videotex. *Telecommunications Policy* 6: 237–241.

Sung Sio Ma. 1983. Strategy for introducing new services. *Telecommunication Journal* 50: 129–133.

Swann, G. M. P. 1987. Industry standard microprocessors and the strategy of second-source production. In *Product Standardization and Competitive Strategy,* ed. H. Landis Gabel. North-Holland.

Tamarin, Christopher. 1988. Telecommunications technology applications and standards. *Telecommunications Policy* 12: 323–331.

Tang, Debra, Michael Anzenberger, Paul Markovitz, and Michael Wallace. 1988. A gateway between MHS (X. 400) and SMTP. In *Computer Standards Evolution.* IEEE.

Tassey, Gregory. 1992. *Technology, Infrastructure and Competitive Position.* Kluwer.

Teece, David J. 1988. Profiting from technological innovation: Implications for integration, collaboration, licensing and public policy. In *Readings in the Management of Innovation,* ed. M. Tushman and W. Moore. Ballinger.

TEEGA Research Consultants. 1985. Telidon and the Standard-Setting Process: Background Study #2 for an Evaluation of Telidon. Ottawa.

Temple, Stephen. 1991. *A Revolution in European Telecommunications Standards Making.* ETSI.

Thomas, Frank. 1988. The politics of growth: The German telephone system. In *The Development of Large Technical Systems,* ed. R. Mayntz and T. Hughes. Campus.

Thompson, George V. 1954. Intercompany technical standardization in the early American automobile industry. *Journal of Economic History* 14: 1–19.

Thompson, James D. 1967. *Technology and Structure: Organizations in Action.* McGraw-Hill.

Tietz, Walter. 1988. Zweite Generation von Standards der Mitteilungs-Übermittlung. *Nachrichtentechnische Zeitschrift* 41: 512–518.

Tietzel, Manfred. 1990. Virtue, vice, and Dr. Pangloss: On the economics of conventions. In *Ordo (Jahrbuch für die Ordnung von Wirtschaft und Gesellschaft),* Band 41, ed. H. Lenel et al. Gustav Fischer.

Tirole, Jean. 1988. *The Theory of Industrial Organization.* MIT Press.

Tsebelis, George. 1990. *Nested Games: Rational Choice in Comparative Politics.* University of California Press.

Tushman, Michael L., and Lori Rosenkopf. 1992. Organizational determinants of technological change: Toward a sociology of technological evolution. *Research in Organizational Behavior* 14: 311–347.

Tversky, Amos, and Daniel Kahneman. 1985. The framing of decisions and the psychology of choice. In *Behavioral Decision Making,* ed. G. Wright. Plenum.

Tversky, Amos, and Daniel Kahneman. 1988. Rational choice and the framing of decisions. In *Decision Making,* ed. D. Bell et al. Cambridge University Press.

Tydeman, John, Hubert Lipinski, Richard P. Adler, Michael Nyhan, and Laurence Zwimpfer. 1982. *Teletext and Videotex in the United States.* McGraw-Hill.

Uhlig, Ronald P., David J. Farber, and James H. Bair. 1979. *The Office of the Future: Communication and Computers.* North-Holland.

Ullmann-Margalit, Edna. 1977. *The Emergence of Norms.* Oxford University Press.

Underdal, Arild. 1991. The outcomes of negotiation. In *International Negotiation,* ed. V. Kremenyuk. Jossey-Bass.

Updegrove, Andrew. 1995. Consortia and the role of the government in standard setting. In *Standards Policy for Information Infrastructure,* ed. B. Kahin and J. Abbate. MIT Press.

Vanberg, Viktor, and James M. Buchanan. 1989. Interests and theories in constitutional choice. *Journal of Theoretical Politics* 1: 49–62.

Vanberg, Viktor, and James M. Buchanan. 1991. Constitutional choice, rational ignorance and the limits of reason. In *Jahrbuch für Neue Politische Ökonomie 10: Systemvergleich und Ordnungspolitik,* ed. E. Boettcher et al. Mohr.

van den Belt, Henk, and Arie Rip. 1987. The Nelson-Winter-Dosi model and synthetic dye chemistry. In *The Social Construction of Technological Systems,* ed. W. Bijker et al. MIT Press.

van Tulder, Rob, and Gerd Junne. 1988. *European Multinationals in Core Technologies.* Wiley.

Verman, Lal C. 1973. *Standardization: A New Discipline.* Archon Books.

Vincenti, Walter G. 1990. *What Engineers Know and How They Know It: Analytical Studies from Aeronautical History.* Johns Hopkins University Press.

Vincenti, Walter G. 1995. The technical shaping of technology: Real-world constraints and technical logic in Edison's electrical lighting system. *Social Studies of Science* 25: 553–574.

Voelzkow, Helmut. 1995. *Private Regierungen in der Techniksteuerung. Eine sozialwissenschaftliche Analyse der technischen Normung.* Campus.

Voelzkow, Helmut, and Volker Eichener. 1992. Techniksteuerung durch Verbände. Institutionelles Arrangement und Interessenberücksichtigungsmuster bei der Harmonisierung technischer Normen in Europa. In *Politische Techniksteuerung,* ed. K. Grimmer et al. Leske + Budrich.

Voelzkow, Helmut, Joseph Hilbert, and Eckard Bolenz. 1987. Wettbewerb durch Kooperation: Kooperation durch Wettbewerb. Zur Funktion und Funktionsweise von Normungsverbänden. In *Dezentrale Gesellschaftssteuerung,* ed. M. Glagow and H. Willke. Centaurus.

von Vignau, Ralph A.1983. Bildschirmtext and the CEPT videotex system. *IEEE Journal on Selected Areas in Communications* 1, no. 2: 254–259.

Wallenstein, Gerd. 1977. Development of policy in the ITU. *Telecommunications Policy* 1: 138–152.

Wallenstein, Gerd. 1990. *Setting Global Telecommunication Standards.* Artech House.

Wärneryd, Karl. 1990. Conventions: An evolutionary approach. *Constitutional Political Economy* 1: 83–107.

Wasserman, Neil H. 1985. *From Invention to Innovation: Long-Distance Telephone Transmission at the Turn of the Century.* Johns Hopkins University Press.

Weingart, Peter. 1984. The structure of technological change: Reflections on a sociological analysis of technology. In *The Nature of Technological Knowledge,* ed. R. Laudan. Reidel.

Weiss, Martin B. H., and Marvin Sirbu. 1990. Technological choice in voluntary standards committees: An empirical analysis. *Economics of Innovation and New Technology* 1: 111–133.

Wenger, Pierre-André. 1989. The future also has a past: The telefax, a young 150-year old service. *Telecommunication Journal* 56: 777–782.

Werle, Raymund. 1990. *Telekommunikation in der Bundesrepublik: Expansion, Differenzierung, Transformation.* Campus.

Werle, Raymund. 1993. Politische Techniksteuerung durch europäische Standardisierung? In *Perspektive Techniksteuerung,* ed. H. Kubicek and P. Seeger. Edition Sigma.

Werle, Raymund. 1995a. Rational Choice und rationale Technikentwicklung. Einige Dilemmata der Technikkoordination. In *Technik und Gesellschaft, Jahrbuch 8,* ed. J. Halfmann et al. Campus.

Werle, Raymund. 1995b. Staat und Standards. In *Gesellschaftliche Selbstregelung und politische Steuerung,* ed. R. Mayntz and F. Scharpf. Campus.

Werle, Raymund, and Gerhard Fuchs. 1993. Liberalization and integration: Pathways to a trans-European network in telecommunications. *Utilities Policy* 3: 187–200.

Wetherington, Joe D. 1983. The story of PLP. *IEEE Journal on Selected Areas in Communications* 1: 267–277.

Williamson, Oliver E. 1975. *Markets and Hierarchies: Analysis and Antitrust Implications.* Free Press.

Williamson, Oliver E. 1985. *The Economic Institutions of Capitalism: Firms, Markets, Relational Contracting.* Free Press.

Willmott, Robert. 1989. Message Handling and Directories. Paper prepared for conference on Telecommunications and the Melbourne Meetings, London.

Wilson, P. A. 1984. *Standards and the Electronic Mailbox.* NCC Publications.

Winner, Langdon. 1993. Upon opening the black box and finding it empty: Social constructivism and the philosophy of technology. *Science, Technology & Human Values* 18, no. 3: 362–378.

Woolgar, Steve. 1987. Reconstructing man and machine: A note on sociological critiques of cognitivism. In *The Social Construction of Technological Systems,* ed. W. Bijker et al. MIT Press.

Woolgar, Steve. 1991. The turn to technology in social studies of science. *Science, Technology & Human Values* 16, no. 1: 20–50.

Young, Oran R. 1992. The effectiveness of international institutions: Hard cases and critical variables. In *Governance without Government,* ed. J. Rosenau and E.-O. Czempiel. Cambridge University Press.

Zimmermann, Rolf. 1979. Problems of international standardization of videotex services. *Displays* 1: 103–109.

Zucker, Lynne G. 1987. Institutional theories of organization. *Annual Review of Sociology* 13: 443–464.

Index